D1476490

WITHDRAWN
CALTECH LIBRARY SERVICES

THE EL NIÑO–SOUTHERN OSCILLATION PHENOMENON

Many widely dispersed climatic extremes around the globe, such as severe droughts and floods and the failure of the Indian monsoon, can be attributed to the periodic warming of the sea surface in the central and eastern equatorial Pacific Ocean, termed the El Niño or Southern Oscillation (ENSO). In the past few decades many advances have been made in our understanding of ENSO. These have led to marked improvements in our ability to forecast its development months or seasons in advance, allowing practical prediction and adaptation to global impacts.

Edward Sarachik and Mark Cane have been key participants in advancing our understanding of ENSO from the modern beginning of its study. The book begins by introducing the basic concepts before moving on to more detailed theoretical treatments. Chapters on the structure and dynamics of the tropical ocean and tropical atmosphere ground the treatment of ENSO in a broader observational and theoretical context. The atmosphere and the oceans are given equal attention. Chapters on ENSO prediction, ENSO past and future, and ENSO impacts introduce the reader to the broader implications of the phenomenon.

This book provides an introduction to all aspects of this most important mode of global climate variability. It will be of great interest to research workers and students in climate science, oceanography and related fields, at all levels of technical sophistication.

EDWARD S. SARACHIK is Emeritus Professor of Atmospheric Sciences at the University of Washington. He has been an active contributor to research on tropical meteorology, tropical oceanography and coupled interactions in the tropics. He has also worked on other problems of oceanography, in particular the global thermohaline circulation. Dr. Sarachik has served on nineteen National Research Council committees, chairing two, in particular the Tropical Ocean Global Atmosphere (TOGA) Panel. He worked to found the International Research Institute for Climate and Society and has chaired IRI advisory committees since its inception. Until his retirement, he co-directed the Center for Science in the Earth System at the University of Washington, a center devoted to the dynamics of climate variability and change and the impacts of such variability and change on the ecology, built environment and people of the Pacific Northwest of the USA. He is a fellow of the American Meteorological Society, the American Geophysical Union, and the American Association for the Advancement of Science.

MARK A. CANE is G. Unger Vetlesen Professor of Earth and Climate Sciences and Director of the Master of Arts Program in Climate and Society at the Lamont Doherty Earth Observatory, Columbia University. With Lamont colleague Steven Zebiak, he devised the first numerical model able to simulate ENSO, and in 1985 this model was used to make the first physically based forecasts of ENSO. Since then the Zebiak–Cane model has been the primary tool used by many investigators to enhance understanding of ENSO. Dr. Cane continues to work on El Niño prediction and its impact on human activity, especially agriculture and health. His efforts over many years were instrumental in the creation of the International Research Institute for Climate and Society at Columbia. His current research interests include paleoclimate problems and the light they shed on future climate change. Dr. Cane has been honoured with the Sverdrup Gold Medal of the American Meteorological Society (1992), the Cody Award in Ocean Sciences from the Scripps Institution of Oceanography (2003), and the Norbert Gerbier–MUMM International Award from the World Meteorological Organization (2009). He is a Fellow of the American Meteorological Society, the American Association for the Advancement of Science, the American Geophysical Union and the American Academy of Arts and Sciences.

THE EL NIÑO–SOUTHERN OSCILLATION PHENOMENON

EDWARD S. SARACHIK

University of Washington, Seattle

MARK A. CANE

Columbia University, New York

CAMBRIDGE
UNIVERSITY PRESS

CAMBRIDGE UNIVERSITY PRESS
Cambridge, New York, Melbourne, Madrid, Cape Town, Singapore,
São Paulo, Delhi, Dubai, Tokyo

Cambridge University Press
The Edinburgh Building, Cambridge CB2 8RU, UK

Published in the United States of America by Cambridge University Press, New York

www.cambridge.org
Information on this title: www.cambridge.org/9780521847865

© Edward S. Sarachik and Mark A. Cane 2010

This publication is in copyright. Subject to statutory exception
and to the provisions of relevant collective licensing agreements,
no reproduction of any part may take place without the written
permission of Cambridge University Press.

First published 2010

Printed in the United Kingdom at the University Press, Cambridge

A catalog record for this publication is available from the British Library

ISBN 978-0-521-84786-5 Hardback

Cambridge University Press has no responsibility for the persistence or
accuracy of URLs for external or third-party Internet websites referred to in
this publication, and does not guarantee that any content on such websites is,
or will remain, accurate or appropriate.

Contents

Preface

This is a book about the set of coupled atmosphere–ocean phenomena known collectively as ENSO (El Niño–Southern Oscillation). While it will concentrate on what is known about ENSO, its mechanism, its effects, and how predictable it is, it will also touch on what is known about the paleohistory of ENSO and what we might expect in the future as mankind puts CO_2 and other radiatively active constituents into the atmosphere. The approach, while theoretical and sometimes necessarily mathematical, will concentrate on observations and on physical principles. Rigor will be acknowledged and appreciated but rarely practiced. When something in the text is stated to be known, but is not explained, the symbol ☼ (usually accompanied by a reference or footnote) will be used. This will be true of all chapters except the Preview (Chapter 1), where much will be arbitrarily stated, subsequently to be explained in the rest of the book.

Because ENSO is an intrinsically coupled ocean–atmosphere process, we will introduce the essentials of both the tropical atmosphere and ocean and explain the unique properties of each medium. Because ENSO is an intrinsically Pacific phenomenon, we will explain the unique aspects of the tropical Pacific and which of its features makes it particularly congenial for the existence of ENSO. We will describe those tropical atmospheric and oceanic mechanisms that ultimately help to explain the mechanism of ENSO. While there is no general agreement about what the ENSO mechanism is, we would expect that a similar book written a decade or so henceforth would contain much of the same material. In pursuit of the ENSO mechanism throughout this book, these themes will recur: the ability of warm sea-surface temperature to anchor regions of persistent precipitation; the ability of regions of persistent precipitation to induce surface westerly wind anomalies to the west of these regions; the tendency of anomalously warm sea-surface temperature anomalies in the Pacific to become warmer by local processes; and the tendency of cold sea-surface temperature anomalies to be associated with shallower thermoclines.

In order to draw the reader into the subject, the book will begin with a Preview which will touch lightly on all the subject matter in the remainder of the book. We recommend that all readers, regardless of sophistication, read the Preview in order to gain a feel for the method and content of the book and to devise a personal plan for reading the subsequent chapters. While not everything in the Preview is explained, the important topics are introduced and, where explanation is complex or requires the kind of mathematical treatment that will be established in a later chapter, a warning will be given that the matter cannot be understood without some additional work.

Each chapter will begin with a short precis which will indicate the broad outlines of the chapter. The book will conclude with a recap which will mirror, but not repeat, the content of the book. It is hoped that, in this way, the reader will be able to read the book in a manner suitable for his or her ability and needs. Essential mathematics will be relegated to the appendices. Some exercises will be interspersed in the chapters in order to give the reader some useful practice in deriving some basic results.

The aim of the authors is to produce a book that can be read on many levels by many audiences, depending on their interests and capabilities. Anyone reading the Preview, the chapter headings, and the final Postview chapter will get a very complete idea of what this book is about. We view our audience as scientists who are at least familiar with the nature of scientific explanation while perhaps not being familiar with the nitty-gritty of fluid mechanics, meteorology or oceanography. We expect that a second-year graduate student in meteorology or oceanography would have enough basic background to work through the entire book.

This book has two authors but many ancestors. Both authors owe a permanent debt to the prime inspiration for our careers in the geosciences, Jule Charney, and it is to his memory that this book is dedicated.

This book, and our approach to the material, arose from a series of lectures addressed to people of diverse backgrounds and abilities. The lecture series was given three times in Fortaleza, Brazil (thanks to the good offices of Antonio Divino Moura and Carlos Nobre, with the cooperation of the Centro de Previsão de Tempo e Estudas Climáticos [CPTEC] and Fundação Cearense de Meteorologiae Recursos Hídricos [FUNCEME]) and twice at the International Centre for Theoretical Physics in Trieste, Italy, with many thanks to J. Shukla and A. D. Moura for setting up the lectures and to Lisa Ianitti for the loving care with which she treated the students, the lecturers and the manuscript. Virginia DiBlasi typed an early version of the draft and provided essential technical support throughout, as well as much-appreciated moral support. Finally, we would like to thank the numerous colleagues and students who did so much to shape our ideas over the years in conversations, seminars and correspondence. Many of their names are scattered throughout this book. We do not educate easily, so we are especially grateful for their perseverance.

We are grateful to Tony Barnston, Mike Halpert, Emilia Jin, Alexey Kaplan, Billy Kessler, Todd Mitchell, Jenny Nakamura and Daiwei Wang for special efforts in providing figures for our use in the book.

For the hitherto thankless job of proofing the initial version of this book we would like to express our profound thanks to Hua Chen, Zhiming Kuang, Eugenia Kalnay and (especially) Peter Gent and Ed Schneider.

ESS was supported throughout the writing of this book by grants from the National Oceanic and Atmospheric Administration (NOAA) Climate Office to the Joint Institute for the Study of the Atmosphere and Ocean (JISAO) Center for Science in the Earth System at the University of Washington and owes special thanks to his Program Managers, Ming Ji and Chet Ropelewski, for their encouragement and forbearance in the (too) long writing of this book. This book was begun on sabbatical leave supported by the University of Washington.

MAC's contributions were supported by the Vetlesen Foundation, by the National Aeronautics and Space Administration (NASA), the National Science Foundation (NSF) and, most importantly, by NOAA's Office of Global Programs. Particular thanks to Mike Hall and Ken Mooney for their inspired and inspiring leadership in enabling so much of the science that forms the content of this book.

Abbreviations

AR4	Fourth Assessment Report of the IPCC
CGCM	coupled general-circulation model
CMZ	Cane–Münnich–Zebiak
CISK	conditional instability of the second kind
COADS	comprehensive ocean–atmosphere data set
CPTEC	Centro de Previsão de Tempo e Estudos Climáticos
ECHAM	European Centre–Hamburg
ECMWF	European Centre for Medium-Range Weather Forecasts
EEP	eastern equatorial Pacific
ENSO	El Niño–Southern Oscillation
EUC	equatorial undercurrent
FUNCEME	Fundação Cearense de Meteorologia e Recursos Hídricos
GCM	general-circulation model
GCOS	global climate observing system
GR	global residual
GTS	global telecommunication system
HadCM3	Hadley Centre coupled climate model
IPCC	Intergovernmental Panel on Climate Change
IPO	Interdecadal Pacific Oscillation
ITCZ	intertropical convergence zone
JISAO	Joint Institute for the Study of the Atmosphere and Ocean
KE	kinetic energy
KPP	K profile parameterization
LCL	lifting condensation level
LGM	last glacial maximum
MJO	Madden–Julian oscillation
MME	multi-model ensemble
MSU	microwave sounding unit

NAO	North Atlantic Oscillation
NASA	National Aeronautics and Space Administration
NCAR	National Center for Atmospheric Research
NCEP (USA)	National Centers for Environmental Prediction
NEC	north equatorial current
NECC	north equatorial counter current
NOAA	National Oceanographic and Atmospheric Administration
NSF	National Science Foundation
OLR	outgoing long-wave radiation
pdf	probability distribution function
PDO	Pacific Decadal Oscillation
PE	potential energy
Q-G	quasi-geostrophic
rms	root-mean-square
SEC	south equatorial current
SLE	St. Louis encephalitis
SLP	sea-level pressure
SO	Southern Oscillation
SODA	simple ocean data analysis
SOI	Southern Oscillation index
SPCZ	South Pacific convergence zone
SS	Schopf and Suarez (and vice versa)
SST	sea-surface temperature
SSTA	sea-surface temperature anomaly
SVD	singular value decomposition
SWE	shallow-water equation
TAO	tropical atmosphere–ocean
TI	trade inversion
TKE	turbulent kinetic energy
TOGA	Tropical Ocean Global Atmosphere
ZC	Zebiak–Cane

1

Preview

This chapter serves as an introduction and preview for the entire book. Topics will be broadly introduced, to be better and more completely explained in the sequel.

1.1 The maritime tropics

It may surprise people living in the midlatitudes that the tropics have such an overwhelming role in the climate of the Earth. Yet it has been shown time and time again that the maritime tropics are the only regions on Earth where changes in the surface-boundary condition, especially sea-surface temperature (SST), have a demonstrable and robust causal correlation with weather effects in midlatitudes. This happens through the ability of warm sea-surface temperature anomalies (deviations of sea-surface temperature from its normal value for that time of year) to organize deep cumulonimbus convection and plentiful rainfall which can then emit large-scale planetary waves which subsequently travel to higher latitudes. The changes of SST, the formation of regions of persistent precipitation, and the resulting forcing of the midlatitude motions by these regions of persistent precipitation, form a set of themes that appear and recur throughout this book.

It is a good rule of thumb (these rules of thumb will be examined in much greater detail in the body of the book), in the tropical Pacific in particular, that regions of persistent precipitation lie over the warmest water, and a good rule of thumb that in the presence of persistent precipitation, the net synoptic motion is upward and the sea-level pressure low. With these rules of thumb, we are in a position to describe the normal conditions over the tropical Pacific, the main region of interest in this book.

1.2 The normal tropical Pacific

The tropical Pacific extends from the coast of South America in the eastern Pacific to the various islands and land masses of Australia and Indonesia that form the

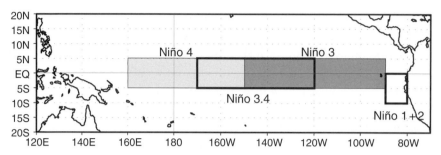

Figure 1.1. The tropical Pacific, including the definition of the four Niño regions. (Courtesy of the NOAA Climate Prediction Center.)

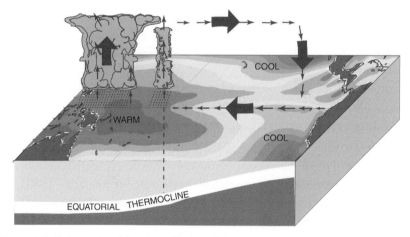

Figure 1.2. Schematic of the normal state of the coupled atmosphere–ocean system in the tropical Pacific during boreal winter. The shading on the surface of the ocean represents sea-surface temperature, warm in the west and cooler to the east and south-west. (Courtesy of the NOAA Climate Prediction Center.)

so-called maritime continent; a somewhat paradoxical idea expressing a collection of land masses without there actually being a land continent present (Figure 1.1). In particular, the equator runs from Ecuador in the east (at 80°W) to Indonesia in the west – the first land the equator crosses in the western Pacific is Halmahera at 129°E and then the more substantial Sulewesi at 120°E. Taking Halmahera as the western boundary of the Pacific gives a total length on the equator of 151 degrees of longitude or 16 778 km, more than one-third of the total distance around the globe.

The climatic state in and over the tropical Pacific is given by a convenient cartoon (Figure 1.2). The surface of the western Pacific is warm and the atmosphere above it is rainy, with the rain coming from deep cumulonimbus clouds. The air rises in the region of the warm water and the rising air is characterized by low pressure at the surface. The winds across the surface of the tropical Pacific blow westward into the region of low pressure, consistent with the westward trade winds. The rising motion in the warm region reaches the tropopause and returns eastward aloft and completes the circuit by descending in the eastern Pacific, leading to higher pressure at the surface.

This tropical Pacific-wide circuit of air proceeding westward at the surface, rising over the (warm) region of persistent precipitation, returning eastward aloft, and descending over the cool eastern Pacific, is called the Walker circulation. Associated with the Walker circulation is the low surface pressure in the western Pacific and the high surface pressure in the east. A measure of the strength of the Walker circulation is the difference in the surface pressure between the east and west – this difference is conventionally called the Southern Oscillation Index (SOI) – we will see below the oscillation to which it refers. When the Walker circulation is strong, the pressure in the west is low and the pressure in the east is high – the SOI is then less negative. When the Walker circulation is weak, the SOI is more negative.

The oceanic part of Figure 1.2 is driven by the westward surface winds; the surface expression of the Walker circulation in the atmosphere. The feature in the ocean called the thermocline is a near-ubiquitous property of the oceans. In the tropics it is a region of such sharp temperature change in the vertical that one may vertically divide the ocean into only two regions, one with warm temperatures and one where the temperatures are cold. The thermocline demarcates the warm-water sphere near the surface from the cold-water sphere below. We will show later that the deeper thermocline in the western Pacific is caused by the westward winds at the surface of the ocean. Thus the stronger the westward surface winds (due to a strong Walker circulation), the deeper the thermocline in the west and the shallower the thermocline in the east. The tilt of the thermocline in the ocean is a measure of the strength of the westward surface winds and, therefore, another measure of the strength of the Walker circulation. The chain of reasoning is continued by noting that the East–West temperature difference, which may be considered to drive the atmospheric motion, is indeed unexpected, since the sun shines equally on the western and eastern Pacific.

The mechanism responsible for the mean East–West sea-surface temperature difference involves both the atmosphere and the ocean. The mean westward surface winds drive ocean motion poleward in both the northern and southern hemispheres within 50 or so meters of the surface very near the equator. Water moving poleward must be replaced by water upwelling on the equator from below. In the eastern Pacific, the thermocline (recall that the thermocline is the demarcation between warm and cold water) is shallower than 50 m, so that cold water is upwelled on the equator, causing the SST to be cold. In the western Pacific, the thermocline lies below 50 m and, while upwelling still occurs, it simply brings up warm water from above the thermocline, allowing the western Pacific SST to remain warm. Heat put into the ocean from the atmosphere counteracts the upwelling influence on SST but it does not win the contest: the eastern Pacific remains cooler than the west.

The cold SST in the eastern Pacific is spread poleward several degrees of latitude by the ocean motions until it encounters another warm region in the northern (but not southern) hemisphere caused by an eastward ocean current. There is again rising motion in the atmosphere above this warm water and a line of deep convection extends pretty much across the entire Pacific at an average latitude of about 6°N. This region of deep convection is called the intertropical convergence zone (ITCZ) and forms the rising tropical branch of a North–South circulation called the Hadley circulation.

Though not mentioned thus far, there is a pronounced seasonal cycle in the tropical Pacific. Unlike midlatitudes, the seasonal extremes are in March–April when the eastern equatorial Pacific is warmest and the ITCZ is closest to the equator, and September–October when the eastern SST is coldest and the ITCZ is furthest north. Since the SSTs in the western Pacific vary only by about 1 °C, the seasonal variations in the East–West gradient co-vary with the eastern Pacific SSTs: weakest in boreal spring, strongest in fall. This annual cycle has a strong influence on the evolution of ENSO phases, which exhibit a marked tendency to be phase-locked to the annual cycle, growing through the (northern) summer and fall to reach a winter peak.

1.3 The phases of ENSO

Superimposed on the normal state of the tropical Pacific is an irregular cycle of warming and cooling of the eastern Pacific with attendant atmospheric and oceanic effects, the panoply of which will be referred to as ENSO. Figure 1.3 shows conditions in and over the tropical Pacific during warm phases of ENSO.

The eastern Pacific warms, and can warm to such an extent that the temperature across the entire tropical Pacific becomes almost uniform. That the temperature reached is that of the western Pacific, rather than that of the eastern Pacific, indicates that the warm phase of ENSO is due to a failure of the eastern Pacific to stay cold. Consistent with this point of view is the relaxation of the westward surface winds (Figure 1.3 shows weaker than normal easterly winds which implies westerly anomalous winds), which produces less upwelling and therefore less cooling. Consistent with weaker westward winds, the thermocline is not as tilted, and any upwelling in the eastern Pacific would bring warmer water to the surface. When the cooling in the eastern Pacific is totally gone and the westward surface winds relaxed to almost zero, the warm phase of ENSO is as strong as it can be, and the temperature over the entire tropical Pacific is uniform and assumes the

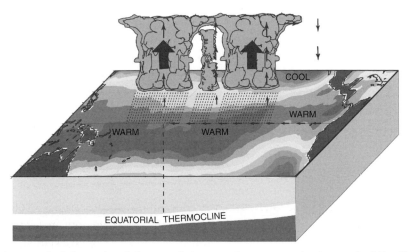

Figure 1.3. Schematic of the coupled atmosphere–ocean in the tropical Pacific during the peak of a warm phase of ENSO during boreal winter. (Courtesy of the NOAA Climate Prediction Center.)

approximate temperature of the western Pacific. This happened in the strong warm phases of ENSO during 1982–3 and 1997–8.

As the eastern Pacific becomes less cold, the region of persistent precipitation that lies over the warmest water expands eastward into the central Pacific. The normally high sea-level pressure (SLP) of the eastern Pacific becomes lower and the sea-level pressure difference between the western and eastern Pacific decreases. Consistent with this decrease is the weakening of the Walker circulation and the relaxation of the normally westward surface winds. As the central and eastern tropical Pacific becomes warm, the ITCZ moves onto the equator and the line of deep convection assumes its southernmost position and the Hadley circulation becomes stronger.

The effect of the warming of the eastern Pacific, and the consequent eastward movement of the region of persistent precipitation, is felt throughout the world (Figure 1.4). In the tropics, the normally rainy western Pacific becomes drier as the region of persistent precipitation moves eastward into the central Pacific. Droughts in Indonesia and in eastern Australia become far more common during the warm phases of ENSO. Rainfall in the normally arid coastal plains of Peru becomes far more likely and warm water spreads north and south along the western coasts of the North and South American continents. The temperature and rainfall in other selected areas of the world (e.g. Zimbabwe, Madagascar) are similarly affected, even though the reasons are either difficult to explain or unknown.

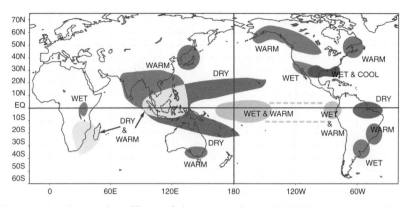

Figure 1.4a. Composite effects of the warm phase of ENSO on global climate during boreal winter. (Courtesy of the NOAA Climate Prediction Center.)

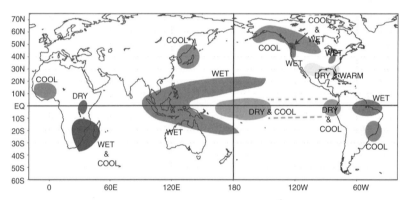

Figure 1.4b. Composite effects of the cold phase of ENSO on global climate during boreal winter. (Courtesy of the NOAA Climate Prediction Center.)

During cold phases of ENSO, the normal cooling of the eastern Pacific becomes even stronger, the surface pressure difference between the eastern Pacific and western Pacific becomes stronger, and the Walker circulation, in general, becomes stronger. Consistent with this, the surface westward winds become stronger, the tilt of the thermocline becomes greater, the stronger westward winds in the eastern Pacific produce even more upwelling and, because the thermocline is closer to the surface, the water upwelled is colder. The regions of warmest water in the western Pacific contract westward under the encroachment of cold water in the east and, with the warm water, the region of persistent precipitation contracts westward onto the maritime continent. Excess rainfall in Indonesia and western Australia becomes far more common during cold phases of ENSO (Figure 1.5).

The SST anomalies (the deviations from the norm) look, in many ways, the obverse of each other (Figure 1.6).

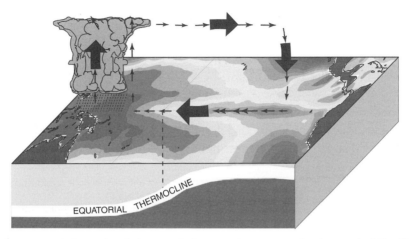

Figure 1.5. Schematic of the coupled atmosphere–ocean in the tropical Pacific during the peak of a cold phase of ENSO during boreal winter. (Courtesy of the NOAA Climate Prediction Center.)

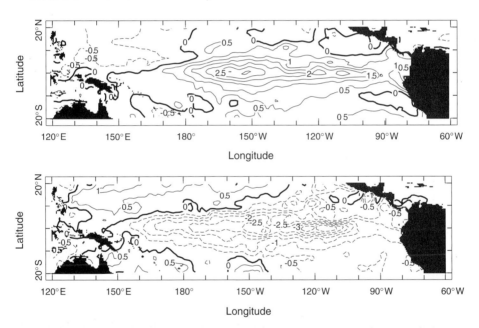

Figure 1.6. Upper panel: SST anomalies for the warm phase of ENSO during December 1991. Lower panel: SST anomalies for the cold phase of ENSO during December 1988. (Downloaded and plotted from www.iridl.ldeo.columbia.edu/ using the Reynolds *et al.*, 2002, updated SST data set.)

It must be kept in mind, however, that in many ways, cold and warm phases are fundamentally different because the quantities that affect the remote atmosphere are not the SST anomalies, but rather the mean location of the regions of persistent precipitation. In the warm phase of ENSO, persistent precipitation extends into the central Pacific, while during the cold phases of ENSO it retreats to the far western Pacific. The SST anomalies can be the inverse of each other but the mean location of the heat source, which drives the response in the low and midlatitudes, is very different. Because the rest of the world is forced by these regions of persistent precipitation and because these regions are in different locations for warm and cold phases of ENSO, there is no expectation that the global effects will be the opposite of each other. Figure 1.4b shows that during cold phases of ENSO, there are some similarities and significant differences in the global response.

1.4 Evolution of phases of ENSO

While we will go into greater detail in later chapters, we simply note here that the phases of ENSO evolve differently each time they appear. The general recurrence time for warm and cold phases is around 4 years, with large variations around this mean. The literature often speaks of an "ENSO band" from 2 to 7 years.

One way of describing ENSO evolution with time is to examine the SST anomalies in various regions of the Pacific defined by Figure 1.1. Figure 1.7 shows the SST anomalies since 1980 in the various regions of the Pacific, and a measure of the strength of the Walker circulation, the SOI. We can infer a number of important properties of the warm and cold phases of ENSO by examining this Figure. First we see that the major phases of ENSO tend to have expression all the

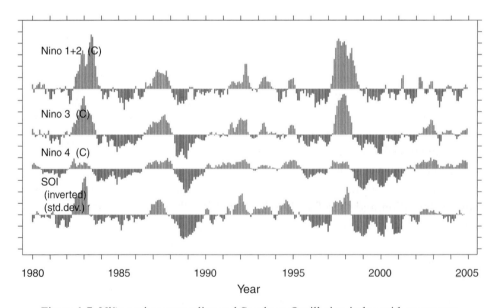

Figure 1.7. Niño region anomalies and Southern Oscillation index with respect to the respective means of 1985–94. (Courtesy of Todd Mitchell. Extended version of Plate 2 of Wallace *et al.*, 1998.)

way across the Pacific, from the coast of South America (Nino $1+2$) to the western Pacific (Nino 4). Second, that the major warm and cold phases tend to set on across the entire Pacific at about the same time. Third, that the larger events seem to start around summer, peak near the end of the year, and end before the next summer, so that the length of warm and cold phases is about a year. Fourth, that there are stretches of time in which not much is happening in the tropical Pacific (the entire 1930s were noted for having no major warm or cold phases – see Figure 1.17) and that these times are punctuated by the appearance of large phases of ENSO. It is worth mentioning that the warm phases in 1982–3 and 1997–8 were the largest of the century.

1.5 Physical ENSO processes

According to what we have seen so far, we need to understand how the SST anomalies characteristic of ENSO are produced, and how the connections of SST with sea-level pressure, precipitation, surface winds, the depth of the thermocline and remote precipitation and temperature are accomplished. Once we have a firm idea of the operation of each of these processes, we will need to know how they fit together to produce ENSO.

1.5.1 The processes that change SST

The temperature of ocean water can change either by directly adding heat (for example, from the sun) or by mixing with water of a different temperature. Because the ocean has no significant internal heat sources, heat can only be added directly at the surface. Heat added at the surface is basically a balance between radiation and evaporation: any net radiation reaching the surface that does not evaporate water is available to cross the ocean surface and heat the ocean water. In general, when water cools, evaporation decreases, and when water warms, evaporation increases. To the extent that the solar radiation reaching the surface is independent of the temperature of the underlying ocean (not entirely true, since the overlying cloudiness can change), warm surface water will have more evaporation and therefore less heat entering the ocean across the surface. Similarly, cooler water will have less evaporation and therefore more heat entering the ocean. Clearly, therefore, the heat entering the surface of the ocean tends to *oppose* the temperature changes.

If we consider some water near the ocean surface, the temperature can change if it is heated by heat entering the ocean through the surface, if it mixes with warm or cold water entering from the sides, or if it is cooled by water entering from below (Figure 1.8).

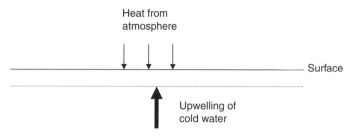

Figure 1.8. Schematic of vertical heat inputs into the tropical ocean mixed layer.

In the eastern Pacific, water at the surface is constantly cooled by water upwelled from below the thermocline and this cooling is opposed by heat entering through the surface. In the western Pacific, the temperature of the water is determined by the interactions with the atmosphere. It is approximately in equilibrium with the atmosphere, and is neither cooled from below nor heated from the atmosphere above. If the upwelling in the east were to decrease to a new, but smaller, steady value, not as much cold water would be brought up from below and the heat entering from the surface would warm the water until it reached a new not as cool temperature – this would be a warm SST anomaly. The evaporation would increase, the heat entering though the surface would decrease, and the water near the surface would reach a new warmer equilibrium; cooled not as much from below and heated not as much from above. The water could also warm if it mixed with warmer water from the west or perhaps from the north. In either case, warm SST anomalies would be associated with more evaporation and therefore less heating of the ocean surface from above.

1.5.2 *The process by which warm SST anchors regions of persistent precipitation*

Warm regions tend to have lighter air above these regions, as the air is warmed by the surface. Warm air is light and since surface pressure is the total weight of the air above it, the surface pressure tends to be lower above warm tropical regions. Air from surrounding higher pressure regions rushes in and is warmed, moistened and raised. Rising air condenses and the heat of condensation raises the air further. If the underlying SST is warm enough, about 28 °C or 29 °C, the clouds can reach to the top of the atmosphere (the tropopause) and regions of deep cumulonimbus convection result. The average amount of rain that falls is equal to the local evaporation plus the amount of moisture that converges into the region. Moisture is confined mostly to the lowest 1 or 2 kilometers of the atmosphere, so it is the low-level moist air that converges into the region that provides the additional moisture for the rainfall. The overall picture may be sketched as in Figure 1.9.

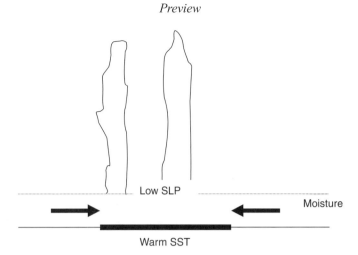

Figure 1.9. Schematic of the convergence of moisture into deep cumulonimbus clouds over regions of warm SST.

1.5.3 *The processes by which regions of persistent precipitation produce surface winds*

It is not an easy matter to understand how regions of persistent precipitation force surface winds, and a full explanation can only come from the deeper considerations in Chapter 5. The problem, however, can be stated relatively straightforwardly.

A region of persistent precipitation is one in which deep cumulonimbus clouds constantly rain in a given area and, therefore, constantly condense heat into the atmosphere: because so much heat is being released into the atmosphere in these regions, regions of persistent precipitation are said to "thermally force" the atmosphere.

Clouds have their base at about 600 m above the ocean surface in the tropics so that any thermal forcing by the cumulonimbus convection occurs only above the cloud base. The problem then is to get the region of thermal forcing to transmit its forcing down below the cloud base to the surface.

There is an alternate mechanism that seems also to affect the winds in the maritime tropics. As we pointed out in the previous section, warm SST tends to have lighter air above it, and cold SST heavier air. The subsequent pressure gradients can drive surface winds into the warm region and these will be in roughly the same direction as those forced by the cumulonimbus convection.

1.5.4 *The processes by which surface winds change thermocline depth*

Since SST changes in the tropical Pacific are due primarily to changes in upwelling of cold water from below, and since the efficacy of this upwelling depends on the

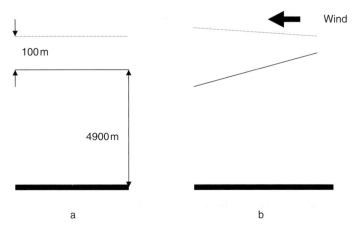

Figure 1.10. Schematic of effect of wind stress on thermocline depth and sea-level height. a) The upper 100 m of ocean is bounded below by the thermocline and above by the sea surface. Without winds, both are flat. b) Response to an easterly wind-stress anomaly. The sea-level height tilts up to the west while the thermocline deepens to the west.

location of the thermocline (the deeper the thermocline, the further from the surface is the cold water), we have to be able to find the depth of the thermocline and how it changes.

Let us assume that the processes that determine the average depth of the thermocline occur on long timescales and, from the point of view of ENSO dynamics, may be considered given. In the absence of any equatorial winds, the depth of the thermocline would be about 100 m and independent of longitude (Figure 1.10a) – below the thermocline is the deep ocean.

In the presence of a westward wind (Figure 1.10b), water is moved westward on the equator and piles up against the western boundary until the westward force (stress) exerted by the wind is balanced by the eastward force exerted by the higher pressure due to the greater weight of the water in the west. If the wind was suddenly removed, the water would flow eastward until the conditions of Figure 1.10a were restored. Across the entire Pacific, the sea level is approximately 40 cm higher in the west than in the east.

When the water piles up in the west, the thermocline moves down in such a way that the total weight of water down to the bottom does not change (if it did change, there would be unbalanced forces in the deep ocean). The water above the thermocline is lighter than the water below by a small amount, so that a large amount of light water is needed to balance the small amount of sea-level rise – the thickness of light water above the thermocline must therefore increase, and the thermocline must descend. Similarly, if the winds were from the west, the sea level would rise and the thermocline would descend in the east.

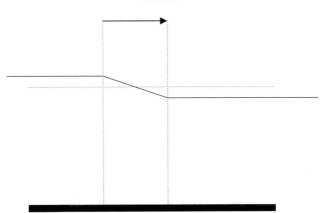

Figure 1.11. Schematic of the response of the thermocline to a westerly wind anomaly (arrow) of limited zonal extent. The thermocline deepens not only at the eastern end of the wind patch, but also everywhere to the east of the wind patch. Similarly, the thermocline rises not only at the western end of the wind patch, but everywhere to the west of the wind patch.

Imagine now that a finite region of *eastward* wind anomaly (i.e. a westerly wind patch) blows over the surface of the equatorial ocean. Superimposed on whatever else is happening would be the picture in Figure 1.11, where only the deviation of the thermocline (and not the small deviation of sea level) is shown.

The Figure shows the final steady stage of the thermocline – it has deepened not only in the east of the region of the winds, but everywhere to the east, and has risen, not only in the western part of the region of the winds, but everywhere to the west. It does this through a time-dependent process of the *adjustment* of the thermocline to the winds. This adjustment takes place thorough a signaling process in which the signals have properties of equatorial waves; in particular, Kelvin waves traveling to the east and Rossby waves traveling to the west. In the presence of real boundaries to the east and west of the wind patch, the signals are reflected and work their way back into the basin.

1.5.5 *The processes by which regions of persistent precipitation affect regions remote from the tropical Pacific*

The remote effects of ENSO arise from the movement of the regions of persistent precipitation, and the subsequent thermal forcing of the atmosphere by the latent heat released in the process of cumulus condensation and precipitation. The air rising at upper levels of the atmosphere eventually diverges and, according to one

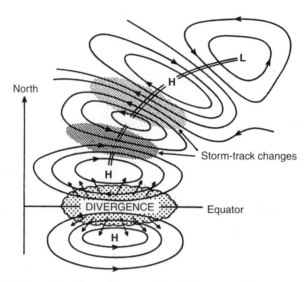

Figure 1.12. Schematic of how the upper-level divergence in regions of persistent precipitation (i.e. in thermal sources) forces a poleward progression of planetary waves at upper levels which subsequently moves the storm tracks. (From Trenberth *et al.*, 1998.)

way of looking at the problem (Figure 1.12), the divergence region acts as a source of planetary waves at the upper levels of the atmosphere that propagate into the mid-latitudes as a series of cyclonic (L in Figure 1.12) and anticyclonic (H) features. Because the high- and low-pressure areas (H and L in Figure 1.12) move the jet streams, the storm paths are moved (Figure 1.13). During warm phases of ENSO, thermal forcing puts a low-pressure area in the Gulf of Alaska. Air blows counterclockwise around the low and brings warm air into the Pacific Northwest. The low-pressure area also moves the storm track southward and brings excess rain to California and Baja California and leaves the Pacific Northwest relatively dry. During cold phases of ENSO, the mechanism of Figure 1.12 produces a high-pressure region in the Gulf of Alaska, cold clockwise flow into the Pacific Northwest, and a northward displacement of the storm track bringing excess precipitation into the Pacific Northwest, while leaving California and Baja California relatively dry.

While Figure 1.12 shows the generation of planetary highs and lows at upper levels only, a more complete theory of generation of planetary waves would show that thermal forcing in the tropics by cumulonimbus convection throughout the atmosphere (not only at upper levels) creates planetary motions that propagate to higher latitudes. Only those motions that are relatively independent of height reach high latitudes and the results of Figure 1.13 then follow.

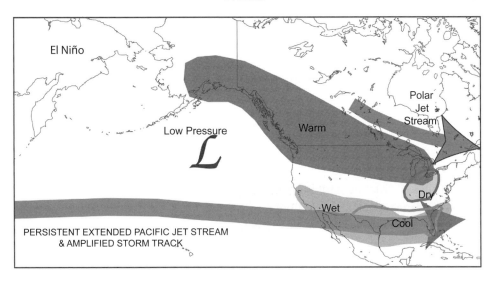

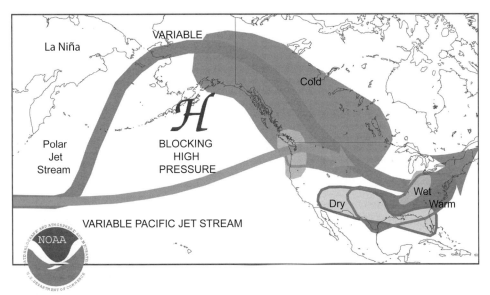

Climate Prediction Center/NCEP/ NWS

Figure 1.13. Effects of warm (Upper panel) and cold (Lower panel) phases of ENSO on blocking and storm tracks in the north-east Pacific. (Courtesy of the NOAA Climate Prediction Center.)

1.6 Modeling ENSO

If the processes introduced in Section 1.5 were complete and accurately portrayed in a coupled atmosphere–ocean model, we would expect that ENSO would be the natural result of the coupling of the atmosphere and the ocean.

The first coupled atmosphere–ocean model of ENSO (the Zebiak–Cane model) took care to represent: the regions of persistent precipitation over the warmest water; the westerly surface winds to the west of the regions of persistent precipitation; the processes that change SST in the surface layer of the ocean; and the correct effects of the winds on the thermocline depth. What the model did not calculate, but rather specified, was the correct annual cycle in the Pacific and it simply calculated the anomalies with respect to this annual cycle. The model produced a recognizable version of the ENSO phenomenon in agreement with reality in important ways.

The spatial structure and amplitude of the warm and cold phases have a good correspondence with nature (see Figure 1.15) and occur irregularly with an average period of about 4 years (Figure 1.14). As in the observations, there are long periods where not much happens, and periods when the events seem to occur relatively regularly. Both warm and cold phases have the correct tendency to peak near the end of the calendar year, and the amplitude of the warm events is greater than that of the cold events. Among the discrepancies with nature, the model events tend to last too long, and set on over the entire eastern Pacific simultaneously more consistently than nature's version.

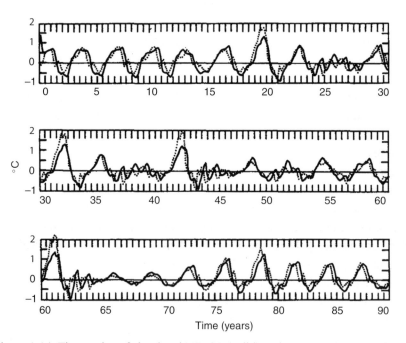

Figure 1.14. Time series of simulated NINO3 (solid) and NINO4 (dotted) indices. (From Zebiak and Cane, 1987.)

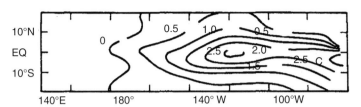

Figure 1.15. December SST anomaly at peak of simulated warm phase of ENSO. (From Zebiak and Cane, 1987.)

1.7 Observing and predicting ENSO

Why would we even suppose that the phases of ENSO could be predicted? There would have to be something of a long timescale that carries the information from the time we started the forecast to the verification time (the time at which we compared the prediction with the actual state of the system). Even if we did not know what this something was (it has to be something in the ocean, since atmospheric timescales are quite short – of the order of a few days to weeks), we would see that the evolution of ENSO phases takes place slowly, the phases beginning around summer, growing to reach a maximum toward the end of the year and decaying into the spring of the next year. If we could, therefore, recognize the characteristic features of SST anomaly growth before the summer, we would be able to make a prediction for the following winter. To do this, we use models – but this requires that we know what the current state of the ocean is. Fortunately, we have an observing system that was designed to tell us just this.

The observing system has combinations of instruments to measure the thermal state of the upper ocean throughout the tropics in the Pacific (in particular the SST and the depth of the thermocline), to measure the surface winds, and to estimate the heat fluxes. The data is telemetered once a day by satellite and made available to everyone.

There are numerous schemes in use for predicting ENSO, many of which have roughly the same skill. (More precisely, in view of the short record used for most forecasts, one cannot say that one is significantly better than another with a high level of confidence.) They can be divided into two classes: statistical methods relying on empirically determined relationships between states in the future and states in the past; and dynamical methods, using numerical models that incorporate equations describing physical laws for the ocean, atmosphere and their interaction. There are also hybrid methods with statistical add-ons to a dynamical model. Our understanding of ENSO puts the long memory of the system in the distribution of upper-ocean heat content, or equivalently, the displacement of the thermocline. This does not mean that all prediction schemes must make explicit use of this field, and in

TOGA in situ ocean-observing system

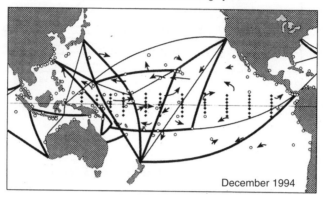

December 1994

Figure 1.16. The ENSO observing system established by the TOGA program at the end of TOGA in December 1994. Solid diamonds are bottom-moored buoys taking both upper-ocean measurements and surface meteorological measurements. Open circles are tide gauge stations, solid lines are ship tracks on which traditional meteorological measurements and some shipboard ocean measurements are taken, and arrows are drifting buoys. (From McPhaden *et al.*, 1998.)

fact most statistical schemes rely solely on SST, the variable with the longest set of reliable observations, and therefore the longest history to use in training an empirical model. There is no contradiction here; if the SST field encodes the essential thermocline information in some way, explicit use of the latter is not necessary.

The dynamical approach casts prediction as what is mathematically an initial-value problem. The model starts from an initial state at a time t_0 and is integrated forward into the future, simulating the evolution of nature. If the model flaws are not too incapacitating, and the initial model state bears sufficient resemblance to the state of nature at time t_0, then that simulation may be realistic enough to yield a good prediction of what nature will do. We say "may" because the climate system is chaotic, which means that its evolution into the future is highly sensitive to the initial state. Since we cannot know all the variables that comprise this state perfectly at all places, some uncertainty in the starting point of our forecast is unavoidable, and hence we cannot be sure which among a range of possible futures will be the one that actually occurs.

We face up to this intrinsic limitation in our ability to forecast the future by running an ensemble of forecasts, each with different plausible initial conditions, which results in a distribution of possible future states. We might, for example, initialize our model with the state of the atmosphere and ocean on successive days. The ocean changes very little over the course of a few weeks, but the atmospheric states (which might be taken from the sophisticated analyses of daily initial states used at a major weather-prediction center) change quite considerably. Any of these daily states is an equally plausible initial-forecast state, but the model forecasts of ENSO a few seasons hence could be quite different. The ensemble of predictions from all these states gives a mean forecast and a range of possibilities: this is the best we can do. Sometimes this range is narrow and the forecast is rather definite; sometimes it is broad and the forecast is highly uncertain. In either case, it could be wrong. Our coupled ocean–atmosphere models are, at present, seriously flawed, and our procedures for creating initial states by combining all available observational data with fields from the model to create a complete best estimate of the initial state for a coupled ocean–atmosphere model – a process known as data assimilation – are still quite primitive in comparison with the comparable state-of-the-art in weather prediction.

Suffice it to say that forecast procedures are in regular operation in a number of places throughout the world and the forecasts are proving to have skill several seasons in advance. The forecasts have some skill but are not perfect. How to use the results of forecasts that have uncertainty is a subject in itself.

1.8 Towards a theory of ENSO

The most widely accepted explanation for ENSO is built upon Jacob Bjerknes' (1969) masterpiece of physical reasoning from observational data. Bjerknes marked the peculiar character of the "normal" equatorial Pacific we noted above: although the equatorial oceans all receive about the same solar insolation, the Pacific is 4–10 °C colder in the east than in the west (see Figure 1.2). The east is cold because of equatorial upwelling, the raising of the thermocline exposing colder waters, and the transport of cold water from the south Pacific. All of these are dynamical features driven by the easterly trade winds. But the winds are due, in part, to the temperature contrast in the ocean, which results in higher atmospheric sea-level pressures in the east than the west. The surface air flows down this gradient. Thus the state of the tropical Pacific is maintained by a coupled positive feedback: colder temperatures in the east drive stronger easterlies which, in turn, drive greater upwelling, pull the thermocline up more strongly, and transport cold waters faster, making the temperatures colder still. Bjerknes, writing in the heyday of atomic energy, referred to it as a "chain reaction." We now prefer "positive feedback" or "instability."

Bjerknes went on to explain the warm El Niño state with the same mechanism. Suppose the east starts to warm; for example, because the thermocline is depressed. Then the East–West SST contrast is reduced, so the pressure gradient and the winds weaken. The weaker winds bring weaker upwelling, a sinking thermocline and slower transports of cold water. The positive feedback between ocean and atmosphere is operating in the opposite sense (see Figure 1.3). Note that this explanation locks together the eastern Pacific SST and the pressure gradient – the Southern Oscillation – into a single mode of the ocean–atmosphere system, ENSO.

Bjerknes' mechanism explains why the system has two favored states, but not why it oscillates between them. That part of the story relies on the understanding of equatorial ocean dynamics that developed in the two decades since he wrote. The key variable is the depth of the thermocline, or, equivalently, the amount of warm water above the thermocline. The depth changes in this warm layer associated with ENSO are much too large to be due to exchanges of heat with the atmosphere; they are a consequence of wind-driven ocean dynamics. While the wind and SST changes in the ENSO cycle are tightly locked together, the sluggish thermocline changes are not in phase with the winds driving them. Every oscillation must contain some element that is not perfectly in phase with the other, and for ENSO it is the tropical thermocline. In particular, it is the mean depth of the thermocline – equivalently, the heat content – in the equatorial region. The most widely accepted account of the underlying dynamics emphasizes wave propagation and is referred to as the "delayed oscillator." Some authors regard the recharge–discharge of the equatorial ocean heat content as the essence of the oscillation. Others emphasize the role of ocean–atmosphere interactions in the western Pacific. One point of view (that of the authors) is that these are different descriptions of what is the same essential physics.

There are two elements in this story: the coupled Bjerknes feedback and the (linear) ocean dynamics, which introduces the out-of-phase element required to make an oscillator. If the coupling is very strong, then the direct link from westerly wind anomaly to deeper eastern thermocline to warmer SST and back to increased westerly anomaly would build too quickly for the out-of-phase signals to ever catch up. There would be no oscillation. If this coupling strength is not quite so strong, then oscillations become possible, as the delayed signal can now catch up and overtake the directly forced component. If the coupling strength is not strong enough, then there can be no oscillations because an initial small disturbance is no longer reinforced and will die out. However, oscillations in this weaker system could be sustained if we add some forcing. This forcing need not be very organized; it could be "weather noise." As one crosses a threshold from self-sustained oscillations to noise-driven oscillations, the characteristics of the oscillations do not change very much; in fact, we are not sure in which regime the real world lies.

The "coupling strength" is determined by a host of physical factors. Among the most important are: how strong the mean wind is, which influences how much wind stress is realized from a wind anomaly; how much atmospheric heating is generated by a given SST change, which will depend on mean atmospheric temperature and humidity; and how sharp and deep the climatological thermocline is, which together determine how big a change in the temperature of upwelled water is realized from a given wind-driven change in the thermocline depth.

In simple linear analyses the ENSO period is determined more by the coupling strength than the time for waves to travel back and forth across the Pacific. In more realistic nonlinear models this general statement still holds, but in contrast to the linear case the periods tend to stay within the 2–7 year band. There is no satisfactory theory explaining why this is so, or more generally, what sets the average period of the ENSO cycle. There is broad disagreement as to why the cycle is irregular; some attribute it to low-order chaotic dynamics, some to noise – weather systems and intraseasonal oscillations – shaking what is essentially a linear, damped system.

It might seem that this distinction is important for the predictability of ENSO, but this is true only in a very limited sense. At present, our predictions are limited by inadequacies in models and data more than limits to predictability intrinsic to the system. The real-world ENSO incorporates a combination of nonlinear effects, climate system noise, and variations in forcing due to, for example, volcanic eruptions and variations in solar radiation.

1.9 The past and future of ENSO

Knowledge of the past history of ENSO will, we expect, lead to an understanding of the mechanisms that led to past changes of ENSO. Knowing these mechanisms might give us some insight into the future of ENSO.

We saw that the ENSO cycle has proceeded, in its irregular manner, for at least the last 25 years (Figure 1.7). Longer instrumental records, though less complete, clearly show that from – at least – the mid nineteenth century, ENSO has had the same character (Figure 1.17).

To extend the record still further back in time, when no instrumental records exist, requires finding proxies that respond to temperature in a consistent way. This is a currently active field of research, and will be taken up in Chapter 9. There is good evidence that ENSO has been a feature of the Earth's climate at least as far back as the last interglacial period (approximately 130 000 years ago). There is some evidence that the ENSO cycle was weaker during the glacial period (before the current Holocene which started 10 000 years ago). It is possible that the weakness of ENSO during the glacial period can be traced to the same mechanisms that produced the glacials themselves, namely the very slow changes in the Earth's orbit, but it may be that the key thing is that the overall colder climate weakened the ocean–atmosphere coupling. For one thing, colder temperatures would mean less evaporation and so weaker heating of the atmosphere for the same wind convergence. We do know that the ENSO cycle was weaker than today for the first 5000 years of the Holocene, and that has been shown to be a consequence of the different phase of the Earth's precession cycle.

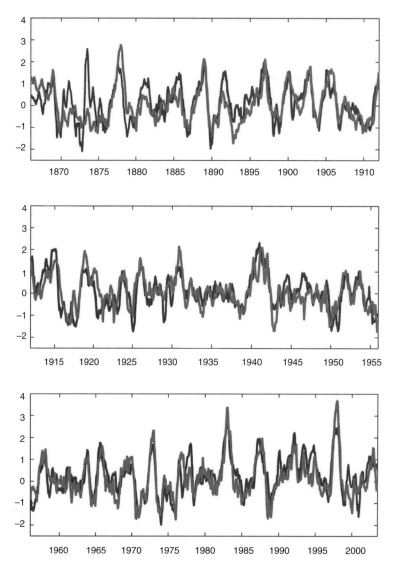

Figure 1.17. Anomalies (relative to the mean from 1865 to 2005) of SST anomalies (gray curve) in the NINO3 area (cf. Figure 1.1) measured in °C and of sea-level pressure (black curve) at Darwin, Australia (measured in hPa). The former is a measure of El Niño and the latter is a measure of the Southern Oscillation. The two are obviously intimately related; both components of a single phenomenon: ENSO. (Courtesy of Alexey Kaplan.)

Unfortunately, models (i.e. the ones in the Fourth Assessment Report of the Intergovernmental Panel on Climate Change; IPCC, 2007) do not agree on how ENSO will change in the future. Arguments have been given that global warming (due to the accumulation of radiatively active gases in the atmosphere) will either increase or decrease the amplitude of the ENSO cycle. Since ENSO has potent effects on temperature and precipitation throughout the world (Figure 1.4), a basic part of the world's future climate cannot currently be predicted with any confidence.

1.10 What is ENSO information good for?

We all instinctively realize that some knowledge about the future is better than no knowledge about the future. The promise of ENSO prediction gives some information about the future probabilities of temperature and precipitation in selected regions of the world. The hope is that such information can be used for public and private benefit in these ENSO-sensitive areas, but the use of such information is turning out to be much more difficult than previously realized. A basic problem is that the system evolves so slowly, there are so few unique forecasts, and one has to live with blown forecasts for such a long time.

One would expect that agriculture, hydrology and water management, energy use and fisheries would be highly influenced by climate variability and would therefore benefit from some information concerning conditions one or two seasons in advance. Those users who understand forecasts that state probabilities of occurrence, and can relate the climate forecast to a forecast of the resource they are most interested in, are in the best position to make use of the climate forecasts to manage their future risks. Chapter 10 will provide some examples of the successful use of forecasts of aspects of ENSO and will indicate the difficulties in making use of this information.

2

The observational basis

This chapter provides an observational survey of the main elements of the tropical atmosphere and ocean needed in the sequel. In particular, the major circulation features in the atmosphere and ocean important for understanding ENSO: SST, SLP, surface winds, surface heat fluxes, the East–West overturning circulation in the Pacific, the Hadley circulation and the depth of the equatorial thermocline. Because the surface plays such a crucial role in atmosphere–ocean interactions, special emphasis will be placed on the fluxes at the surface, in particular the wind stresses, the latent heat flux and the net heat flux into the ocean. The annual cycle of the crucial quantities needed to define the climatology of the tropics: SST, SLP, precipitation, winds and thermocline depth, will be presented. "Anomalies," including those characteristic of ENSO, can be defined relative to this climatology.

The major features of ENSO and the evolution of ENSO as we now know them will be presented, with some discussion of how typical an ENSO event is likely to be. Some effects of ENSO on the globe, especially tropical temperature and precipitation, Atlantic hurricane landings and monsoon rainfall will be described. Some observations of both higher frequencies (periods less than a year) and lower frequency (especially decadal variability) will be introduced.

2.1 The nature and source of climate observations relevant to ENSO

It would be valuable to have an accurate picture of the Earth's atmosphere and ocean throughout the temporal evolution of climatic variability but, unfortunately, the measurement of variables important for climate has a relatively short history. While a few individual records of temperature extend back hundreds of years, the instrumental record adequate to measure the temperature of the extent of the Earth's surface is generally taken to have begun around 1880 with at-sea shipboard measurements, although a reasonable global description of the Earth's surface was not complete until the 1950s and a full global description had to await the development of satellite observations in the 1980s. The global upper-air network of radiosondes and rawinsondes used for weather prediction began in the 1950s and, while sporadic ship-based measurements of the surface and depths of the ocean have been going on for at least 150 years, the systematic measurements of the state of the top 1500 meters of the ocean is just now getting underway at the beginning of the twenty-first century. Observations taken for other purposes (weather prediction, agriculture, water resources, etc.) have then been used for defining the climate system, but even today there is no observing system adequate for climate; i.e. a system of measurements which satisfies the internationally agreed-upon principles of climate measurement (GCOS, 2004) and which defines the basic variables of the climate system to sufficient accuracy. Because the observational records are short, and the long-term accuracy of each record cannot be assured, the climate record is uncertain and incomplete.

The surface of the ocean is the site of the interaction between the atmosphere and the ocean. This interaction is mediated by the exchange of heat and momentum fluxes between the atmosphere and ocean through the ocean surface. For records longer than a very few decades, there is only one source of information on these fluxes, namely the shipboard-based meteorological observations taken routinely by many voluntary observing ships. These observations include: temperature at ship level; humidity at ship level; winds at ship level or winds at sea level by proxy observations of sea state; sea-surface temperature as measured by the temperature of a bucket of surface water or by the water temperature at the ship's engine intake; and approximate cloud cover in eighths (oktas) as estimated by shipboard observers. The prime compilation of these records is the comprehensive ocean–atmosphere data set (COADS) containing over 30 million reports since 1880 (Woodruff *et al.*, 1987, 1993); from these reports, fields of surface fluxes can be constructed. A number of atlases have been compiled from this freely available data, in particular by Oberhuber (1988) and Josey *et al.* (1998). The data distribution within COADS depends on where ships have traditionally gone: it is quite good in the North Atlantic and quite poor in the tropical Pacific. COADS data was the major source of historical information about ENSO until the deployment since 1995 of the system of 70 bottom-moored buoys that make measurements both of the surface meteorology and also the thermal state of the upper ocean. The data is telemetered to satellites and made freely available within 24 h of measurement (McPhaden *et al.*, 1998) at www.pmel.noaa.gov/tao/realtime.html.

There is another source of surface observations and this arises from the many-times-a-day model analyses produced by the weather centers and their re-analysis over the entire record of observations using a single model and the best current data-assimilation procedure. A recent atlas (Kållberg *et al.*, 2005) produced at the European Centre for Medium-Range Weather Forecasts (ECMWF), based on data from 1957–2002, contains a dynamically consistent climatology of the surface fluxes and concomitant upper-atmosphere fields and uses satellite data, where available, for much of this length of time. The re-analysis methods use weather data assimilated into a numerical atmospheric model (designed for weather prediction) with fields of SST as the only oceanic input. Because not all the original data taken was available rapidly enough to meet the stringent time requirements for weather prediction, and therefore had to be set aside and not used for the weather analysis, the re-analyses offer the possibility of using more input data than was originally available for the real-time weather analyses. More importantly, improvements in the models and advances in the techniques for assimilating data into models make these re-analyses superior to what was possible in the past.

While there have been no direct comparisons of the surface fluxes from the two different methods, we would expect that the climatology data (see Section 2.3) would be better for the re-analysis methods, while the shipboard method would be necessary when longer-term records are required and no other measurements are available. Our presentation of the annual cycle will therefore rely on re-analysis data, especially from the ECMWF compilation, and our description of ENSO evolutions and its longer-term evolution will necessarily rely on shipboard observations.

2.2 Solar forcing and fluxes at the surface

We begin by tracing solar radiation from the top of the atmosphere to the surface to give an idea of the magnitude of the various terms of the global energy budget.

The solar flux at the mean position of the Earth is $1367\,W/m^2$ and, because the area of the Earth $(4\pi R^2)$ is four times the disc intercepted (πR^2), the average solar flux at the top of the atmosphere is $342\,W/m^2$. The albedo of the Earth is about 0.3, so $107\,W/m^2$ is reflected back to space, partly by the clouds and aerosols in the atmosphere and partly by the bottom surface. An additional $67\,W/m^2$ is absorbed by the atmosphere so that, on average, $168\,W/m^2$ of direct solar radiation makes it to the surface. The net radiation at the surface is the difference between the solar radiation reaching the surface and the net infrared radiation leaving the surface. Since the mean temperature of the Earth is $15\,^\circ C$ ($288\,K$), σT^4 is $390\,W/m^2$ (see Appendix 1), while the back radiation from the radiatively active gases in the atmosphere (mostly water vapor and carbon dioxide) is $324\,W/m^2$. The net infrared radiation at the surface is therefore upward and has the value $66\,W/m^2$. The net radiation at the surface is therefore $102\,W/m^2$.

The net radiation at the surface can generally do three distinct things: it can evaporate water from the surface; it can warm the atmosphere by transferring heat from the surface to the atmosphere; or it can warm the ocean by transferring heat from the surface to the interior of the ocean. Because Figure 2.1 represents

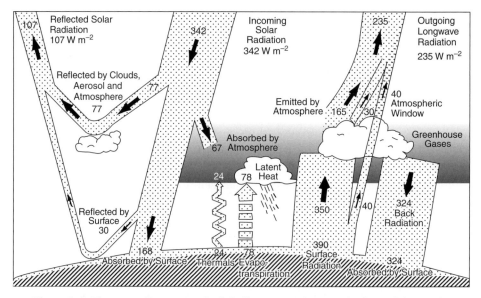

Figure 2.1. The annually averaged, globally averaged energy budget of the Earth. (From Kiehl and Trenberth, 1997.)

the entire Earth averaged over the year, the ocean neither heats nor warms over the course of the year, and the heat flux into the ocean is zero. (In reality, the anthropogenic addition of greenhouse gases into the atmosphere means that the Earth system is slightly out of equilibrium and, in particular, the ocean is warming slightly, with a current net input of about 0.5 W/m^2. The top of the atmosphere heat balance is also out of equilibrium, with a net of about 0.8 W/m^2 less net outgoing infrared radiation at the top of the atmosphere than net solar incoming radiation, thereby heating the Earth system – the approximately 0.3 W/m^2 difference between this number and the amount entering the ocean goes into melting ice and evaporating water. These numbers are characteristic of 2006 and will be different as time passes, since the emission of greenhouse gases continues and, indeed, seems to be accelerating.)

The net radiation at the surface therefore either warms the atmosphere by direct transfer of sensible heat or evaporates water from the surface: 78 W/m^2 evaporates 2.7 mm/d of water (see Appendix 1). The remainder of the surface heat budget, 24 W/m^2, goes as sensible heat from the surface of the Earth to the atmosphere, where it helps warm the atmosphere. In equilibrium, the surface heat budget must balance, the top of the atmosphere budget must balance, and the total heat absorbed by the atmosphere by radiation, latent heating and sensible heating must also balance. Careful inspection of Figure 2.1 (highly recommended!) indicates that the Earth system depicted by this Figure is indeed in equilibrium.

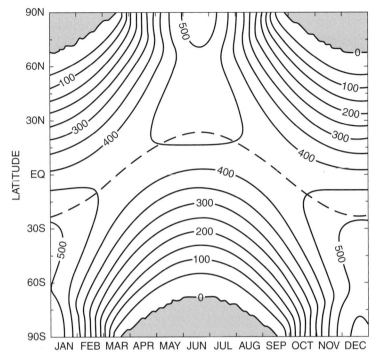

Figure 2.2. The net solar radiation at the top of the atmosphere, zonally averaged, as a function of latitude and month, W/m². Contour interval 50 W/m². (From Hartmann, 1994.)

There are, of course, spatial variations to the solar input at the top of the atmosphere, so that if we do not average over the entire area of the Earth and over the entire year, the situation becomes more complicated. Figure 2.2 shows the incoming solar radiation at the top of the atmosphere averaged around latitude bands as a function of month and latitude. On the equator, the solar radiation has only a few percent variation, with a mean value of about 425 W/m² – note that the sun is overhead twice a year. Some of the solar radiation is directly reflected back to space. Only about 325 W/m² is available after reflection as input to the Earth – this is the net solar radiation at the top of the atmosphere – see Figure 2.4.

The solar radiation reaching the ground depends on the intervening clouds and aerosols. From Figure 2.3, we see that rather than 325 W/m², the values in the tropical Pacific range from 225–250 W/m² in the eastern Pacific, which as we will see is relatively clear, to something like 50 W/m² less than this under the heavy clouds in the western Pacific and in the ITCZ, 5–10 degrees north of the equator.

Because the tropics is not in radiative equilibrium, some of the excess absorbed solar radiation (as well as some of the absorbed infrared radiation from the surface of the Earth) is diverged out of the tropics into midlatitudes. Figure 2.4 shows the

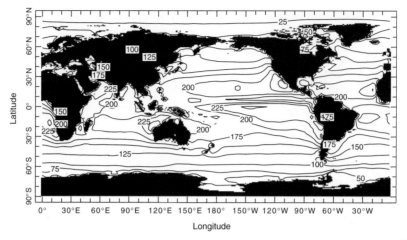

Figure 2.3. Annually averaged solar radiation reaching the sea surface in W/m². Downward solar radiation taken as positive. (Plotted and downloaded, with permission of ECMWF, from the ECMWF ERA-40 data set [Uppala *et al.*, 2005] at www.ingrid.ldeo.columbia.edu/sources/.ecmw/.ERA-40/.)

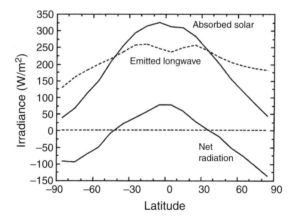

Figure 2.4. The net annual radiation balance at the top of the atmosphere as a function of latitude. (From Hartmann, 1994.)

annually averaged top of atmosphere difference between the net incoming solar radiation and net outgoing infrared radiation.

The tropics is radiatively heated while the higher latitudes are radiatively cooled. Since this is an annual average, we assume the atmosphere and ocean are in equilibrium (subject to the same proviso given earlier in this section), and there is, therefore, no net annual averaged heat storage. To remain in equilibrium, the net radiative heating of the tropics has to be balanced by net divergence of heat to higher latitudes by both the atmosphere and the ocean and, conversely, the radiative net cooling of the higher latitudes is balanced by the net convergence of heat by the atmosphere and the ocean. It may be noted that the net emitted long-wave radiation in Figure 2.4 is relatively flat with latitude because infrared radiation emitted to space is emitted from the top-unit optical depth of the radiating atmosphere, which is due to water vapor. Since the tropics are warm, they have much more water vapor than midlatitudes and one optical depth is higher in the atmosphere. As we move poleward, the amount of water vapor decreases, so one optical depth is lower in the atmosphere. The net effect is to emit infrared radiation at roughly similar temperatures at all latitudes, thus accounting for the relative flatness of the infrared profile.

The local net radiation at the surface can now do three distinct things. As in the discussion of the global budgets in Figure 2.1, it can leave the surface as sensible heat into the atmosphere or as latent heat into the atmosphere. But now, because we are not averaging over the entire globe, the net radiation at the surface that does not become sensible or latent heating of the atmosphere is also available to enter into the ocean as sensible heat Q:

$$F_{Net} = F_{Solar} + F_{IR} = S + LE + Q. \qquad (2.1)$$

Note that in Equation 2.1 we have defined fluxes as positive when upward, and negative when downward. The solar flux reaching the surface is therefore downward and negative. A common alternate convention is taking the solar flux as positive, which makes the latent heat into the atmosphere negative.

The net infrared flux at the surface is the difference between the emitted blackbody radiation at the temperature of the surface and the downward infrared radiation received at the surface from the rest of the radiating atmosphere. This is relatively constant in the surface of the tropical oceans at a value of about $50\,W/m^2$ (not shown – see Figure B6 of Kållberg *et al.*, 2005). The sensible heat from the ocean surface is also relatively spatially constant and has a small value of about $10\,W/m^2$. The net flux available at the surface for LE and Q in Equation 2.1 is therefore in the range of 165 to 190 W/m^2. Figure 2.5 shows the annual amount of latent heat due to the evaporation of water vapor leaving the surface.

The evaporation from the surface is of the order 3–5 mm/d in the eastern Pacific (Appendix 1 indicates that $29\,W/m^2$ evaporates 1 mm/d of water) except in the cold tongue of the eastern Pacific, where the evaporation is 1–2 mm/d. Since the net radiation at the surface does not have the spatial dependence of the cold tongue while the evaporation does, the net heat flux into the ocean will be largest where the evaporation is least and, therefore, will also have the spatial dependence of the cold tongue. Figure 2.6 shows the net heat flux into the ocean.

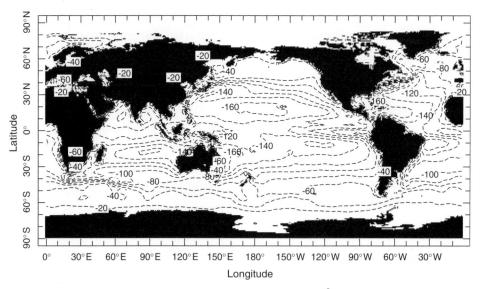

Figure 2.5. Net annually averaged latent heating, in W/m^2, at the surface. In this Figure, upward latent heating is taken as negative. (Plotted and downloaded, with permission of ECMWF, from the ECMWF ERA-40 data set [Uppala *et al.*, 2005] at www.ingrid.ldeo.columbia.edu/sources/.ecmwf/.ERA-40/.)

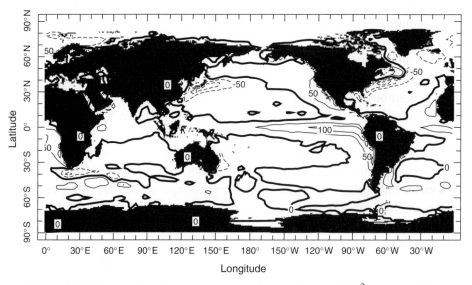

Figure 2.6. Net heat flux into the ocean at the surface, in W/m^2 (here positive downward). (Plotted and downloaded, with permission of ECMWF, from the ECMWF ERA-40 data set [Uppala *et al.*, 2005] at www.ingrid.ldeo.columbia. edu/sources/.ecmwf/.era-40/.)

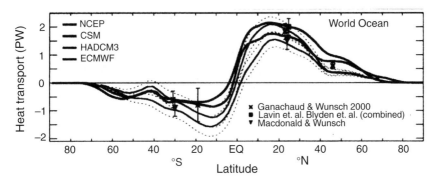

Figure 2.7. Annually averaged heat transport in the ocean from a number of model analyses. (From Trenberth and Caron, 2001.)

We see that heat enters the equatorial Pacific in the cold tongue and leaves the ocean at higher latitudes in the regions of the warm Gulf Stream and Kuroshio. The net heat into the ocean is far greater on the eastern side of the tropical Pacific than it is in the west (over $100\,\mathrm{W/m^2}$ versus less than $10\,\mathrm{W/m^2}$) primarily because the ocean is far cooler in the east than in the west (we will offer a more dynamical argument when we discuss the maintenance of the sea-surface temperature in Chapter 5). There is also heat into the tropical oceans in other cool regions: in the equatorial Atlantic and in the upwelling regions in the south-east Pacific and south-east Atlantic, and north of the equator in the upwelling regions off Africa and Central America. The heat transport implied by the heat flux at the surface is shown for a number of atmospheric models in Figure 2.7, along with estimates based on ocean estimates available at a few latitudes.

Since all the model heat-flux analyses shown in Figure 2.7 were obtained from surface fluxes determined by *atmospheric* models with sea-surface temperatures as their bottom boundary conditions, the differences between them are indicative of uncertainties in the ability of such models to accurately generate surface fluxes – these differences are probably mostly due to differences in simulated cloud distributions. The oceans transport a maximum of 2 petawatts (PW; 2×10^{15} W), with the maximum within ±20 degrees of latitude of the equator. The total heat transport in both the atmosphere and ocean from the top of the atmosphere radiative balance in Figure 2.4 is about 6 PW, with maxima at about ±30 degrees of latitude. The ocean therefore carries a significant part of the total needed heat transport, and a majority of the heat transport, equatorward of ±20 degrees of latitude.

2.3 The annually averaged tropical Pacific

The annual mean near-surface temperature, sea-level pressure and precipitation are shown in Figure 2.8.

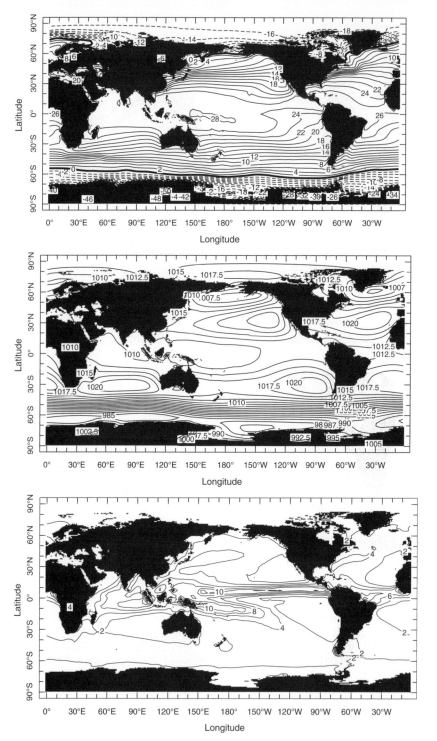

Figure 2.8. Annual means. Upper panel: Near-surface air temperature at 2 m which, over the oceans, is almost the sea-surface temperature (SST) in °C, negative values dashed. Middle panel: The sea-level pressure (SLP) in hPa. Lower panel: Precipitation at the surface in mm/d. (Plotted and downloaded, with permission of ECMWF, from the ECMWF ERA-40 data set [Uppala *et al.*, 2005] from www.ingrid.ldeo.columbia.edu/sources/.ecmwf/.ERA-40/.)

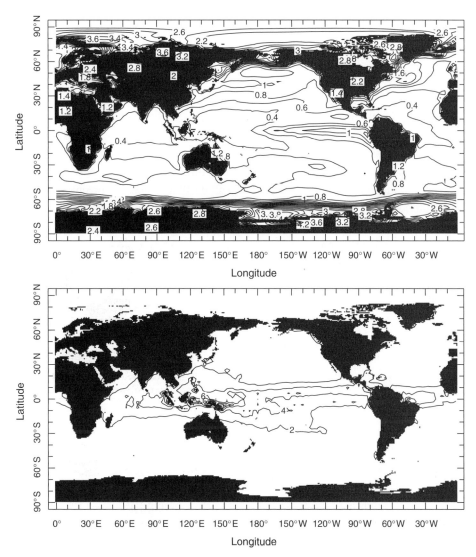

Figure 2.9. Monthly rms variability of annual means. Upper panel: 2 m air temperature, in °C. Lower panel: Precipitation at the surface in mm/d. (Plotted and downloaded, with permission of ECMWF, from the ECMWF ERA-40 data set [Uppala *et al.*, 2005] at www.ingrid.ldeo.columbia.edu/sources/.ecmwf/.ERA-40/.)

The annual mean of a quantity is, by definition, the average over a year. Figure 2.8 shows the average of the individual annual averages over all the years from 1957 to 2002 to produce a climatological annual mean. (Hence this particular climatology is relative to 1957–2002.). Figure 2.9 shows the monthly variance of the SST and precipitation to identify those regions of the ocean that are associated with large variability. By comparison with Figure B3 of Kållberg *et al.* (2005), we see that most of the variability is, in fact, interannual variability.

We see from Figure 2.8 (Upper) that the western tropical Pacific is warmer than the eastern tropical Pacific and, from Figure 2.8 (Lower), that the warm pool in the western Pacific is a locus of heavy precipitation. The warm-pool precipitation extends westward into the warm Indian Ocean, south-eastward into the south Pacific as the South Pacific convergence zone (SPCZ) and eastward into the Pacific north of the equator at about 7°N as a linear feature, the ITCZ. The ITCZ lies over a band of warm water which extends from the warm pool eastward into the eastern Pacific and coincides with a warm eastward ocean current, the north equatorial counter current (NECC). The dry zone in the south-east Pacific, off the coast of Peru and Chile, is extremely dry and, as indicated in Chapter 1, lies in the downward branch of the zonal atmospheric circulation that is upward over the warm pool. This is the Walker circulation.

The dominant variability of surface temperature over the tropical oceans occurs over the eastern tropical Pacific, as seen in Figure 2.9 (Upper). As we will see in Section 2.4, this is the interannual variability of SST connected to the evolution of the ENSO phenomenon. The greatest variability of precipitation (Figure 2.9 Lower) not surprisingly occurs where the precipitation is greatest. The meridional thickening of the variance compared to the annual mean indicates that the ITCZ moves meridionally on an interannual basis. Again, as we will see in Section 2.4, this occurs because the interannual warming and cooling of the eastern equatorial Pacific moves the warmest water meridionally. When the equatorial region is warm (the warm phase of ENSO), the ITCZ moves onto the equator; when cold, it moves to its northward position.

The annual mean sea-level pressure is shown in Figure 2.8 (Middle). The western Pacific, which is warm at the surface and has heavy precipitation, is a region of low mean pressure, consistent with the rising motion that accompanies the heavy precipitation. The eastern Pacific has relatively high pressure which accompanies downward motion and lack of precipitation. The subtropics are dominated by the subtropical anti-cyclones (surface high-pressure areas) and the north Pacific is dominated by the surface expression of the Aleutian Low.

Figure 2.10 shows the mean surface winds in the tropical Pacific and their relation to the mean SST. While the winds are generally to the west (towards the region of low SLP over the maritime continent) they have a distinct southerly component across the equator in the eastern Pacific, where they converge into the ITCZ which lies over the warm water centered at about 6°N. The mean winds at the center of the ITCZ are weak because they are constantly mixed vertically by storms moving through the region. Further west, the center of the ITCZ is a confluence zone for the winds, with southerly components to the south of the ITCZ and northerly

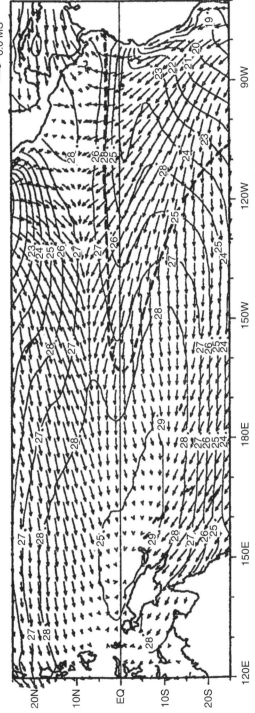

Figure 2.10. The annually averaged winds (arrows) superimposed on the annually averaged SST. (The heavy dashed line corresponds to the region in which the annually averaged SST in the eastern Pacific is less than the zonal average at that latitude – the reason why the region is called the "cold tongue.") (From Wang, 1994.)

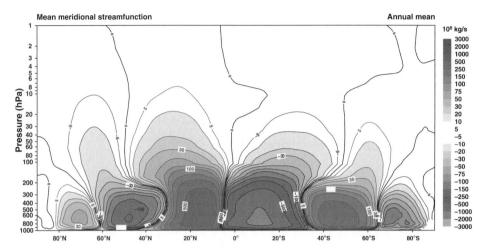

Figure 2.11. Mean meridional streamfunction. (Downloaded from the ECMWF ERA-40 Atlas [Kållberg *et al.*, 2005] site www.ecmwf.int/research/era/era-40_ atlas/docs/index.html, and used with permission of ECMWF.)

components to the north. The surface winds over the warm pool are also weak and disorganized, presumably due to the vertical mixing by storms and the lack of substantial gradients in SLP needed to drive an organized wind system.

That the annual mean precipitation over the tropical Pacific is north of the equator implies mean upward motion in the ITCZ region. The zonally averaged mean circulation is shown by Figure 2.11. This is the streamfunction averaged around the entire zonal band. There is rising motion north of the equator and sinking into the respective hemispheres in the subtropics. The two cells connected by rising in the ITCZ are expressions of the Hadley circulation.

Figure 2.12 shows the annual mean vertical velocity in mid-troposphere. The maximum upward motion lies over the maritime continent with descent into the south-east Pacific off Peru and Chile. This East–West circulation over the Pacific, rising in the west Pacific and descending in the east Pacific, forms the Walker circulation.

The vertical structure of the tropical atmosphere is basically layered (Figure 2.13). The subcloud layer is well mixed and extends up to cloud base. Above a transition layer, the shallow cloud layer extends to 2–3 km above the sea surface and is populated by puffy non-precipitating trade cumulus clouds. Deep clouds extend from cloud base to the tropopause.

Figure 2.14 shows the temperature structure of the ocean near the equator beneath the surface of the ocean. Across the entire ocean, there is a region of sharp vertical gradients ("the thermocline") centered at the 20 °C isotherm. Below the thermocline lies cold water and above the thermocline warm water. As we will see, the

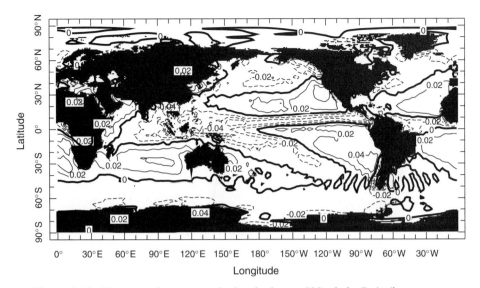

Figure 2.12. The annual mean vertical velocity at 500 mb in Pa/s (in pressure coordinates, dashed contours indicate upward vertical velocity). (Plotted and downloaded, with permission of ECMWF, from the ECMWF ERA-40 data set [Uppala *et al.*, 2005] at www.ingrid.ldeo.columbia.edu/SOURCES/.ECMWF/. ERA-40/.)

thermocline is deep in the west because the mean winds along the equator are westward. The shoaling of the thermocline in the east means that cold water is closer to the surface in the east and the Ekman divergence on the equator guarantees that upwelling brings the cold water to the surface.

The currents at the surface and below the surface are shown in a cross-sectional diagram (Figure 2.15). Although the details will vary at different longitudes, the surface currents are generally in the direction of the wind near the equator (the south equatorial current, SEC) and against the winds north of the equator (the NECC). The NECC flows from the warm western Pacific to the eastern Pacific, where it keeps the SST warm. Above this warm water lies the heavy convective region, the ITCZ.

Below the surface of the ocean, and under the westward-moving SEC, lies a rapidly moving (currents of the order of 1 m/s) current to the east, opposite to the direction of the winds, the equatorial undercurrent (EUC). The EUC is *in* the thermocline and, while the picture makes it look tubular (like bucatini), it is really ribbon-like (more like lasagna), being of the order of 200 km wide and 100 meters deep. The heavy arrows below the surface of the ocean show upwelling on the equator, poleward Ekman divergence on either side of the equator, and equatorward replenishment of the surface-diverging water at a few hundred meters depth.

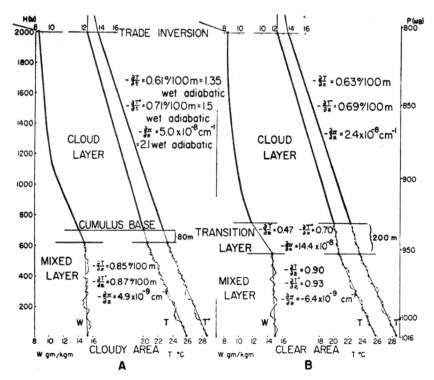

Figure 2.13. Structure of the trade wind moist layer showing the characteristic subcloud (well-mixed) and shallow cloud layers in regions of deep convection (A) and in clear areas (B). In this diagram, *w* is the water-vapor mixing ratio. (From Malkus, 1958.)

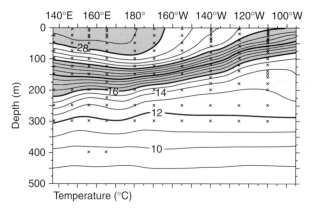

Figure 2.14. Temperature section across the equatorial Pacific averaged from 2°S to 2°N and 1980–1996 from the TAO array. (From McPhaden *et al.*, 1998.)

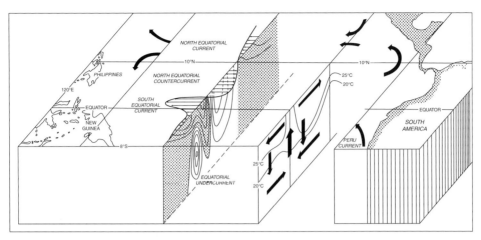

Figure 2.15. Cross-section of currents in the upper ocean of the tropical Pacific. (From Philander, 1990.)

2.4 The annual cycle in the tropical Pacific

The annual variation of the tropical SST is shown in Figure 2.16. While the midlatitude ocean is warmest in (northern) summer and coldest in winter, the tropics has March–April as its warmest period and September–October as its coolest. Along the equator, the deviations from the annual mean clearly propagate westward (Figure 2.17) with the majority of annual amplitude confined to the eastern third of the tropical Pacific. Note that in the western Pacific, the wind anomaly is westerly in November and December, a marker of the monsoonal winds in that region.

During boreal winter, the northern hemisphere component of the Hadley circulation is largest and descends into the winter hemisphere. During boreal summer, the southern hemisphere component is largest and again descends into the winter (southern) hemisphere (Figure 2.18).

The thermocline hardly varies annually in the middle third of the basin on the equator (Figure 2.19), but is shallowest in the east when the SST is coldest and deepest when the SST is warmest.

2.5 The evolution of ENSO

We saw in Figure 2.9 (Upper) that the locus of variability of SST was in the eastern Pacific. The evolution of the SST component of ENSO was first examined in a classic paper by Rasmusson and Carpenter (1982), who used a compositing technique to gain enough data to define the signal. The signal they define is that of a

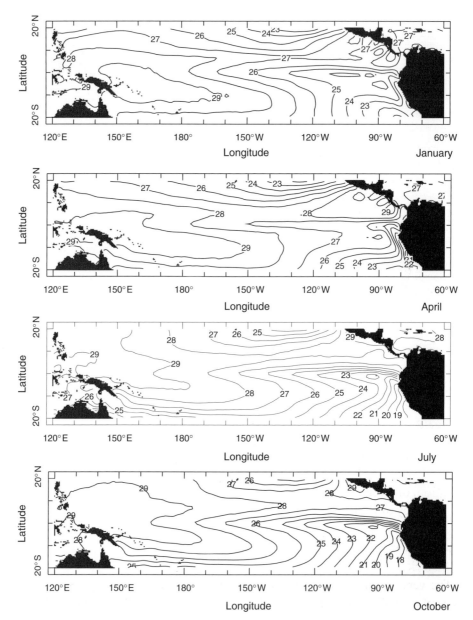

Figure 2.16. Climatological snapshots of SST (January, April, July, and October) in °C in the tropical Pacific. (Plotted and downloaded from www.iridl.ldeo.columbia. edu/ using the Reynolds *et al*., 2002 data set.)

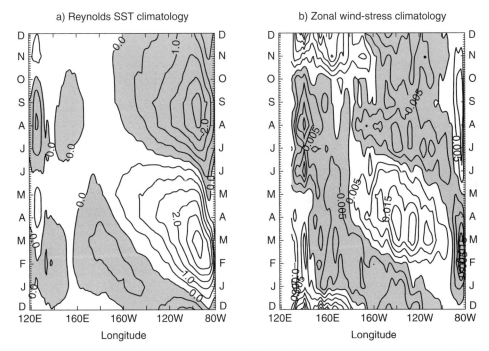

Figure 2.17. Monthly deviations of a) SST, and b) Zonal wind stress from their annual means along the equator. Note that time proceeds upward. Contour interval for SST is 0.5 °C and for wind stress 0.01 N/m². (From Yuan, 2005.)

"canonical" El Niño event; any individual event will have some idiosyncrasies. The basic idea is that there is enough similarity between the various individual phases of ENSO that occur in different years that one can define the various stages and use data from different years to define a composite stage. This compositing is aided by the tendency of warm and cold phases of ENSO to peak around December. The year in which ENSO peaks is usually called year (0), and the year before and after year (−1) and year (+1), respectively.

A quick look at the evolution of the SST on the equator is shown in Figure 2.20.

The very warm phase of ENSO during 1997–8 (i.e. the warm phase peaked in December 1997) is apparent, as are the cold phases during 1988–9 and 1998–2001. Note that the thermocline in the east deepened a few months *before* the peak of the SST and started to shallow, while the SST was still anomalously warm, again by a few months. An alternate way of looking at long series of ENSO phases is through indices such as the widely used indices of equatorial SST anomalies NINO 1 + 2, NINO 3, NINO 4 and NINO 3.4. The indices are defined as monthly averages of anomalies of SST from their annual march in the regions defined by Figure 1.1.

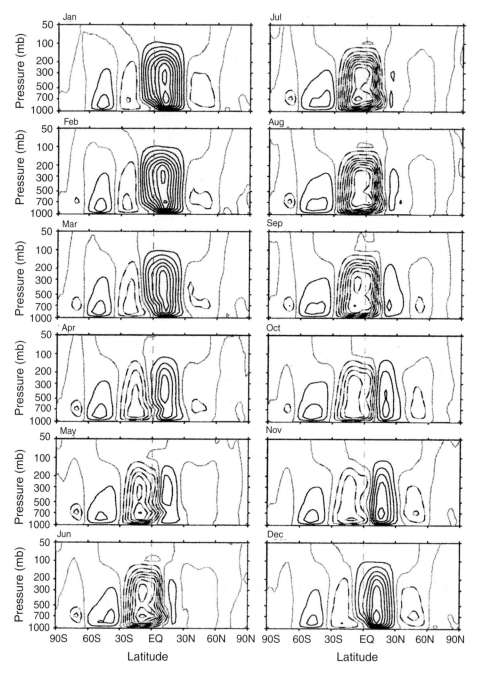

Figure 2.18. Monthly variation of the Hadley circulation. (From Dima and Wallace, 2003.)

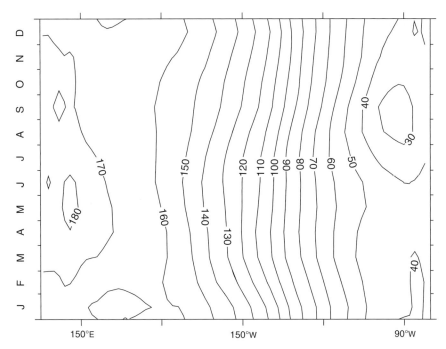

Figure 2.19. The annual variation of depth of the 20° isotherm (a measure of thermocline depth) on the equator in the Pacific based on a compilation of XBT, TAO mooring and Argo CTD data for the period 1980–2008, according to the method of Smith (1995). Contour interval 10 m. (Courtesy W.M. Kessler.)

Figure 2.21 shows the SOI, defined as the difference between Tahiti and Darwin sea-level pressure anomalies. Clearly the two series move in opposition and the difference makes a more robust index: negative values of the SOI are characteristic of warm ENSO phases and positive values of the SOI are characteristic of cold ENSO phases.

The temperature indices are averaged over the four Niño regions (Figure 1.1) and are shown in Figure 2.22. It is clear from this Figure that the peaks of NINO 3, 3.4 and 4 are relatively coincident in time, consistent with the flatness of the SST anomalies with time (i.e. lack of propagation) in Figure 2.20. The coastal SST, NINO 1 + 2, exhibits somewhat different behavior, especially when no large event prevails.

To see how exactly the anomaly evolves, we turn to the composite analysis, Figure 2.23, which gives the evolution of the composite warm phase of ENSO.

The first hint of warming occurs in April and May of year (0). There is some growth of the warm SST anomaly from May to July of year (0) but after that time, the SST anomaly has reached its full westernmost extent in the tropical Pacific, and

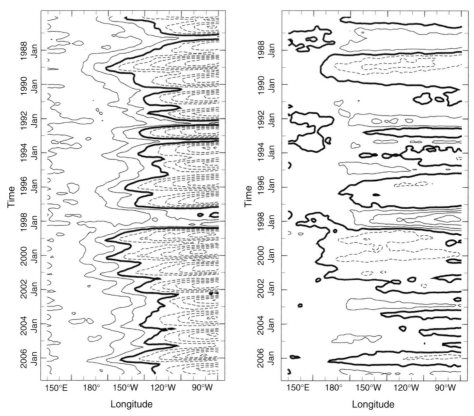

Figure 2.20a. Evolution of SST (left panel) and SST anomalies (right panel) on the equator (averaged from 2°S to 2°N) from 1986–2007. The contour interval is 1°C. In the left panel, the heavy line is 27 °C and temperatures less than 27 °C are dashed. In the right panel, the heavy line is 0 °C, and negative SSTAs are dashed. (Courtesy of Alexey Kaplan. Produced from Hadley Centre SST products.)

from that time to its maximum in December of year (0), the SST anomaly simply grows in place. The warm SST anomaly dies in place in the early part of year (1) and is essentially gone by April of year (1). During those months that the SST anomaly is strong in the eastern Pacific, there are warm anomalies in the Indian Ocean and cold anomalies in the north Pacific. The zonal wind anomalies on the equator are westerly at the western flank of the warm SST anomaly, and westerly anomalies also exist in the north Pacific (Figure 2.24).

A very useful cartoon of the composite is shown in Figure 2.25. Although precipitation was not part of the analyses, the region of persistent precipitation normally lying over the warm pool in the western Pacific moves eastward into the central Pacific. This tends to produce high SLP anomalies in the west and low SLP anomalies in the central and eastern Pacific, where it is now raining. At the peak of

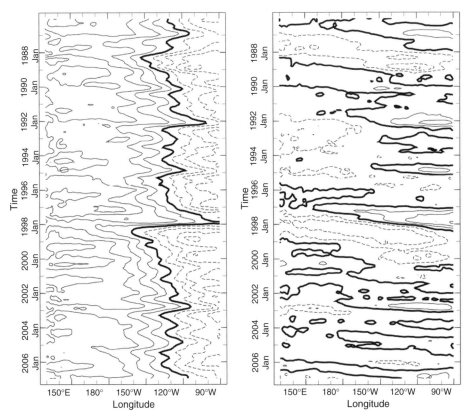

Figure 2.20b. Evolution of the mean (left) and anomalous (right) thermocline depth (as measured by the 20 °C isotherm) on the equator (averaged from 2°S to 2°N) from 1986–2007. The contour interval is 20 m. In the left panel, the heavy line is 100 m and depths less than 100 m are dashed. In the right panel, the heavy line is 0 m, and negative depths are dashed. (Courtesy of Alexey Kaplan. Produced from the SODA [Simple Ocean Data Analysis] product, which incorporates a variety of ocean data.)

the warm phase, the ITCZ collapses onto the equator, where the need for moisture convergence indicates that the meridional wind anomalies near the equator become equatorward in both hemispheres and indicate low-level moisture convergence to feed the anomalous rainfall on the equator. The zonal wind anomalies exist to the west of the SST anomalies and over the north Pacific. There are easterly anomalies in the Indian Ocean consistent with the anomalous divergence, i.e. the absence of convergence in the western Pacific region, where the usual persistent precipitation is no longer present, having moved into the central Pacific.

Lastly, we should comment on the role of heat fluxes into the ocean. We already saw, in Figure 2.6, that on an annual basis there is heat flux into the tropical ocean where the tropical ocean is coldest. Barnett *et al.* (1991) pointed out that this also

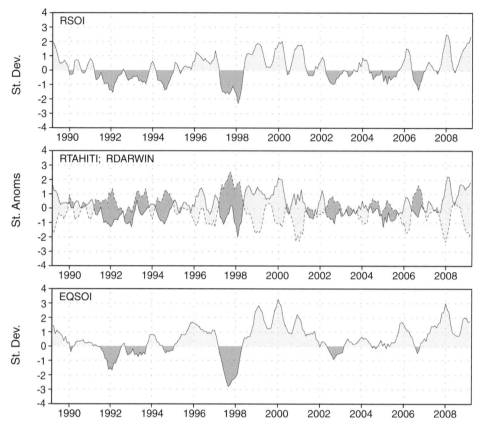

Figure 2.21. Southern Oscillation index (SOI). Upper panel: SOI (Pressure anomalies Tahiti–Darwin) from NCEP Reanalysis. Middle panel: Pressure anomalies Tahiti–Darwin. Lower panel: SOI from Tahiti–Darwin in middle panel. (Downloaded from www.cpc.noaa.gov/products/analysis_monitoring/ bulletin/ figt2.gif.)

applies in an anomaly sense: the heat flux into the ocean tends to counteract the anomaly, i.e. acts as a negative feedback on the SST. Thus, warm phases of ENSO have less heat flux into the equatorial Pacific and cold phases have more. The value of this negative feedback is about 40 W/m^2 per degree C in the western Pacific and 10 W/m^2 per degree C in the eastern Pacific.

It should not be presumed that the phases of ENSO evolve the same way every time. Figure 2.26 shows a number of different warm phases of ENSO as a function of time.

We see that there is some variation in the evolution of the warm phases, but by and large, the warm phases grow during the spring of year (0), peak toward the end of year (0) and decay during the spring of year (+1).

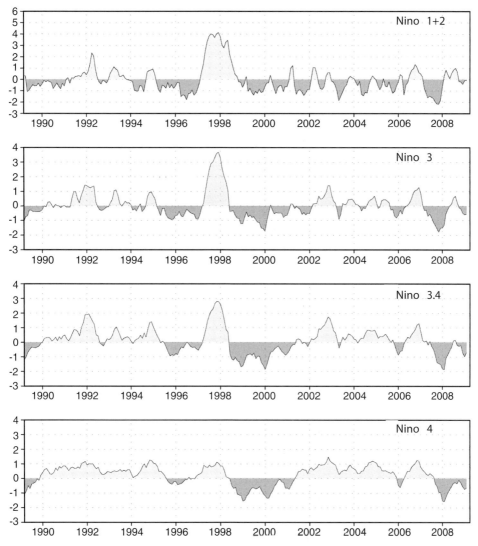

Figure 2.22. Time series of SST anomalies in the ENSO regions from 1986–2009.
(Downloaded from www.iridl.ldeo.columbia.edu/maproom/.ENSO/.)

2.6 ENSO effects

Some effects of ENSO on the rest of the globe will be noted here, but without
much detail.

Figure 1.4 is a version of the well-known diagram of the global influence of an
ENSO warm event (after Ropelewski and Halpert, 1987; all of the relationships
discussed below may be found in that paper or in Ropelewski and Halpert, 1996).

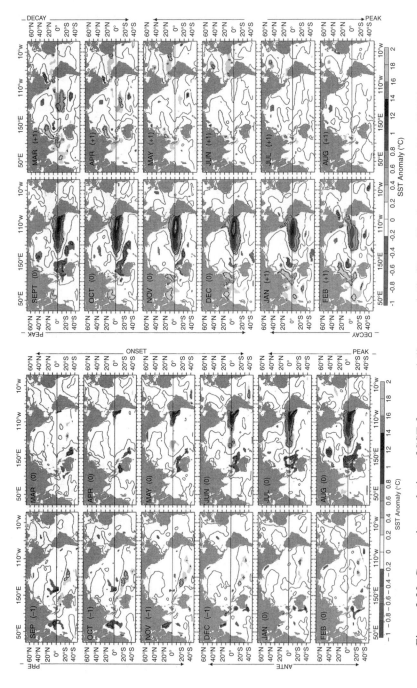

Figure 2.23. Composite evolution of SST from year (−1) to year (+1). (From Harrison and Larkin, 1998.)

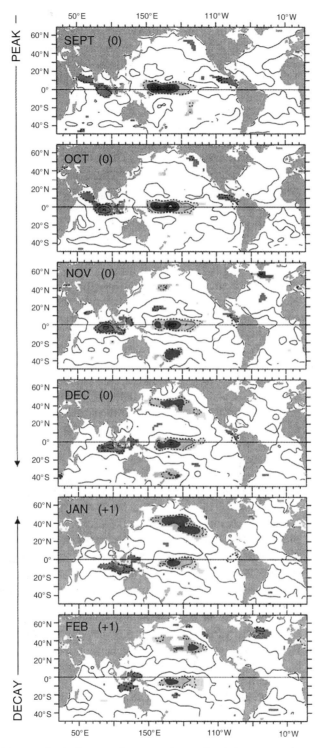

Figure 2.24. Zonal wind anomalies near the peak of the warm phase of ENSO. (From Harrison and Larkin, 1998.)

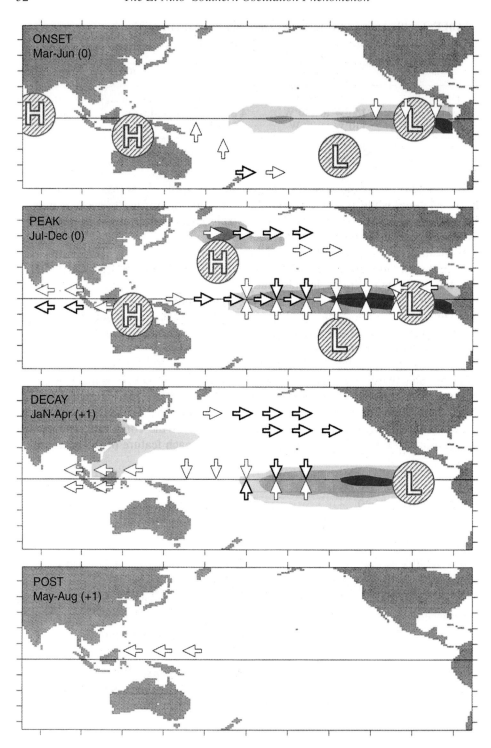

Figure 2.25. Diagram of SST and wind anomalies throughout the warm phase of ENSO. L and H represent regions of low and high sea-level pressure anomalies, respectively. (From Harrison and Larkin, 1998.)

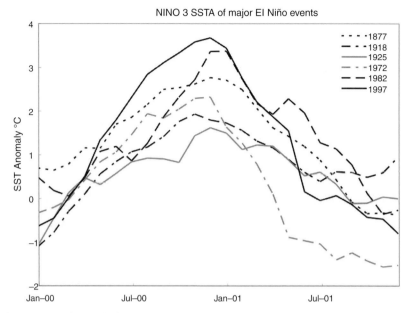

Figure 2.26. The evolution of several different warm phases of ENSO averaged over the NINO3 region, normalized by the standard deviation over the period 1964–94. The numbers refer to the different years having warm phases during this period.

As a crude first approximation, one may say that ENSO cold events have the opposite effects, but there are significant exceptions. As a general rule, the effects of an ENSO event are strongest and most reliable in the tropical Pacific genesis region and on contiguous continents. When there is a warm phase of ENSO, one can be fairly certain of heavy rains in Peru, drought in parts of Indonesia, and fewer typhoons in the western Pacific Ocean. Typical consequences are somewhat less reliable in the global tropics, but still highly likely. We will deal with four additional effects of ENSO: the effects of ENSO on rainfall in the western Pacific; the effects of ENSO on the Indian monsoon; the variation of temperature of the entire tropical atmosphere (up to the tropopause) with ENSO; and the effect of the phases of ENSO on hurricanes in the tropical Atlantic.

There is no question that ENSO has an influence in extratropical latitudes, but the response is less certain than in the tropics. Other factors may intervene, and the extratropical atmosphere is characteristically more chaotic, and thus less determined by SSTs. In these latitudes an ENSO event should be thought of as putting a probable bias in the system, rather than as a certain cause. With warm phases of ENSO (El Niño), heavy rains in the Great Basin region of the USA are more likely, and with cold (La Niña) events, midwestern US drought (1988, for example) and lower corn yields are more likely. Certain patterns are more likely to persist, altering the paths of hurricanes, typhoons and winter storms.

Another way of saying that not all ENSO connections are equally strong and reliable is the more general statement that the global impacts of each ENSO event are different. Not every warm event is accompanied by the same global variations, nor is the magnitude of what variations there are simply related to the strength of the event. Understanding of these differences is limited; they have hardly been classified satisfactorily, let alone explained in physical terms. A corollary is that the differences between events are not well predicted.

There are a number of reasons why this might be so. Surely, in some cases, failed forecasts are a consequence of the intrinsic limits to the predictability of the climate system. In other cases it may be that the prediction schemes fail to respond to the idiosyncrasies of each event such as the subtle (and not so subtle) differences in the pattern of its SST anomalies. It is known that the global response is sensitive to the location and strength of the atmospheric heating in the tropics (for example, see Hoerling *et al.*, 1997), but our understanding of what features truly matter is very limited.

A brief global tour through the historical record and the events of 1997–9 will illustrate the range of possibilities in ENSO impacts. In Indonesia and New Guinea it is virtually certain that warm ENSO years (El Niño years) are drought years and cold ENSO years (La Niña years) bring excess rain (Figure 2.27a). The 1997–8 forest fires in Indonesia and famine-inducing drought in Papua New Guinea fit the pattern, as does the greater than average rainfall that occurred during the 1998–9 La Niña. In Australia the expected rainfall anomalies are in the same direction, but are not nearly as reliable (Figure 2.27b). Drought in Australia during the 1997–8 El Niño was not as severe as the size of the event would have suggested. In Zimbabwe there is a very strong connection in the same direction between ENSO and rainfall and an even stronger connection to the corn crop, which integrates rainfall and temperature effects (Cane *et al.*, 1994). However, the relationship is not entirely reliable or straightforward: 1992 was the most severe drought year in at least the last 150 years in southern Africa, but produced only a moderate El Niño.

Figure 2.28 shows the relationship between ENSO and a measure of the intensity of the Indian monsoon, the All-India rainfall index. It is obvious that poor monsoons are generally associated with El Niño events, and excess rain with La Niña events, but the connection is far from perfect. Sometimes El Niño-year rainfall is average, and sometimes there is a poor monsoon without an El Niño event. Based on this history, if one had been asked early in 1997 what sort of monsoon to expect, the forecast would have been that a poor monsoon was likely. Indeed, two of the best atmospheric general-circulation models used for global prediction (the models of the National Centers for Environmental Prediction [NCEP] and the European Centre–Hamburg [ECHAM]) model) predicted significantly below average June to September rainfall for India. In the event, the rainfall turned out to be indistinguishable from the climatological norm.

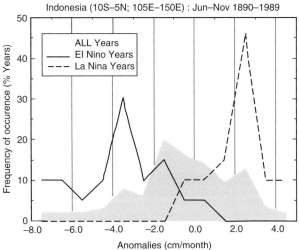

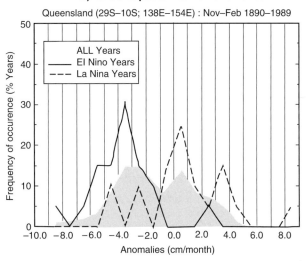

Figure 2.27. Rainfall anomalies a) Over Indonesia, and b) Over Queensland, Australia during warm ENSO years (solid) and cold ENSO years (dashed). The average over all years is shaded. (Courtesy of the IRI.)

Figure 2.29 shows the zonally and vertically averaged temperature anomaly for various latitude bands. The tropical troposphere (here defined as 20°S to 20°N) warms of the order of one degree during warm phases of ENSO with a lag of one or two seasons after the eastern Pacific SST anomalies characteristic of ENSO (the lag is not obvious from the Figure).

The pattern of the warming in relation to the precipitation is illustrated in Figure 2.30.

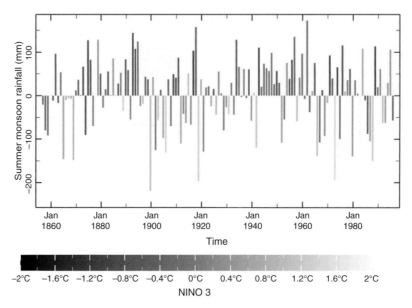

Figure 2.28. Annual anomaly of monsoon rainfall, where shading indicates average value of NINO3 index.

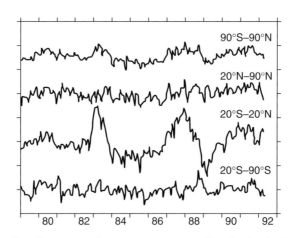

Figure 2.29. Zonally averaged temperature anomalies in the indicated latitude bands from the microwave sounding unit (MSU) vertically averaged over the atmosphere. The vertical tick interval is 0.5 K. (From Yulaeva and Wallace, 1994.)

Figure 2.30 shows that the anomalous precipitation (negative outgoing long-wave radiation [OLR] corresponding to the high cold tops of precipitating cumulonimbus clouds) during warm phases of ENSO moves into the central Pacific. The anomalous temperature pattern that goes with this anomalous precipitation pattern is shown in Figure 2.30b and consists of warm centers on the poleward flanks of the region of anomalous precipitation. The upper-level pressure field mirrors the

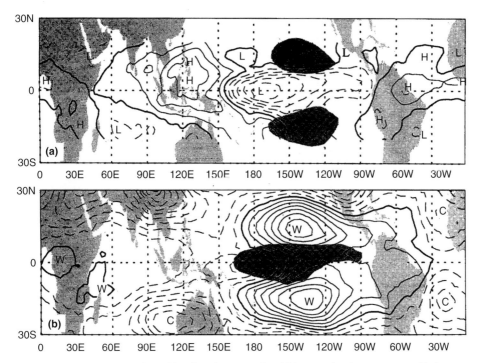

Figure 2.30. The combined leading SVD mode of (a) outgoing long-wave radiation (OLR) and (b) residual temperature (i.e. temperature anomalies with domain mean subtracted) from the microwave sounding unit (MSU). The shaded region in each quantity indicates the position of the extreme of the other. (From Yulaeva and Wallace, 1994.)

temperature patterns (higher mean temperatures imply greater thickness, and therefore anomalous highs coincident with the warm temperature centers) and therefore anticyclones. Interpretations of this pattern will be given in Chapter 5 when we discuss the atmospheric response to regions of persistent precipitation.

Figure 2.31 indicates that during cold phases of ENSO there are more hurricanes hitting the Atlantic and Gulf coasts of the USA. A detailed histogram giving the probability of a given number of hurricanes hitting the US coastline (Bove *et al.*, 1998) verifies that the probabilities are minimum during warm phases of ENSO and maximum during cold phases. As a logical concomitant, hurricane damages (normalized for increased coastal development and population over time) are also greater during cold phases of ENSO (Pielke and Landsea, 1999).

2.7 Variability at periods of less than a year

There is continuous variability occurring at periods of less than a year, notably including the inevitable turbulent gusts lasting seconds or minutes, organized wave

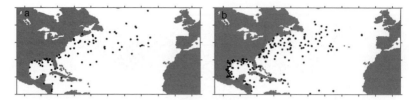

Figure 2.31. Hurricane positions on the last day that they exhibit hurricane-force winds (> 64 knots) during the (Left panel) 25 warmest and (Right panel) 25 coldest years in terms of sea-surface temperature in the equatorial cold-tongue region (6°N–6°S, 180–90°W) during the period of record 1870–2007. (Courtesy of Todd Mitchell, constructed from National Center for Atmospheric Research data sets at www.dss.ucar.edu/datasets/ds824.1/.)

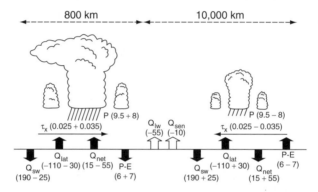

Figure 2.32. Schematic of the effects of the MJO on the surface fluxes (here, fluxes are positive downward). The short-wave (sw), long-wave (lw), sensible (sen), latent (lat) and net heat flux are in W/m^2, precipitation and evaporation in mm/d and zonal stress τ_x in N/m^2. (From Zhang, 2005.)

motions in the ITCZ of periods of a few days (see Section 5.2), and intraseasonal oscillations, the so-called Madden–Julian oscillations (MJOs) (see Zhang, 2005, for a complete review).

The MJO is an eastward-propagating, global wavenumber-1 disturbance that seems to arise from the western Indian Ocean and work its way eastward across the Indian Ocean and the Pacific. It has a locus of convergence that becomes apparent in convective regions, where it enhances local precipitation and has westerly surface-wind anomalies to the west of the convergence and easterly surface-wind anomalies to the east. The MJO has considerable local influence on surface fluxes and travels slowly enough that its surface winds last long enough to have a role in the ENSO story.

Figure 2.32 shows a synthesis of the effect of the MJO on surface fluxes. In the deep convective regions, which cover a small area compared to the rest of the

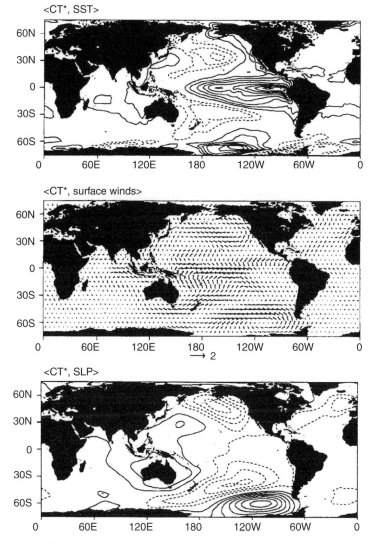

Figure 2.33. Global decadal patterns derived from regressing an equatorial Pacific index with interannual variability removed, with SST, surface winds and SLP throughout the world. (From Garreaud and Battisti, 1999.)

oscillation, the short-wave radiation reaching the surface decreases, both the precipitation and evaporation increase and the westerly stress increases.

The importance of the effects of the anomalous-MJO winds may be understood as follows. Say the MJO has surface-wind anomalies of ± 3 m/s. The effect of these winds on the ocean will depend on the pre-existing local mean winds, since the stress is proportional to the square of the *total* wind. If the mean wind is westerly, say $+1$ m/s, then the westerly phase of the MJO will give a westerly stress of relative magnitude 16, while the easterly phases will give an easterly stress of magnitude 4. If the mean wind is easterly, say -1 m/s, the MJO will give an easterly stress of relative magnitude 16, and westerly stress of magnitude 4. The MJO therefore nonlinearly promotes the pre-existing mean wind stress. These considerations will figure in the early stages of the growth of the phases of ENSO.

2.8 Decadal variability

A glance at Figure 2.20a indicates that, because the evolution of warm and cold phases of ENSO are different each time, there must exist a longer-term variation to ENSO. Indeed, if the interannual variability of ENSO is removed from an eastern Pacific SST index, the resulting index, called the global residual (GR) represents the effects of the non-interannual part of ENSO. Regressing this GR index, which has clear decadal variability, on global SST, SLP and surface winds gives a representation of the global decadal variability co-varying with the tropical Pacific (Figure 2.33). The decadal pattern is "ENSO-like" in having a signature in the tropical Pacific that looks like the ENSO pattern, but has considerably wider meridional extent. The north and south Pacific SST varies out of phase, and the Indian Ocean varies in phase with the tropical Pacific. The phase relations are similar to those of interannual ENSO and seem to indicate that the global effects may simply be due to the longer-term variation of ENSO, although other interpretations are possible (e.g. Vimont, 2005).

3

The equations of motion and some simplifications

In this chapter we introduce the equations of motion for both the atmosphere and ocean and develop some simplifications for later use. While the atmosphere and ocean are both fluids, and therefore, despite their difference in density, obey the same basic fluid equations, there are some essential differences that make their treatment and simplification very different. We will derive the equations of motion on a rotating sphere and show how the equations can be written on an f-plane tangent to the rotating sphere. The basic simplifications of hydrostatic and geostrophic balance will be motivated and introduced and the Boussinesq approximations, where differences of density are important only when coupled to gravity, are introduced for both the atmosphere and ocean. For the ocean, the existence of standing vertical modes leads to a profoundly useful simplification, the shallow-water equations (SWEs). The SWEs turn out to be an effective model for the atmosphere as well, though the interpretation is not straightforward and there are a number of different ideas about why it works as well as it does, as discussed in Chapter 5.

The material in this chapter, familiar to those with a background in atmospheric or ocean dynamics, is a necessary prerequisite for the mathematical treatments that follow. Aside from a few idiosyncrasies, we claim no great originality or excitement here and those who know this material are invited to skip it. The reader should recognize that necessary notation, concepts and derivations are collected here.

3.1 Equations governing the ocean and atmosphere

There are a number of similarities and differences between the atmosphere and oceans that are dynamically important and should be kept in mind as we develop the equations.

(a) Motions of interest in both the atmosphere and ocean may be considered shallow. In particular, if H is the characteristic scale of vertical motions, L is the characteristic scale of horizontal motions and a is the Earth's radius, then motions are shallow when $H \ll L, a$.

61

(b) Both the atmosphere and the ocean may be considered to be rapidly rotating. If U is a characteristic horizontal velocity and Ω is the rotation rate of the Earth, rapid means that the deviation from solid-body rotation is small: $U \ll \Omega a = 463$ mls.

(c) Both the atmosphere and the ocean are stratified fluids (usually stably stratified with lighter fluid on top of heavier fluid). The implication is that both gravity and buoyancy are important.

It may be noted that properties (a), (b), and (c) are general characteristics of geophysical fluid dynamics.

(d) Both the atmosphere and the ocean have significant bottom topography.

(e) The ocean has sidewall boundaries while the atmosphere does not.

(f) The ocean has a definite top while the atmosphere does not. The implication is that vertically standing modes exist in the ocean whereas outgoing radiation boundary conditions for the atmosphere generally imply that such modes do not exist for the atmosphere (there is more about this in Chapter 5).

(g) The atmosphere is driven primarily by thermal forcing instigated at its lower boundary; the ocean is driven primarily by wind stresses at its surface. Topographic forcing at the bottom of the atmosphere and heat fluxes at the surface of the ocean are not unimportant, however.

(h) The atmosphere has significant diabatic heating in its interior; in particular, latent heat release in clouds and the absorption and emission of radiation, while the oceans, by and large, do not have internal heat sources. There is some geothermal heating at the bottom of the ocean and some internal heating by radioactive decay but, for our purposes, these are small and can be neglected. In some places the water is quite clear, allowing blue–green solar radiation to penetrate many tens of meters into the ocean, while in other places abundant phytoplankton absorb it close to the surface. The difference can affect the distribution of sea-surface temperature. For didactic reasons, we will take solar radiation to be absorbed entirely at the surface.

(i) Both the atmosphere and the ocean are primarily two-component systems with the oceans being composed of water and salt and the atmosphere composed of air and water (in convertible solid, liquid and vapor forms). The atmosphere also has a number of constituents that are dynamically minor but are radiatively major; in particular carbon dioxide, methane, nitrous oxide and the chlorofluorocarbons, and in addition, radiatively active aerosols. The dominant constituents of the atmosphere, air and water vapor, are ideal gases, while the constituents of the ocean satisfy a highly complex equation relating density to concentrations of water and salt as a function of pressure.

(j) The atmosphere is a compressible gas, the ocean a nearly incompressible fluid. This difference turns out to be rather unimportant for most of the motions we will be interested in (but see Chapter 5, where some geophysical motions have the speed of sound waves, which depend essentially on compressibility).

(k) The ocean is dense, with a large heat capacity and large inertia. For a unit-area column of atmosphere and ocean, 10 meters of ocean has the same weight as the entire atmospheric column extending from the surface to the outer reaches of the atmosphere. Since the thermal heat capacity of water is four times the thermal heat capacity of air (for equivalent weights), the heat capacity of the entire atmospheric column is the same as the heat capacity of 2.5 meters of the ocean column. Clearly, most of the thermal heat capacity of the entire climate system resides in the ocean.

(l) The ocean is data-poor, especially for times before the last few decades, while the atmosphere has long had an observing system for weather prediction that is constantly being analyzed. Remote sensing from satellites has added global coverage of the atmosphere and of the ocean surface, but has added nothing directly about the subsurface ocean, since electromagnetic radiation barely penetrates below the surface.

3.1.1 Equations of motion on a rotating sphere

We begin with Newton's law, $\mathbf{F} = m\,\mathbf{A}$, where $\mathbf{A} = \dfrac{d\mathbf{V}}{dt}$ and $\mathbf{V} = \dfrac{d\mathbf{r}}{dt}$. We consider space dimensions as well as time and take a Eulerian approach. The independent variables are: $(\mathbf{r}, t) = (x, y, z, t)$. The dependent variables (i.e. the properties of the fluid) are: $\mathbf{V}, p, \rho, T, \ldots \theta$ where $\mathbf{V} = (u, v, w)$ gives the horizontal and vertical velocities, p, ρ and T are pressure, density and temperature, $\ldots$ denotes any additional properties (e.g. salinity), and θ is potential temperature.

Rewrite Newton's law in the form

$$\frac{d\mathbf{V}}{dt} = \frac{1}{\rho}\mathbf{F}' = \mathbf{F}, \tag{3.1}$$

where $\mathbf{F}'$ is now force/unit volume and $\mathbf{F}$ is force/unit mass. Note that

$$\frac{d\mathbf{V}}{dt}(x, y, z, t) = \frac{\partial \mathbf{V}}{\partial t} + \frac{\partial \mathbf{V}}{\partial x}\frac{dx}{dt} + \frac{\partial \mathbf{V}}{\partial y}\frac{dy}{dt} + \frac{\partial \mathbf{V}}{\partial z}\frac{dz}{dt}$$

$$= \frac{\partial \mathbf{V}}{\partial t} + u\frac{\partial \mathbf{V}}{\partial x} + v\frac{\partial \mathbf{V}}{\partial y} + w\frac{\partial \mathbf{V}}{\partial z} = \frac{\partial \mathbf{V}}{\partial t} + \mathbf{V} \cdot \nabla \mathbf{V}.$$

We want to take (x, y) fixed on the Earth's surface, i.e. in a rotating frame of reference, whereas Equation 3.1 is appropriate to an inertial frame. First consider uniform rotation $\mathbf{\Omega}$ about an unvarying axis, i.e $\dfrac{d\mathbf{\Omega}}{dt} \equiv 0$ (Figure 3.1a).

We will use the following notation for any vector $\mathbf{w}$:

$\left(\dfrac{d\mathbf{w}}{dt}\right)_a$ is the absolute rate of change of $\mathbf{w}$ and $\left(\dfrac{d\mathbf{w}}{dt}\right)_r$ is the rate of change of $\mathbf{w}$ relative to the rotating system. Note that $\left(\dfrac{d\mathbf{w}}{dt}\right)_a$ is perpendicular to both $\mathbf{\Omega}$ and $\mathbf{w}$ and that

$$\left(\frac{d\mathbf{w}}{dt}\right)_a = \left(\frac{d\mathbf{w}}{dt}\right)_r + \mathbf{\Omega} \times \mathbf{w}.$$

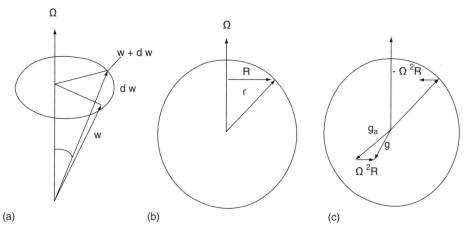

Figure 3.1. (a) Change of arbitrary vector in an absolute rotating system. (b) Definition of the vector **R**. (c) Modification of gravity **g** by centrifugal force.

Equation 3.1 is $A_a \equiv \left(\dfrac{d}{dt}\left(\dfrac{d\mathbf{r}}{dt}\right)_a\right)_a = \mathbf{F}_a$ but we want an equation for

$\mathbf{A}_r \equiv \left(\dfrac{d}{dt}\left(\dfrac{d\mathbf{r}}{dt}\right)_r\right)_r$.

Now $\left(\dfrac{d\mathbf{r}}{dt}\right)_a = \left(\dfrac{d\mathbf{r}}{dt}\right)_r + \boldsymbol{\Omega} \times \mathbf{r}$ so that

$$\mathbf{A}_a = \frac{d}{dt}\left[\left(\frac{d\mathbf{r}}{dt}\right)_r + \boldsymbol{\Omega} \times \mathbf{r}\right]_a = \left(\frac{d^2\mathbf{r}}{dt^2}\right)_r + \boldsymbol{\Omega} \times \left(\frac{d\mathbf{r}}{dt}\right)_r + \frac{d}{dt}(\boldsymbol{\Omega} \times \mathbf{r})_r + \boldsymbol{\Omega} \times (\boldsymbol{\Omega} \times \mathbf{r})$$

$$= \left(\frac{d^2\mathbf{r}}{dt^2}\right)_r + 2\boldsymbol{\Omega} \times \left(\frac{d\mathbf{r}}{dt}\right)_r + \boldsymbol{\Omega} \times (\boldsymbol{\Omega} \times \mathbf{r}).$$

Dropping the subscripts in the relative (rotating) system gives:

$$\mathbf{A}_a = \mathbf{A} + 2\boldsymbol{\Omega} \times \mathbf{V} + \boldsymbol{\Omega} \times (\boldsymbol{\Omega} \times \mathbf{r}). \tag{3.2}$$

The second term in Equation 3.2 is the Coriolis force and the third is the centrifugal force.

Take the origin of the coordinate system to be the center of the Earth; with the vector **R** as the perpendicular from the axis of rotation to **r** (Figure 3.1b). Then

$$\boldsymbol{\Omega} \times (\boldsymbol{\Omega} \times \mathbf{r}) = \boldsymbol{\Omega} \times (\boldsymbol{\Omega} \times \mathbf{R}) = -\Omega^2 \mathbf{R}$$

and

$$\frac{d\mathbf{v}}{dt} + 2\mathbf{\Omega} \times \mathbf{V} = \mathbf{F}_a + \mathbf{\Omega}^2\mathbf{R}.$$

One important force is gravity $\mathbf{g}_a$. We can write the force as $\mathbf{F}_a = \mathbf{F}' + \mathbf{g}_a$ where $\mathbf{g}_a$ is true gravity. Parcels of fluid feel true gravity plus the centrifugal force so that we can combine the terms and define the apparent gravity $\mathbf{g}$ as

$$\mathbf{g}_a + \mathbf{\Omega}^2\mathbf{R} = \mathbf{g} \equiv -\nabla\Phi,$$

where Φ is the geopotential (Figure 3.1c). The Earth's surface is approximately a surface of constant Φ and so is not precisely spherical. Now, with the radius of the Earth denoted as a,

$$\frac{\Omega^2 R}{g} \simeq \frac{\Omega^2 a}{g} \approx \frac{(7.29 \times 10^{-5}s^{-1})^2(6.37 \times 10^6 m)}{9.81 ms^{-2}} \approx \frac{1}{300},$$

so the difference between $\mathbf{g}$ and $\mathbf{g}_a$ is small. The difference between the Earth's radius at the equator and at the pole is about 214 km so we may take the Earth's surface (and other Φ = constant surfaces) to be approximately spherical. (Veronis, 1973, gives a very thorough discussion of the effects of the ellipticity of the Earth.)

We will also take $\Phi = gz$ with g = constant and $z = r - a$ so $z = 0$ at mean sea level. This is a good approximation for $\sim$100 km above sea level or 5 km below it – except that it ignores tidal forces. (Alternately, $z \equiv g^{-1}[\Phi - \Phi(sfc)]$.)

Since we are concerned with fluids, the pressure force, $\rho^{-1}\nabla p$, is also important. We now write the momentum equation as

$$\frac{d_3\mathbf{V}}{dt} + 2\mathbf{\Omega} \times \mathbf{V} = \rho^{-1}\nabla_3 p - \nabla\Phi + \mathbf{F}_3. \tag{3.3}$$

The subscript 3 explicitly recognizes the three dimensions; $\mathbf{F}_3$ represents frictional forces. Air and water are Newtonian fluids, but if we wish to consider only large-scale motions, $\mathbf{F}_3$ may stand for the effects of smaller-scale turbulent motions. In the latter case the proper form for $\mathbf{F}_3$ is not immediately obvious.

We now seek the equations for the velocity *components* corresponding to Equation 3.3. Introduce the spherical coordinates: λ, θ are longitude and latitude, respectively, and $z = r - a$ is the altitude above mean sea level.

Let $\mathbf{i}$, $\mathbf{j}$ and $\mathbf{k}$ be unit vectors in the direction of increasing λ, θ and z, and let the velocity components be $u = \mathbf{V} \cdot \mathbf{i}$, $v = \mathbf{V} \cdot \mathbf{j}$ and $w = \mathbf{V} \cdot \mathbf{k}$ (see Figure 3.2) so that we can write

$$\mathbf{V} = u\mathbf{i} + v\mathbf{j} + w\mathbf{k}.$$

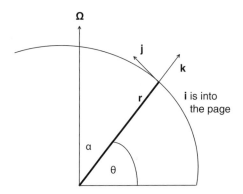

Figure 3.2. Sketch of relation between $\mathbf{i}, \mathbf{j}, \mathbf{k}$ and the rotation axis.

Exercise 3.1: Determine $\dfrac{d\mathbf{i}}{dt}, \dfrac{d\mathbf{j}}{dt}, \dfrac{d\mathbf{k}}{dt}$ in terms of $\mathbf{i}, \mathbf{j}, \mathbf{k}, u, v, w$. Also determine the components of $\boldsymbol{\Omega} \times \mathbf{V}$. (The relations $\mathbf{k} = \dfrac{\mathbf{r}}{|r|}, \mathbf{i} = \dfrac{\boldsymbol{\Omega} \times \mathbf{r}}{|\boldsymbol{\Omega} \times \mathbf{r}|} = \dfrac{\boldsymbol{\Omega} \times \mathbf{k}}{|\boldsymbol{\Omega} \times \mathbf{k}|}, \mathbf{j} = \mathbf{k} \times \mathbf{i}$ may be helpful.)

Exercise 3.2: Derive Equations 3.4 below from Equation 3.3 using the results of the previous exercise.

The momentum equations in component form are :

$$\frac{d_3 \mathbf{u}}{dt} + \left[f + \frac{u\tan\theta}{a+z} \right] \mathbf{k} \times \mathbf{u} + \frac{w}{a+z}\mathbf{u} + w \cdot 2\Omega \cos\theta\, \mathbf{i} = -\frac{1}{\rho}\nabla p + \mathbf{F} \qquad (3.4\text{a})$$

and

$$\frac{d_3 w}{dt} - \frac{u^2 + v^2}{a+z} - u \cdot 2\Omega \cos\theta = -\frac{1}{\rho}\frac{\partial p}{\partial z} - g + F_r, \qquad (3.4\text{b})$$

where $f = 2\Omega \sin\theta$ is the vertical component of the rotation vector $\boldsymbol{\Omega}$ and

$$\mathbf{u} = (u, v) \text{ and } \nabla \equiv \mathbf{i}\frac{1}{(a+z)\cos\theta}\frac{\partial}{\partial\lambda} + \mathbf{j}\frac{1}{(a+z)}\frac{\partial}{\partial\theta}.$$

Equations 3.4 have an angular-momentum principle

$$\frac{dM}{dt} = -\frac{1}{\rho}\frac{\partial p}{\partial\lambda} + r\cos\theta F_\lambda, \text{ where } M = (\Omega r\cos\theta + u)r\cos\theta. \qquad (3.5)$$

Exercise 3.3: Derive Equation 3.5 from Equations 3.4.

Since $z \ll a$, one is tempted to replace $a+z$ by a in Equations 3.4. The resulting equations do not have an angular-momentum principle. (A detailed discussion appears in Phillips, 1968, and references therein; also see Veronis, 1973.) The difficulty arises from the terms $2\Omega cos\theta w$ and uw/a in Equation 3.4a. A way out is given in the following exercise.

Exercise 3.4: Starting from the vector-invariant form of Equation 3.3

$$\frac{\partial}{\partial t}\mathbf{V} + \nabla_3\left(\frac{1}{2}V^2\right) - \mathbf{V} \times \nabla \times (\mathbf{V} + 2\Omega rcos\theta\mathbf{i}) = \frac{-1}{\rho}\nabla_3 p - \nabla_3\Phi + \mathbf{F},$$

use the approximate relations

$$x = acos\theta\lambda; \quad u = \frac{dx}{dt} = acos\theta\frac{d\lambda}{dt},$$

$$y = a\theta; \quad v = \frac{dy}{dt} = a\frac{d\theta}{dt}$$

and

$$z = r - a; \quad w = \frac{dz}{dt},$$

to obtain the component equations

$$\frac{d_3}{dt}u - \left(f + \frac{utan\theta}{a}\right)v = -\frac{1}{\rho}\frac{\partial p}{\partial x} + F_\lambda, \tag{3.6a}$$

$$\frac{d_3}{dt}v - \left(f + \frac{utan\theta}{a}\right)u = -\frac{1}{\rho}\frac{\partial p}{\partial x} + F_\theta \tag{3.6b}$$

and

$$\frac{d_3}{dt}w = -\frac{1}{\rho}\frac{\partial p}{\partial z} - g + F_r, \tag{3.6c}$$

and show that Equations 3.6 have an angular-momentum principle with

$$M \equiv (\Omega acos\theta + u)acos\theta,$$

(i.e. M evaluated as if the parcel of fluid is at $z = 0$).

Equations 3.6 (i.e. with $2\Omega cos\theta$ terms neglected) may be derived by a scaling argument that relies on the stratification being strong enough to inhibit vertical motions. The vertical direction – the direction determined by gravity – has been singled out (rather than the direction of Ω). From here on we will work with the momentum equations in the form of Equations 3.6, and consistently replace r by a as a coefficient. The terms proportional to $1/a$ arise because of the curvature of the Earth; we will do much of our analysis on planes where they may be neglected. Note that all the equations in Equations 3.6 are nonlinear.

3.1.2 The continuity equation and equation of state

The continuity equation is:

$$\frac{\partial \rho}{\partial t} + \nabla_3 \cdot (\rho \mathbf{V}) = 0$$

or, alternately,

$$\frac{1}{\rho}\frac{d_3\rho}{dt} + \nabla_3 \cdot \mathbf{V} = 0.$$

In spherical coordinates, with the same approximations as above

$$\frac{1}{\rho}\frac{d_3\rho}{dt} + \frac{1}{a\cos\theta}\frac{\partial u}{\partial \lambda} + \frac{1}{a\cos\theta}\frac{\partial}{\partial \theta}(v\cos\theta) + \frac{\partial w}{\partial z} = 0 \qquad (3.7)$$

(a term $2w/a$ has been neglected).

The equation of state has the form:

$$F(p,\rho,T;\mathbf{c}) = 0, \qquad (3.8)$$

where T = temperature and $\mathbf{c}$ stands for an array of other constituents. For the Earth's atmosphere the most important example of $\mathbf{c}$ is some measure of water-vapor content such as specific humidity q. For the ocean $\mathbf{c}$ is S, the salinity. For other planets and other situations there are other possibilities.

Dry air may be treated as a perfect gas:

$$p = \rho RT; \quad R = const = 287\,\mathrm{Jkg^{-1}K^{-1}}. \qquad (3.9)$$

The equation for moist air may be put in the same form by replacing T by T_v, the virtual temperature (see Section 5.1, Thermodynamic quantities).

The equation of state for sea water is generally written in the form $\rho = \rho_o F(T, S, P)$; and is linearized as

$$\rho = \rho_o[-\alpha T + \beta S + \gamma p]. \qquad (3.10)$$

In general α, β and γ are not constant, but for near-surface work the approximation α, β constant and $\gamma = 0$ is acceptable. For temperatures well above freezing $\alpha > 0$ in accord with the intuitive expectation that density is greater in water that is colder (or saltier; $\beta > 0$).

3.1.3 Constituent equations

In general, a constituent c (e.g. water vapor, salinity, carbon dioxide etc.) per unit mass is governed by an equation of the form

$$\rho \frac{d_3 c}{dt} = Sources - Sinks + Diffusion. \tag{3.11}$$

For water vapor q, a source would be evaporation and a sink would be condensation of liquid water; for salinity S the source–sink is evaporation–precipitation and is localized to the ocean surface. (Note that salt is not actually added to the ocean [except perhaps in river runoff] but the concentration of salt in water changes by the addition and subtraction of fresh water.) Active chemicals can have very complicated right-hand sides to their equations.

3.1.4 The first law of thermodynamics

The entropy form of the first law is

$$\frac{d\eta}{dt} = \frac{J}{T}, \tag{3.12}$$

where η = entropy per unit mass, and J is the rate of heating per unit mass by irreversible processes.

For a dry atmosphere, the ideal gas law is $p - \rho RT$ so that

$$d\eta = c_p \frac{d\theta}{\theta} = c_p \frac{dT}{T} - \frac{Rdp}{p},$$

where $\theta = T(p_0/p)^{\frac{R}{c_p}}$ is the potential temperature, i.e. the temperature a parcel of air of temperature T and pressure p would have if brought adiabatically to pressure p_0 (= 1000 mb). Hence

$$\frac{d\theta}{dt} = \theta \frac{J}{c_p T}. \tag{3.13}$$

Generally the heating term J includes radiation, latent and sensible heating, and diffusion and conduction of heat.

3.1.5 Boundary conditions

Equations 3.6, 3.7, 3.10 and 3.8 are six equations in the variables u, v, w, ρ, p, T and S or q. Equation 3.11 provides an equation for S or q. Let us ignore these for the present – assume dry air and S = constant. Then we have six equations in six unknowns if the friction and heating terms are specified externally or as functions of the calculated variables.

For a solution to these equations, we need boundary conditions.

For the atmosphere these are:

(a) At the lower surface, denoted by $z_0 = z_0(x, y)$, $w = \mathbf{u} \cdot \nabla z_o$ so there is no normal flow at the surface. Near the surface, friction becomes far more important than in the free atmosphere. We will consider this more carefully in Chapter 4.
(b) At the top of the atmosphere (or at least at such a low pressure that there is not much more atmosphere above that point) we impose radiation conditions: energy fluxes (wave energy) are outward. However, numerical models have tops and often impose $w = 0$ at the top. As we will see in Chapter 5, this may lead to spurious vertically standing modes.

For the ocean, the boundary conditions are:

(a) At the bottom of the ocean $w = \mathbf{u} \cdot \nabla z_o$ or some frictional form.
(b) At the top surface of the ocean, a kinematic condition: either $w = 0$ (rigid lid) or $w = \dfrac{dz_{top}}{dt}$, which can be made into a condition on the pressure. Also continuity of pressure across the interface $p(0) = P_{atm}$ (dynamic condition).
(c) At the side boundaries of the ocean, the boundary condition depends on whether the fluid is taken as inviscid or frictional.

Frictionless: $\mathbf{u} \cdot \mathbf{n} = 0$ where $\mathbf{n}$ is the vector normal to the boundary.

Frictional: no slip: $u = 0$; free slip: $\mathbf{u} \cdot \mathbf{n} = 0$ and tangential stress vanishes: $\dfrac{\partial}{\partial n}(\mathbf{k} \times \mathbf{u}) = 0.$

(Note that friction requires more boundary conditions.)

(d) At the interface between the atmosphere and the ocean, the stresses across the interface are equal: $\tau_{wind} = \tau_{water}$. Stresses are not directly observed and are generally derived from near-surface velocities using a bulk formula: $\tau = \rho c_D |\mathbf{u}| \mathbf{u}$ where c_D is the drag coefficient (see Chapter 4). c_D is the same for air and water so that, since $\rho_{water} \approx 1000 \times \rho_{air}$, $u_{air} \approx 30 \times u_{water}$.

3.2 The f-plane and the beta-plane

A major simplification of the equations on a rotating sphere is the f-plane approximation.

Our equations are

$$\frac{d_3}{dt}\mathbf{u} - \left[f + \frac{u\tan\theta}{a}\right]\mathbf{k} \times \mathbf{u} = -\frac{1}{\rho}\nabla p + \mathbf{F}_2, \tag{3.14a}$$

$$\frac{d_3}{dt}w = -\frac{1}{\rho}\frac{\partial p}{\partial z} - g + F_r, \tag{3.14b}$$

and

$$\frac{\partial \rho}{\partial t} + \nabla_3 \cdot (\rho \mathbf{V}_3) = 0, \tag{3.14c}$$

where $\mathbf{u} = (u,v)$, $\mathbf{V}_3 = (u,v,w)$, $f = 2\Omega \sin\theta$ and

$$\frac{d_3}{dt} \equiv \frac{\partial}{\partial t} + u\frac{\partial}{\partial x} + v\frac{\partial}{\partial y} + w\frac{\partial}{\partial z},$$

with

$$\nabla \equiv \mathbf{i}\frac{\partial}{\partial x} + \mathbf{j}\frac{\partial}{\partial y},$$

and

$$\nabla_3 \equiv \mathbf{i}\frac{\partial}{\partial x} + \mathbf{j}\frac{\partial}{\partial y} + \mathbf{k}\frac{\partial}{\partial z} - \frac{\tan\theta}{a}\mathbf{j}.$$

The equations for an f-plane follow by taking $a \to \infty$ while y remains finite, thereby eliminating curvature terms. Taking $f = $ constant defines an *f-plane*; a plane rotating at a rate $\frac{1}{2}f$. Alternately, view the f-plane as a piece of the sphere where the relevant length scale L is small enough so that we may ignore the curvature of the Earth *and* the variation of f.

We may retain the simpler geometry of the plane while variation of rotation rate with latitude is taken into account by putting

$$f = f_0 + \beta y,$$

where $f_0 = f(y = y_0)$ and $\beta = \dfrac{df(y_0)}{dy}$ is the local variation of f with latitude. This configuration is known as the beta-plane (or β-plane).

3.3 The hydrostatic approximation

Consider a part of a motionless column of fluid with density ρ, thickness Δz and area A. The mass of the parcel is being pulled down by the force of gravity: $g\rho A\Delta z$. The parcel stays in place because this force is balanced by the pressure difference between the bottom and the top of the parcel:

$$[p(z) - p(z + \Delta z)]A = \rho g A \Delta z;$$

or

$$\frac{dp}{dz} = -\rho g.$$

For obvious reasons this relation is known as "hydrostatic balance." It turns out that, to a good approximation, it holds for large-scale motions in the atmosphere and ocean.

3.3.1 The hydrostatic equations, with formalities

We now proceed to formally justify the hydrostatic approximation. We will focus on the atmosphere and follow Phillips (1973). Our starting equations are the inviscid forms of Equations 3.4 and Equation 3.7:

$$\frac{d\mathbf{u}}{dt} + \left[f + u \frac{tan\theta}{a+z} \right] \mathbf{k} \times \mathbf{u} + \frac{w}{a+z}\mathbf{u} + 2\Omega cos\theta w \mathbf{i} = -\frac{1}{\rho}\nabla p, \tag{3.15a}$$

$$\frac{dw}{dt} - \frac{(u^2+v^2)}{a+z} - 2\Omega cos\theta u = -\frac{1}{\rho}\frac{\partial p}{\partial z} - g, \tag{3.15b}$$

and

$$\frac{d\rho}{dt} + \rho \left[\nabla \cdot \mathbf{u} + \frac{\partial w}{\partial z} + \frac{2w}{a+z} \right] = 0, \tag{3.15c}$$

where

$$\frac{d}{dt} \equiv \frac{\partial}{\partial t} + \frac{u}{a+z}\frac{1}{cos\theta}\frac{\partial}{\partial \lambda} + \frac{v}{a+z}\frac{\partial}{\partial \theta} + w\frac{\partial}{\partial z},$$

and

$$\nabla \cdot \equiv \frac{1}{a+z}\left[\mathbf{i}\frac{1}{cos\theta}\frac{\partial}{\partial \lambda} + \mathbf{j}\frac{1}{cos\theta}\frac{\partial}{\partial \theta}(cos\theta) \right].$$

First write

$$p = \bar{p}(z) + p^*(x, y, t, z)$$

and

$$\rho = \bar{\rho}(z) + \rho^*(x, y, t, z),$$

where $\frac{\partial \bar{p}}{\partial z} = -\bar{\rho}g$, so the mean fields are defined to be hydrostatic. Note that only p^* enters dynamically (i.e. in Equations 3.15). For the motions to be hydrostatic, the *variable* parts p^*, ρ^* must be; it is not enough that the mean part is hydrostatic.

We introduce characteristic scales for the motions to be considered:

L = horizontal length scale of the motion (1/4 wavelength, say)
H = vertical length scale of the motion

U = magnitude of horizontal velocity u, v

W = magnitude of vertical velocity w

τ = timescale of motions.

Using these scalings, we can scale:

$$\frac{\partial}{\partial x} \sim \frac{\partial}{\partial y} \sim \frac{1}{L}; \frac{\partial}{\partial z} \sim \frac{1}{H}; \frac{\partial}{\partial t} \sim \frac{1}{\tau}$$

(it could be that $\tau = \dfrac{L}{U}$, the advective timescale).

To start off we assume:

(a) $L \leq a$.

(b) $H \leq L$.

(c) It may be that $H \ll L$ or $L \ll a$ but in any case we assume that $H \ll a$.

As a consequence of (c) we may replace $a + z$ by a since $a + z \leq a\left[1 + \frac{H}{a}\right]$; we will do so from here on without further comment. This, if done properly, also eliminates w and other metric terms and $2\Omega \cos\theta$ terms, resulting, as before, in Equations 3.6.

Now consider Equation 3.15a. There are two cases to consider:

(1) Slow motion: $\tau^{-1} \ll f$ or $f^{-1} \ll \tau$.

In this case $\dfrac{du}{dt} \sim \dfrac{U}{\tau} \ll fU$ so that

$$p^* \sim \rho LfU. \tag{3.16}$$

(2) Fast motions: $\tau^{-1} \geq f$ so that

$$\frac{p^*}{\rho} \sim P = L\frac{U}{\tau}. \tag{3.17}$$

Now consider the continuity equation and use the adiabatic relation $dp = c_s^2 d\rho$ (where $c_s = \sqrt{\gamma RT}$ is the speed of sound in the atmosphere) and Equations 3.9 and 3.13 with $J = 0$ to obtain

$$\frac{1}{\rho c_s^2}\frac{dp}{dt} + \nabla \cdot \mathbf{u} + w_z = 0$$

or

$$\frac{1}{\rho c_s^2}\frac{dp^*}{dt} - \frac{g}{c_s^2}w + w_z + \nabla \cdot \mathbf{u} = 0. \tag{3.18}$$

Define the length $D = c_s^2/g$ and assume

(d) $D \geq H$.

The order of magnitude of each of the terms in Equation 3.18 is respectively:

$$\frac{1}{\rho c_s^2}\frac{p^*}{\tau}+\frac{w}{D}+\frac{w}{H}+\frac{U}{L}\sim 0 \quad . \tag{3.19}$$

$$I \qquad\qquad II \quad III$$

Since $H \leq D$ we must have $\dfrac{W}{D}\leq\dfrac{W}{H}$.

(a) Suppose the mean balance is $I+II\approx 0$ so that $W\sim HP/(c_s^2\tau)$.

Then in Equation 3.15b

$$\frac{dw}{dt}\sim\frac{W}{\tau}\sim\frac{HP^*}{c_s^2\tau^2}\quad\text{while}\quad\frac{1}{\rho}\frac{\partial p^*}{\partial z}\sim\frac{P^*}{H}.$$

Hence

$$\frac{dw}{dt}\ll\frac{1}{\rho}\frac{\partial p^*}{\partial z}\quad\text{if}\quad\tau^2\gg\left(\frac{H}{c_s}\right)^2$$

In the atmosphere, $\dfrac{H}{c_s}\sim\dfrac{10^4\,\mathrm{m}}{300\,\mathrm{ms^{-1}}}\sim 30\,\mathrm{s}$ and in the ocean, $\dfrac{H}{c_s}\sim\dfrac{5\times 10^3\,\mathrm{m}}{1500\,\mathrm{ms^{-1}}}\sim 3\,\mathrm{s}$. Our interest is in motions with timescales of many days, so this condition is clearly satisfied. In the case where τ is the advective timescale $\tau\sim L/U$; $\tau\gg\dfrac{H}{c_s}$ is equivalent to $\dfrac{c_s}{U}\gg\dfrac{H}{L}$, which is certainly true if $H/L\ll 1$.

(b) Suppose the term I in Equation 3.19 is small so the balance is $II+III\approx 0$. Then

$$W\sim\frac{H}{L}U\quad\text{so}\quad\frac{dw}{dt}\sim\frac{H}{L\tau}U.$$

If Equation 3.17 holds, i.e. fast motions, then $\dfrac{1}{\rho}p_z^*\sim\dfrac{L}{H}\dfrac{U}{\tau}$ so

$$\frac{dw}{dt}\ll\frac{1}{\rho}\frac{\partial p^*}{\partial z}\quad\text{if}\quad\left(\frac{H}{L}\right)^2\ll 1\quad\text{or}\quad L\gg H.$$

If Equation 3.16 holds, i.e. slow motions, then $P\sim LfU$ so

$$\frac{dw}{dt}\ll\frac{1}{\rho}\frac{\partial p^*}{\partial z}\quad\text{if}\quad\frac{HU}{L\tau}\ll\frac{L}{H}fU\text{ or }(H/L)^2\frac{1}{f\tau}\ll 1.$$

In this case of slow motion $(f\tau)^{-1}\ll 1$. So in either case

$$\frac{dw}{dt}\ll\frac{1}{\rho}\frac{\partial p^*}{\partial z}\quad\text{if}\quad\frac{H}{L}\ll 1.$$

(c) If the balance is $I+III\approx 0$ then $W\sim\dfrac{H}{L}U$ and the above arguments hold.

The Coriolis term in the equation for w, $2\Omega cos\theta u \sim fU$. Now in either Equation 3.16 or 3.17, $P \geq LUf$ so $\frac{1}{\rho}p_z \geq \frac{L}{H}fU$. Hence the Coriolis term is clearly negligible compared to $\frac{1}{\rho}p_z^*$ if $H/L \ll 1$.

It is easy to show that if Equation 3.15b is replaced by the hydrostatic equation, then $2\Omega cos\theta w$ is negligible in Equation 3.15a. The shallow approximation neglects this and all terms in $\frac{w}{a}$, $\frac{u^2 + v^2}{a}$.

There are two major implications of the use of the hydrostatic equation instead of Equation 3.15b. The first is that there is no longer a prognostic equation for the vertical velocity w. The vertical velocity becomes a diagnostic quantity: it takes on whatever values it needs to assure hydrostatic balance. In any case, for large-scale problems, dw/dt is too small to compute. In this sense, hydrostatic balance is forced on us. This does not mean $dw/dt = 0$, only that it is negligible compared to other terms in the force balance. The second implication is that buoyancy oscillations are ruled out. Static instability is no longer automatically handled by the equations because the modified equations have built in that the horizontal scale of motions is much greater than the vertical scale $(L \gg H)$, which eliminates convective motions in which $L \approx H$. Such convective motions are "sub-grid scale" and modelers must parameterize them in terms of variables at the larger scales the models do allow.

With $H/L \ll 1$ and $\tau \gg \frac{H}{c_s}$, the full set of approximate equations become:

$$\frac{d}{dt}\mathbf{u} + \left[f + u\frac{tan\theta}{a} \right] \mathbf{k} \times \mathbf{u} + \frac{1}{\rho}\nabla\rho = 0, \qquad (3.20a)$$

$$\frac{\partial p}{\partial z} = -\rho g, \qquad (3.20b)$$

and

$$\frac{1}{\rho}\frac{d\rho}{dt} + \nabla \cdot \mathbf{u} + \frac{\partial w}{\partial z} = 0, \qquad (3.20c)$$

where

$$\frac{d}{dt} \equiv \frac{\partial}{\partial t} + \frac{u}{acos\theta}\frac{\partial}{\partial\lambda} + \frac{v}{a}\frac{\partial}{\partial\theta} + w\frac{\partial}{\partial z}.$$

We also have the first law of thermodynamics; in its adiabatic form, in terms of potential temperature:

$$\frac{d\ln\theta}{dt} = \frac{d\ln\theta^*}{dt} + \frac{N^2}{g}w = 0 \quad \text{where} \quad N^2 = g\frac{\partial\ln\bar{\theta}}{\partial z}. \tag{3.20d}$$

3.3.2 Boussinesq equations

Since the density variations of water – e.g. in the ocean – are very small, it is attractive to ignore the slight variations in the inertia of the fluid and take ρ to be constant in the momentum balance. The variations of density in the continuity equation are similarly small, so it is a good approximation to take $\rho = $ constant, which effectively renders the fluid incompressible. However, we do not want to remove the important influence of stratification, so we must allow the variations in buoyancy and thus cannot simply say $\rho = $ constant. The Boussinesq equations accomplish all these desirable goals.

While the formal derivation of the Boussinesq approximation is complicated and will not be given here, the basic idea of this approximation is simple: $\rho \approx \rho_0 = constant$ except where ρ is coupled to gravity.

Therefore, with subscripts representing differentiation:

$$\frac{1}{\rho}p_x \approx \frac{1}{\rho_0}p_x = P_x,$$

where $P = \dfrac{p}{\rho_0}$ is the dynamic pressure and since

$$\frac{d\rho}{dt} + \rho\nabla_3 \cdot \mathbf{V}_3 = 0,$$

taking ρ to be constant yields

$$\nabla \cdot \mathbf{V}_3 = 0,$$

which means that the fluid acts like an incompressible fluid. *But $\rho - \rho_0$ is important when coupled to gravity:* i.e. let $p = p_0 + p'$ where $\dfrac{\partial p_0}{\partial z} = -g\rho_0$ so that $\dfrac{\partial p'}{\partial z} = -g(\rho - \rho_0)$.

Define $P \equiv \frac{1}{\rho_0}p'$. Then

$$\frac{\partial p}{\partial z} = -g\frac{\rho - \rho_0}{\rho_0} \equiv b,$$

where b is the buoyancy force (or in more common parlance, the buoyancy).

The first law may now be written as

$$\frac{d\rho}{dt} = Q$$

where Q is the rate of heating (in density units). Therefore

$$\frac{db}{dt} - \frac{g^2}{c_s^2} w = B \quad \text{where} \quad B = -\frac{(gQ)}{\rho_o},$$

or, since $\dfrac{g}{c_s^2} \approx \dfrac{1}{200 \text{ km}}$ for the ocean, the first law simply becomes:

$$\frac{db}{dt} = B.$$

We note that this is certainly a good approximation in the upper ocean. We also note that the Boussinesq approximation does not require hydrostatic balance and is often used to study convection.

3.3.3 Hydrostatic balance and pressure coordinates

We have shown that the hydrostatic relation

$$\frac{\partial p}{\partial z} = -\rho g \tag{3.21}$$

holds for the large-scale motions in the atmosphere and oceans. When Equation 3.21 holds, the equations for the atmosphere may be simplified by changing to pressure coordinates.

First define the geopotential $\Phi = gz$ so that the hydrostatic relation becomes

$$\frac{\partial \Phi}{\partial p} = -\frac{1}{\rho} = -\frac{RT}{p}. \tag{3.22}$$

In pressure coordinates p replaces z as the vertical coordinate. To derive the new equations let $s'(x, y, z(x, y, p, \tau), \tau) = s(x, y, p, \tau)$ and note that for any function $s(x, y, p, \tau)$:

$$\left(\frac{\partial s}{\partial x}\right)_{y,p,t} = \left(\frac{\partial s'}{\partial x}\right)_{y,z,t} + \left(\frac{\partial s'}{\partial z}\right)_{x,y,t} \left(\frac{\partial z}{\partial x}\right)_{y,p,t}$$

and similarly for y and t. In the vertical

$$\frac{\partial s}{\partial p} = \frac{\partial s'}{\partial z} \cdot \frac{\partial z}{\partial p} = -\frac{1}{\rho g} \frac{\partial s'}{\partial z}$$

and with $\nabla_p \equiv$ two-dimensional ∇ on constant p surfaces we have:

$$\nabla_p s = \nabla_z s + \frac{\partial s}{\partial z} \nabla_p z.$$

In particular

$$0 = \nabla_p p = \nabla_z p + \frac{\partial p}{\partial z} \nabla_p z = \nabla_z p - \rho g \nabla_p z$$

so that

$$\frac{1}{\rho} \nabla_z p = \nabla_p \Phi.$$

This makes sense in that, while there is no pressure gradient on a constant pressure surface to accelerate the flow, there is a gravitational force because the surface is not at a constant height. In other words, the fluid tends to flow downhill under the influence of gravity.

The equations in pressure coordinates, with w replaced by $\omega \equiv dp/dt$, are then:

$$\frac{d\mathbf{u}}{dt} + f\mathbf{k} \times \mathbf{u} = -\nabla_p \Phi + \mathbf{F}, \tag{3.23a}$$

$$\frac{\partial \Phi}{\partial p} = -RT/p = -1/\rho, \tag{3.23b}$$

$$\nabla_p \cdot \mathbf{u} + \frac{\partial \omega}{\partial p} = 0, \tag{3.23c}$$

and

$$c_p \frac{dT}{dt} - \frac{RT}{p} w = Q, \tag{3.23d}$$

where

$$\frac{d}{dt} \equiv \frac{\partial}{\partial t} + \mathbf{u} \cdot \nabla_p + \omega \frac{\partial}{\partial p}.$$

Note that in this form of the equations the pressure gradient term is linear and the continuity equation is like the incompressible one. Hence, it is formally like the Boussinesq equations applicable to the ocean. Note also that the $2\Omega cos\theta w$ term is gone.

The disadvantage of this set is that the lower boundary condition is now quite complicated. At the ground, $gw = g\mathbf{u} \cdot \nabla z_o = \mathbf{u} \cdot \nabla \Phi_0$ and

$$gw = g\frac{dz}{dt} = \frac{d\Phi}{dt} = \left(\frac{\partial \Phi}{\partial t} + \mathbf{u} \cdot \nabla \Phi \right)_p + \omega \frac{\partial \Phi}{\partial p}.$$

Also, $\dfrac{\partial \Phi}{\partial p} = -\dfrac{1}{\rho}$ so

$$\omega = \rho(\frac{\partial \Phi}{\partial t} + \mathbf{u} \cdot \nabla \Phi)_p - \mathbf{u} \cdot \nabla \Phi_o \quad \text{at} \quad p = p_o.$$

But p_o changes in time (and space), which is awkward. So most models with topography follow Phillips (1957), and use $\sigma = p/p_o$ as the vertical coordinate. At the ground $\sigma = 1$ and $\dfrac{d\sigma}{dt} = 0$. The equations get messier – extra terms appear – but computers are undaunted by this extra complexity.

3.3.4 Ocean dynamic height

The hydrostatic relation may be written as

$$\frac{\partial \Phi}{\partial p} = -\alpha$$

where $\alpha = \rho^{-1}$ is specific volume.

Using this form of the hydrostatic relation, differences of geopotential can be written as

$$\Phi_1 - \Phi_2 = -\int_{p_1}^{p_2} \alpha dp \,.$$

Oceanographers usually work with specific volume anomalies; i.e. write $\alpha = \delta\alpha + \alpha_{35,o,p}$ where the last term is the reference specific volume at $T = 0\,°\mathrm{C}$ and salinity $\mathrm{S} = 35‰$ (parts per thousand). Now define

$$D = \int_{p_1}^{p_2} \delta\alpha dp \tag{3.24}$$

as the dynamic height anomaly. To see its meaning consider that

$$\nabla_p D = \nabla_p \Phi_1 - \nabla_p \Phi_2.$$

Now suppose the geopotential surface Φ_2 is "flat," i.e. $\nabla_p \Phi_2 = 0$. Then D determines the pressure gradient force at level 1. Typically, level 2 is taken as a reference level such as 1000 decibars (approximately 1000 m) and D is then the dynamic height (anomaly) relative to 1000 decibars. The concept is useful because the flow at 1000 m (or 500 m for that matter) is so weak compared to the near-surface flow that

$$\nabla \Phi_2 \ll \nabla \Phi_1,$$

so D is an excellent approximation to the pressure forces.

3.4 Geostrophy

The quasi-geostrophic (Q-G) approximation is central to much of oceanography and meteorology. This is especially true for midlatitudes, but the concept is also essential for understanding equatorial dynamics. In recognition of its importance, derivations, formal and informal, can be found in virtually any textbook for ocean or atmosphere dynamics. Our recommendation for a formal derivation bundled with valuable insights and interesting history is the original Charney (1948) paper on geostrophic scaling. We will restrict ourselves here to a heuristic discussion. For simplicity – and for a change – we will work with the atmosphere in pressure coordinates. Everything easily carries over to a Boussinesq ocean, and just as easily to a compressible ocean or atmosphere in z-coordinates.

We begin by *defining* the geostrophic velocity (u^g, v^g) to be the velocity that would balance the Coriolis force against the geopotential height gradient:

$$fu^g \equiv -\frac{\partial \Phi}{\partial y} \quad \text{and} \quad fv^g \equiv \frac{\partial \Phi}{\partial x}, \quad \text{or} \quad f\mathbf{k} \times \mathbf{u}^g = -\nabla_p \Phi. \tag{3.25}$$

Now the question is whether the actual velocity is equal to the geostrophic velocity to a good approximation. Comparing to the full-momentum balance (Equation 3.23a), we see that in regions where friction is small (i.e. away from boundaries in the ocean or atmosphere), this equality demands that the acceleration $d\mathbf{u}/dt$ be small compared to the Coriolis term. If the characteristic scales for f, velocity, temporal variations and spatial variations are f_0, U, T and L, respectively, then this translates into the conditions

$$T^{-1} \ll f_o \text{ and } U/L \ll f_o$$

or

$$f_o^{-1} \ll T \text{ and } R_0 \equiv U/f_o L \ll 1;$$

that is, the timescale of the motions of interest must be long compared to an inertial period, and the "Rossby number," R_0 must be small. Another interpretation of the Rossby number being small is that the relative vorticity, $\zeta = \partial v/\partial x - \partial u/\partial y$, must be small compared to the planetary vorticity, in particular, $\zeta \approx R_0 f$. The reader is invited to plug in typical numbers for synoptic scale motions in the atmosphere or ocean and verify that these conditions are met; the flows at these scales are in geostrophic balance. The system is then also hydrostatic.

An important equation is obtained by differentiating the geostrophic relation with respect to p and using the hydrostatic relation (Equation 3.22) to obtain:

$$f\mathbf{k} \times \frac{\partial \mathbf{u}}{\partial p} = -\frac{\partial}{\partial p} \nabla_p \Phi = -\nabla_p \left[\frac{\partial \Phi}{\partial p}\right] = -\nabla_p \frac{1}{\rho}. \tag{3.26}$$

For the atmosphere the equation of state may be written as

$$p = \rho RT \text{ or } 1/\rho = RT/p, \text{ so}$$

$$f\mathbf{k} \times \frac{\partial \mathbf{u}}{\partial p} = \frac{R}{p} \nabla_p T. \tag{3.27}$$

This relation, known as the "thermal wind equation," clearly shows the dependence of geostrophic wind shear on quasi-horizontal temperature gradients.

Another important concomitant of geostrophy is that the horizontal velocity field is approximately non-divergent. Taking the (vertical component of the) curl of the geostrophic relation,

$$\frac{\partial}{\partial x}(fu) - \frac{\partial}{\partial y}(-fv) = \frac{\partial}{\partial x}\left(-\frac{\partial \Phi}{\partial y}\right) - \frac{\partial}{\partial y}\left(-\frac{\partial \Phi}{\partial x}\right) = 0$$

or

$$f\nabla \cdot \mathbf{u} + \beta v = 0.$$

If the scales of motion are not too large, so that $f_0 \gg \beta L$, then this means the velocity must be nearly non-divergent. Even if $f_0 \approx \beta L$, the divergence $\delta \approx R_0 \zeta$, and so is smaller by a factor of a Rossby number than the vorticity. Note that if f is constant $\beta = 0$, so geostrophy then implies exact non-divergence. It is also worth noting that geostrophy actually requires sufficient stratification so that vertical motions will be small, and in accord with the continuity Equation 3.23c, the flow can then be horizontally non-divergent (see, e.g. Charney, 1948).

Returning to Equation 3.27, we can use the continuity Equation 3.23c to write it as

$$\frac{df}{dt} = \beta v = f\frac{\partial \omega}{\partial p},$$

a relation (one of two) known as the Sverdrup relation. This states that the advection of planetary vorticity is balanced by vortex stretching. Note that the relative vorticity ζ of a fluid parcel is neglected relative to the planetary vorticity f since by our scaling, $\zeta \approx R_0 f$.

Geostrophy is a *balance*: a static diagnostic relation between the velocity and the geopotential height (or pressure, if we were in z-coordinates) gradient. It cannot tell us how either evolves. It implies nothing about cause: it does not mean that the velocity causes the height gradient or vice versa. It only states that, somehow, the two have mutually adjusted to be in geostrophic balance. ("Geostrophic adjustment" is a fascinating topic with a large literature.)

There is an evolution equation for geostrophically balanced flows, the "quasi-geostrophic (Q-G) potential vorticity equation." Formally, it may be derived by expanding the equations in powers of a small parameter, the Rossby number R_0. (As we saw above, the geostrophic scaling gives $f{:}\zeta{:}\delta = 1{:}R_0{:}R_0{}^2$.) The leading order is the geostrophic relation and the next order is an evolution equation for the Q-G potential vorticity. Less formally, if we take the curl of the horizontal momentum (Equation 3.23a), the leading terms, the Coriolis terms and the geopotential height gradient are eliminated, and the resulting time-dependent equation for the relative vorticity allows the evolution of the flow to be calculated:

$$\frac{d_g\varsigma^g}{dt} + \beta v^g + f\nabla{\cdot}u^g = \text{ lower order terms } (\textit{l.o.t.})$$

where

$$\frac{d_g}{dt} \equiv \frac{\partial}{\partial t} + \mathbf{u}^g{\cdot}\nabla,$$

and the superscript or subscript "g" indicates geostrophic quantities. From the continuity Equation 3.23c we may replace $\nabla{\cdot}u^g$ with $-\partial\omega/\partial p$ and then use the energy Equation 3.23d (with $Q = 0$ for simplicity) in a Q-G variant,

$$\frac{d_gT}{dt} + \left[\frac{\partial\bar{T}}{\partial p} - \frac{1}{\bar{\rho}c_p}\right]\omega = \textit{l.o.t.}$$

to eliminate ω in favor of T:

$$\frac{d_g\varsigma^g}{dt} + \beta v^g + f\frac{\partial}{\partial p}\left\{\left[\frac{\partial\bar{T}}{\partial p} - \frac{1}{\bar{\rho}c_p}\right]^{-1}\frac{d_gT}{dt}\right\} = \textit{l.o.t.}.$$

Since the geostrophic velocity is approximately non-divergent, we may define a streamfunction ψ such that

$$\frac{\partial\psi}{\partial x} = v^g; \frac{\partial\psi}{\partial y} = -u^g; \text{ so } \nabla^2\psi = \varsigma^g \text{ and } \frac{\partial\psi}{\partial p} = \frac{-R}{fp}T,$$

with the last relation being a consequence of the thermal wind relation (Equation 3.27). Substituting these relations in the equation above yields the Q-G potential vorticity equation in terms of the single variable ψ:

$$\frac{d_g}{dt}\left\{\nabla^2\psi + f + \frac{\partial}{\partial p}\left(S\frac{\partial\psi}{\partial p}\right)\right\} = \textit{l.o.t.} \tag{3.28}$$

where the stability factor $S = \dfrac{-f_0^2 p}{R}\left[\dfrac{\partial\bar{T}}{\partial p} - \dfrac{1}{\bar{\rho}c_p}\right]^{-1}$, f_0 is the mean value of f and the overbars are the horizontal means so that $\bar{T}(p)$ and $\bar{\rho}(p)$ vary only in the vertical. In deriving this equation we swapped the order of $\dfrac{d_g}{dt}$ and $\dfrac{\partial}{\partial p}$; to do so we used the fact that, because of the thermal wind relation, there is no advection of temperature by thermal wind (the vertical shear of the geostrophic wind that appears in the thermal wind equation); i.e.

$$\frac{\partial \mathbf{u}}{\partial p} \cdot \nabla_p T = \frac{\partial \mathbf{u}}{\partial p} \cdot \frac{p}{R} f\mathbf{k} \times \frac{\partial \mathbf{u}}{\partial p} = 0.$$

The quasi-geostrophic equations are a filtered system, allowing the evolution through a succession of geostrophically balanced states while eliminating the modes of motion that bring about these balances. For quasi-geostrophy it is inertia-gravity waves that make the adjustments, so the large-scale slow motions stay in geostrophic balance. It is remarkable that it is possible to use the conservation of quasi-geostrophic potential vorticity to evolve the flow without having to worry about the details of how inertia-gravity waves make it happen.

3.5 Simple layered models of the ocean

3.5.1 Shallow-water equations

A simple system arises from the primitive equations when we apply them to a homogeneous fluid ($\rho = const$), shallow enough so we may approximate $\partial u/\partial z = 0$. Since it is hydrostatic

$$p_z = -\rho g \Rightarrow p_a - p(z) = -\int_z^h \rho g dz \Rightarrow p(z) = \rho g(h-z) + p_a,$$

where p_a is the pressure of the overlying atmosphere at the top surface of the fluid which we will take to be spatially uniform. Hence

$$\frac{1}{\rho}\nabla p = g\nabla h \quad \text{for all} \quad z$$

and

$$\frac{d\mathbf{u}}{dt} + f\mathbf{k} \times \mathbf{u} = -g\nabla h + \mathbf{F}, \tag{3.29a}$$

where $\mathbf{F}$ stands for all frictional forces, including surface-wind stress.

The continuity equation is the incompressible one:

$$w_z + \nabla \cdot u = 0.$$

Integrate from $z=0$ where $w=0$ to $z=h$:

$$0 = w(h) + \int_o^h \nabla \cdot \mathbf{u} \, dz = w(h) + \nabla \cdot \int_o^h \mathbf{u} \, dz - \mathbf{u}(h) \cdot \nabla h$$

$$= w(h) - \mathbf{u} \cdot \nabla h + \nabla \cdot (h\mathbf{u}).$$

But

$$w(h) \equiv \frac{dh}{dt} = \frac{\partial h}{\partial t} + \mathbf{u} \cdot \nabla h.$$

Therefore:

$$\frac{\partial h}{\partial t} + \nabla \cdot (h\mathbf{u}) = 0. \tag{3.29b}$$

Equations 3.29 form the shallow-water equations. They are clearly nonlinear.

We can linearize the shallow-water equations by writing $h = h' + H$; $h' \ll H$ and $\mathbf{u} = \bar{\mathbf{u}} + \mathbf{u}'$ with $\bar{\mathbf{u}} = 0$. Then, dropping the primes gives the linear shallow-water equations:

$$u_t - fv + gh_x = F^{(x)}, \tag{3.30a}$$

$$v_t + fu + gh_y = F^{(y)}, \tag{3.30b}$$

$$h_t + H\nabla \cdot \mathbf{u} = 0. \tag{3.30c}$$

An alternate form is obtained by letting the geopotential $\Phi = gh$ and $c^2 = gH$:

$$u_t - fv + \Phi_x = F^{(x)}, \tag{3.31a}$$

$$v_t + fu + \Phi_y = F^{(y)}, \tag{3.31b}$$

$$\Phi_t + c^2\nabla \cdot \mathbf{u} = 0, \tag{3.31c}$$

where c is the gravity wave speed on a fluid of depth H.

Exercise 3.5: Suppose the bottom is not flat and has the form $D(x, y)$. What are the corresponding shallow-water equations?

3.5.2 Transport equations

Assuming that the Rossby number $R_0 = U/fL \ll 1$, where, as in Section 3.4, U and L are characteristic velocity and length scales, respectively, we can define the vertically integrated transport as $\mathbf{U} = \int_{-D}^h \mathbf{u} \, dz$ where we integrate from the assumed flat bottom at $z = -D$ to the surface at $h+H$, where h is the perturbed surface height:

$$h + H = \int_{-D}^h \rho \, dz.$$

The boundary conditions are $w = 0$ at $z = -D$ and $w = \frac{dh}{dt} \approx \frac{\partial h}{\partial t}$ at $z = h$. The governing equations are now:

$$\frac{\partial \mathbf{U}}{\partial t} + f\mathbf{k} \times \mathbf{U} + gH\nabla h = \boldsymbol{\tau}_0 - \boldsymbol{\tau}_D$$

and

$$\frac{\partial h}{\partial t} + \nabla \cdot \mathbf{U} = 0,$$

where $\boldsymbol{\tau}_0$ and $\boldsymbol{\tau}_D$ are the stresses at the top and bottom, respectively. If we now assume quasi-geostrophy (Section 3.4) $f_0 U \approx -gHh_y$ and $f_0 V \approx gHh_x$ and define a transport streamfunction, $U = -\psi_y$ and $V = \psi_x$ then $h = \frac{f_0}{gH}\psi$. Now the vorticity $\zeta = V_x - U_y = \nabla^2 \psi$. Taking the curl of the transport equations yields:

$$\frac{\partial \zeta}{\partial t} + \beta V - f_0 \frac{\partial h}{\partial t} = \nabla \times \boldsymbol{\tau}_0 - \kappa \zeta,$$

where the last term is a usual linear form for the bottom stress. In terms of the stream-function, the vorticity equation becomes:

$$[\nabla^2 - \lambda^{-2}]\psi_t + \beta\psi_x = \nabla \times \boldsymbol{\tau}_0 - \kappa\nabla^2\psi$$

where $\lambda = \frac{\sqrt{gH}}{f_0}$ is the Rossby radius of deformation. We will see later that this vorticity equation and its generalization will give the planetary-wave equations.

3.5.3 1½-Layer model

Consider an ocean of depth H for which the shallow-water equations apply, i.e. the depth is small compared to the characteristic scale of the motion.

Since ρ is constant within the layer, it follows that $\partial u/\partial z = 0$ so that u is independent of depth. Consider, for example, the thermal wind relation – but we must also assume that there are no stresses within the layer which allow a velocity shear. Integrating through the layer gives the non-forced version of Equations 3.30 where h is the deviation of the upper surface from its flat value:

$$u_t - fv = -gh_x, \tag{3.32a}$$

$$v_t + fu = -gh_y, \tag{3.32b}$$

$$h_t = -H(u_x + v_y). \tag{3.32c}$$

These are the shallow-water equations of a shallow fluid of depth H.

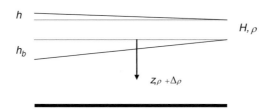

Figure 3.3. Schematic structure of the ocean with the lower layer at rest.

Now consider a shallow ocean consisting of a layer of the same undisturbed depth H and density $\rho - \Delta\rho$ on top of a *motionless* layer of density ρ. Let the deviations of the upper surface be h and the deviations of the lower surface be h_b as shown in Figure 3.3.

The pressure at a point z below the *undisturbed* layer is calculated hydrostatically as

$$p(z) = \rho g H + (\rho + \Delta\rho)gz. \tag{3.33}$$

When there are deviations of the depth of the upper layer, the pressure at the point z is

$$p(z) = \rho g(h + H + h_b) + g(\rho + \Delta\rho)(z - h_b). \tag{3.34}$$

Because the bottom layer is motionless, no change of upper-layer depths can change the hydrostatic pressure in the motionless layer or else the lower layer would move. Therefore, the hydrostatic pressure given by Equations 3.33 and 3.34 must be the same. This gives:

$$h_b = \frac{\rho}{\Delta\rho} h, \tag{3.35}$$

which means that the lower surface must move by the large factor of $\dfrac{\rho}{\Delta\rho}$ more than the upper surface and in the opposite direction in order to make sure the lower layer remains motionless. For the upper layers of the tropical ocean, $\rho/\Delta\rho \sim 500$. Hence, in the upper-layer momentum equation we can write Equations 3.32 in terms of h_b as

$$u_t - fv = -g'h_{bx}, \tag{3.36a}$$

$$v_t + fu = -g'h_{by}, \tag{3.36b}$$

$$h_{bt} = -H_e(u_x + v_y), \tag{3.36c}$$

where $g' = \dfrac{\Delta\rho}{\rho} g$ is known as the reduced gravity. The wave speed in this system is $c = \sqrt{g'H}$ and, as an alternate to reduced gravity, we can write the equations in exactly the same form as Equations 3.32 but with $H_e = \dfrac{\Delta\rho}{\rho} H$ replacing H in Equation 3.32c so that the new (equivalent) depth is about 500 times smaller than the original depth of the active layer if $H = 200$ m, $H_e = 40$ cm and $c = 2$ m/s.

We see that simply by putting the layer of depth H on a motionless denser fluid, the effective depth becomes H_e, a much smaller equivalent depth, and the effective gravity becomes much smaller, g'. The undulations of the interface layer become large compared with those of the free surface.

3.5.4 2-Layer model

As a natural extension of the above, we now allow the lower layer to move. Consider a two-layer ocean with an upper layer of mean depth H_1, density ρ_1, lying over a layer of mean depth H_2, density ρ_2. Let h_1 and h_2 be the deviations of the surface and interface, respectively, measured from the bottom of the ocean.

The linear equations for the upper layer are

$$\frac{\partial \mathbf{u}_1}{\partial t} + f\mathbf{k} \times \mathbf{u}_1 + \frac{1}{\rho_1}\nabla p_1 = \frac{\boldsymbol{\tau}_s - \boldsymbol{\tau}_I}{H_1} \tag{3.37a}$$

and

$$\frac{\partial h_1}{\partial t} + H_1 \nabla \cdot \mathbf{u}_1 = 0. \tag{3.37b}$$

For the lower layer:

$$\frac{\partial \mathbf{u}_2}{\partial t} + f\mathbf{k} \times \mathbf{u}_2 + \frac{1}{\rho_2}\nabla p_2 = \frac{\boldsymbol{\tau}_I - \boldsymbol{\tau}_B}{H_2} \tag{3.38a}$$

$$\frac{\partial h_2}{\partial t} + H_2 \nabla \cdot \mathbf{u}_2 = 0, \tag{3.38b}$$

where $\boldsymbol{\tau}_s$, $\boldsymbol{\tau}_I$ and $\boldsymbol{\tau}_B$ are stresses at the top, interface and bottom, respectively.

Assuming the pressure above the fluid may be taken as zero and integrating the hydrostatic relation $p_z = -\rho g$ from the surface at $h_1 + h_2$ to a depth z, in layer 1 yields:

$$p(z_1) = \rho_1 g[h_1 + h_2 - z_1].$$

Hence

$$\frac{1}{\rho_1}\nabla p_1 = g\nabla[h_1 + h_2]. \tag{3.39}$$

Similarly:

$$p_2 = g\rho_1[h_1 + h_2 - h_2] + g\rho_2[h_2 - z_2];$$

$$\frac{1}{\rho_2}\nabla p_2 = g\frac{\rho_1}{\rho_2}\nabla h_1 + g\nabla h_2 = g\nabla[h_1 + h_2] - g\delta\nabla h_1, \tag{3.40}$$

$$\text{where } \delta = \frac{\rho_2 - \rho_1}{\rho_2} = \frac{\Delta\rho}{\rho} \ll 1.$$

We can recast the 2-layer Equations 3.37 and 3.38 into the SWE form. Define $\mathbf{u} = a_1\mathbf{u}_1 + a_2\mathbf{u}_2$ where a_1, a_2 are constants, and form a_1 (Equation 3.37a) + a_2 (Equation 3.37b):

$$\frac{\partial \mathbf{u}}{\partial t} + f\mathbf{k} \times \mathbf{u} + g\nabla\{[a_1 + a_2(1 - \delta)]h_1 + [a_1 + a_2]h_2\} = a_1\boldsymbol{\tau}_s/H_e. \tag{3.41}$$

(For simplicity we take $\boldsymbol{\tau}_I = \boldsymbol{\tau}_B = 0$.) Now let $h = b_1h_1 + b_1h_2$ and form b_1 (Equation 3.37b) + b_2 (Equation 3.38b):

$$\frac{\partial h}{\partial t} + \nabla \cdot [b_1 H_1 \mathbf{u}_1 + b_2 H_2 \mathbf{u}_2] = 0. \tag{3.42}$$

For Equations 3.41 and 3.42 to take the form of the SWE with velocity $\mathbf{u}$, mean depth H_e, and variable depth h requires that

$$[a_1 + a_2(1 - \delta)]h_1 + [a_1 + a_2]h_2 = h = b_1h_1 + b_2h_2, \tag{3.43}$$

$$H_e\mathbf{u} = H_e[a_1\mathbf{u}_1 + a_2\mathbf{u}_2] = b_1 H_1\mathbf{u}_1 + b_2 H_2\mathbf{u}_2. \tag{3.44}$$

Equations 3.43 and 3.44 must hold for all $\mathbf{u}_1$, $\mathbf{u}_2$ and h_1, h_2 so the coefficients of each of these four variables must be equal:

$$a_1 + a_2(1 - \delta) = b_1 \quad ; \quad a_1 + a_2 = b_2,$$

$$a_1 H_e = H_1 b_1 \quad ; \quad a_2 H_e = H_2 b_2.$$

So

$$H_1 a_1 + H_1 a_2(1 - \delta) = b_1 H_1 = H_e a_1;$$

$$H_2 a_1 + H_2 a_1 = b_2 H_2 = H_e a_2;$$

$$\begin{pmatrix} \dfrac{H_1}{H_2} & \dfrac{H_1(1-\delta)}{H_2} \end{pmatrix} \begin{pmatrix} a_1 \\ a_2 \end{pmatrix} = H_e \begin{pmatrix} a_1 \\ a_2 \end{pmatrix}, \tag{3.45}$$

which is an eigenvalue problem for H_e, the equivalent depth. H_e satisfies

$$(H_1 - H_e)(H_2 - H_e) - H_1 H_2 (1 - \delta) = 0$$

or

$$H_e = \frac{H_1 + H_2 \pm [(H_1 + H_2)^2 - 4\delta H_1 H_2]^{1/2}}{2}.$$

Since (for the ocean) $\delta \ll 1$, the two solutions are approximately

$$H_e^+ \approx H_1 + H_2 - \frac{\delta H_1 H_2}{H_1 + H_2} \approx H_1 + H_2 \tag{3.46a}$$

and

$$H_e^- \approx \frac{\delta H_1 H_2}{H_1 + H_2}. \tag{3.46b}$$

For the + mode,

$$\frac{a_1}{a_2} \approx \frac{H_1}{H_2};$$

e.g. $H_e \mathbf{u}^+ \approx H_1 \mathbf{u}_1 + H_2 \mathbf{u}_2$ is the total transport of the water column. This mode is called the barotropic mode. The currents are approximately equal in the two layers, the equivalent depth of the fluid is equal to the actual mean depth and the barotropic mode basically acts as if the fluid were of constant density from top to bottom.

For the − mode,

$$a_1 \approx -a_2 \left[1 - \frac{\delta H_1}{H_1 + H_2} \right],$$

so the motions in the two layers are opposite. This is called the baroclinic mode and has approximately zero net transport. The two modes are sketched in Figure 3.4.

Note that if $H_2 \gg H_1$ then $H_e \approx \delta H_1$ and $a_1 \approx -a_2$ are in the baroclinic mode. The current is much stronger in the upper layer: zero transport $\Rightarrow |u_1/u_2| = |H_2/H_1| \gg 1$. Suppose that layer 2 is infinitely deep. There can, therefore, be no motion in this layer – or there would be infinite energy. If $\mathbf{u}_2 = 0$ then $\nabla p_2 = 0$. Hence, from Equation 3.40,

$$\nabla(h_1 + h_2) - \delta \nabla h_1 = 0. \tag{3.47}$$

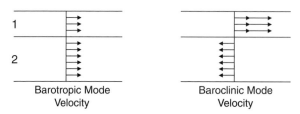

Figure 3.4. Sketch of vertical velocity structure of horizontal velocities for barotropic mode (Left) in which velocities are substantially the same in the two layers, and the baroclinic mode (Right) in which the modes are in opposite directions in such a way that the net (vertically averaged) transport is zero.

Now $\eta = h_1 + h_2$ is sea level; hence the slope of the interface between the two layers is δ^{-1} times the sea-level slope. It is easy to show that as $H_2 \to \infty$ the equations for the upper layer are just the equations for the 1½-layer model of the previous section:

$$\frac{\partial \mathbf{u}_1}{\partial t} + f\mathbf{k} \times \mathbf{u}_1 + g'\nabla h_1 = \frac{\tau_s - \tau_I}{H_1}, \tag{3.48a}$$

$$\frac{\partial h_1}{\partial t} + H_1\nabla \cdot \mathbf{u}_1 = 0, \tag{3.48b}$$

where $g' = g\delta$ is reduced gravity.

Exercise 3.6: We have ignored τ_I and τ_B. Suppose $\tau_I = \kappa_1(\mathbf{u}_1 - \alpha\mathbf{u}_2)$ and $\tau_B = \kappa_B\mathbf{u}_2$. What values can κ_I, κ_B and α have to allow a separation into baroclinic and barotropic modes?

3.6 Vertical ocean modes in a continuously stratified fluid

We saw that, in layer models, the barotropic and baroclinic modes were properties of the difference of density in the various layers. We can extend the concept of equivalent depths and standing modes to the case of continuously varying density in the ocean. Because the atmosphere has no top, these considerations do not apply and we will delay consideration of the modes (or lack of them) for the atmosphere to Chapter 5.

The linearized Boussinesq equations in a stratified ocean are:

$$\frac{\partial u}{\partial t} - fv = -\frac{1}{\rho_0}\frac{\partial p}{\partial x}, \tag{3.49a}$$

$$\frac{\partial v}{\partial t} + fu = -\frac{1}{\rho_0}\frac{\partial p}{\partial y}, \tag{3.49b}$$

$$\frac{\partial p}{\partial z} = -g\rho, \tag{3.49c}$$

$$\frac{\partial u}{\partial x} + \frac{\partial v}{\partial y} + \frac{\partial w}{\partial z} = 0, \tag{3.49d}$$

$$\frac{\partial \rho}{\partial t} + w\frac{d\overline{\rho_0}}{dz} = 0, \tag{3.49e}$$

where the basic stratification is assumed to be a function of z only: $\overline{\rho_0} = \overline{\rho_0}(z)$ and the boundary conditions are taken to be $w = 0$ at the bottom of the ocean, $z = -D$ and $w = \dfrac{\partial \eta}{\partial t}$ at the top, where η is sea level. It is also assumed that the basic state has zero velocities. We can define a buoyancy $b(z) = g\left(\dfrac{\overline{\rho_0} - \rho}{\rho_0}\right)$, and in terms of buoyancy, the hydrostatic and continuity equations become:

$$\frac{1}{\rho_0}\frac{\partial p}{\partial z} = b \tag{3.49c'}$$

and

$$\frac{\partial b}{\partial t} + N^2 w = 0, \tag{3.49e'}$$

where the Brunt–Väisälä or buoyancy frequency is given by

$$N^2 = \frac{-g}{\rho_0}\frac{\partial \overline{\rho_0}}{\partial z} = \frac{\partial b_0}{\partial z}.$$

We will look for solutions via separation of variables into functions of the form

$$(u, v, \frac{p}{\rho_0}) = (\hat{u}(x, y, t), \hat{v}(x, y, t), g\hat{h}(x, y, t))A(z)$$

in view of Equations 3.49a and 3.49b we assume the same vertical structure for u, v, P/ρ_0. On the other hand, because of Equation 3.49d, w must have a different z structure:

$$w = \widehat{w}(x, y, t)B(z).$$

We can eliminate b between Equations 3.49c' and 3.49e':

$$\frac{1}{\rho_0}\frac{\partial}{\partial t}\frac{\partial p}{\partial z} = \frac{\partial b}{\partial t} = -N^2 w$$

so that

$$\frac{\partial}{\partial t} g \hat{h}(x, y, t) \left[\frac{d}{dz} A(z) \right] = -[\widehat{w}(x, y, t)] N^2(z) B(z). \qquad (3.50)$$

Taking the z derivative of Equation 3.49e′ using Equation 3.49d

$$\left[\frac{\partial}{\partial x} \hat{u}(x, y, t) + \frac{\partial}{\partial y} \hat{v}(x, y, t) \right] \frac{d}{dz} A(z) = -\widehat{w}(x, y, t) \frac{d^2}{dz^2} B(z). \qquad (3.51)$$

Divide Equation 3.50 by Equation 3.51 to yield:

$$\frac{g \dfrac{\partial}{\partial t} \hat{h}(x, y, t)}{\dfrac{\partial}{\partial x} \hat{u}(x, y, t) + \dfrac{\partial}{\partial y} \hat{v}(x, y, t)} = \frac{N^2(z) B(z)}{\dfrac{d^2}{dz^2} B(z)}. \qquad (3.52)$$

Since the left-hand side of Equation 3.52 is a function of *(x,y,t)* but not z, and the right-hand side is a function of z only, the only way they can be equal everywhere is if both sides are equal to a constant. Call this constant $-gH_e$. Then Equation 3.52 becomes the two equations:

$$\frac{d^2}{dz^2} B(z) + \frac{N^2}{gH_e} B(z) = 0 \qquad (3.53)$$

and

$$\frac{\partial}{\partial t} \hat{h}(x, y, t) + H_e \nabla \cdot \hat{\mathbf{u}}(x, y, t) = 0. \qquad (3.54)$$

Equations 3.49a and 3.49b can now be written as

$$\frac{\partial \hat{\mathbf{u}}}{\partial t} + f\mathbf{k} \times \hat{\mathbf{u}} + g\nabla \hat{h} = 0. \qquad (3.55)$$

Equations 3.54 and 3.55 are the shallow-water equations for a fluid of depth H_e. For obvious reasons H_e is called the "equivalent depth." But remember, it is just a separation constant – in certain circumstances, the equivalent depth can be negative – atmospheric tides are a prime example (see Lindzen, 1967).

The "vertical structure" is determined by Equation 3.53 (called the vertical-structure equation) plus boundary conditions. For a flat-bottomed ocean with bottom at $z = -D$, $w = 0$ or equivalently, $B = 0$ at $z = -D$.

At the top, $z = 0$, $p(z = 0) = \rho_0 g\eta$ where η, the sea level, deviates only slightly from $z = 0$. Therefore

$$\frac{1}{g\rho_0} \frac{\partial}{\partial t} p(z = 0) = \frac{\partial \eta}{\partial t} \approx \frac{d\eta}{dt} = w(z = \eta) \approx w(z = 0)$$

or

$$\frac{\partial \hat{h}}{\partial t} A(z) = \hat{w} B(z).$$

Then, from Equations 3.49e′ and 3.54

$$\frac{\partial \hat{h}}{\partial t} A(z) = -H_e [\nabla \cdot \hat{\mathbf{u}}] A(z) = H_e \hat{w} \frac{dB(z)}{dz},$$

or

$$H_e \frac{dB}{dz} = B \text{ at } z = 0. \tag{3.56a}$$

To make things still simpler, one often assumes a "rigid lid" at $z = 0$ so $w(z = 0) = 0$ or

$$B = 0 \text{ at } z = 0. \tag{3.56b}$$

As an example, we can return to Equation 3.53 and take $N^2 =$ constant. The solutions are then

$$B = \alpha \cos mz + \beta \sin mz \text{ with } m = \frac{N}{\sqrt{gH_e}} \tag{3.57}$$

The bottom boundary condition, $B = 0$, implies a relationship between α and β which we can write as

$$B = \gamma \sin m(z + D). \tag{3.58}$$

The top boundary condition, either Equation 3.56a or Equation 3.56b, determines the eigenvalues H_e (via m). If the boundary condition is Equation 3.56b then we must have

$$\sin mD = 0 \Rightarrow mD = n\pi \text{ where } n = 1, 2, 3 \dots$$

so that

$$H_e = \frac{N^2 D^2}{g(n\pi)^2} = \frac{D^2}{H_\rho (n\pi)^2} \quad \text{for } n = 1, 2, 3, \dots \tag{3.59}$$

where $H_\rho = \left(\frac{N^2}{g}\right)^{-1} = \left(\frac{1}{\rho_o} \frac{\partial \rho_o}{\partial z}\right)^{-1}$ is the scale height for ocean density. In the real ocean, $H_\rho \approx 200$ km so that

$$\frac{D}{H_\rho} = \frac{N^2 D}{g} \ll 1$$

and H_e is small. In fact, $H_e < 1$ m in the real ocean.

Since H_e is small – in particular, small compared to the vertical scale of $B(z)$ – it follows that the solution (Equation 3.59) with the rigid-lid condition (Equation 3.56b) is also the approximate solution with the free-surface condition (Equation 3.56a). Note that all these modes are internal, or baroclinic modes. They owe their existence to the stratification, and the vertical-structure function A for u, v, p changes sign with depth. Note that the continuity equation implies that $A \sim \dfrac{dB}{dz}$.

In addition, however, the free surface allows an additional mode, the external or barotropic mode. Substituting the expression for $B(z)$ given by Equation 3.58 into Equation 3.56a yields the eigenvalue equation

$$\tan(mD) = mH_e. \tag{3.60}$$

In addition to the internal modes with H_e small, Equation 3.60 has a solution with H_e large. If $mD \ll 1$, then

$$H_e = \frac{\tan(mD)}{m} \approx D.$$

Exercise 3.7: Verify that with $H_e = D$, it is true that $mD \ll 1$.
Then with $mD \ll 1$ it follows from Equation 3.58 that $w(z) \sim z + D$ and $A(z)$ is approximately independent of depth.

The rigid-lid approximation, that $w = 0$ at the surface, does allow an external mode of sorts. The requirement that there be no net divergence at any point in the water column eliminates the possibility of inertia-gravity waves, where the restoring force is gravity acting on the variations in surface elevation. But it does not preclude non-divergent motions; in particular, non-divergent Rossby waves are allowed.

3.7 The shallow-water equations on a sphere and equatorial beta-plane

Earlier we saw that the linear motions in a two-layer fluid could be described by two sets of shallow-water equations: one for the barotropic mode and one for the baroclinic mode. The equations for each mode are identical in form, differing only in the equivalent depth, h, in each set. One may derive the same form for a fluid with n layers, in which case n sets of shallow-water equations are obtained for n modes, each with its characteristic equivalent depth. For a continuously stratified ocean; $\bar{\rho} = \bar{\rho}(z)$, one may separate the motions into an infinite set of vertically standing modes, with characteristic equivalent depths, each of which is governed by the shallow-water equations. Hence much insight into the motions in the ocean – and into models of the ocean – may be derived from an analysis of the shallow-water equations:

$$\mathbf{u}_t - f\mathbf{k} \times \mathbf{u} + \nabla p = 0, \tag{3.61a}$$

$$p_t + gh\nabla \cdot \mathbf{u} = 0. \tag{3.61b}$$

These are the equations governing a shallow homogeneous ocean of depth h or the horizontal structure of a vertical mode with equivalent depth h. As we have seen, in forced problems h may be negative and the idea of waves on a shallow ocean of negative depth must be abandoned. For the free modes we consider here, $h > 0$.

Equations 3.61 would describe the motions on an f-plane if $f = $ constant or a non-rotating fluid if $f = 0$. To investigate the free oscillations on a sphere, we first write

$$(u, v, p) = (U(\theta), V(\theta), 2\Omega a P(\theta)) \exp[i(s\lambda - 2\Omega\omega t)] \tag{3.62}$$

to obtain

$$-i\omega U - sin\theta V + is(cos\theta)^{-1}P = 0, \tag{3.63a}$$

$$-i\omega V + sin\theta U + dP/d\theta = 0, \tag{3.63b}$$

and

$$\frac{1}{cos\theta}\left[isU + \frac{d}{d\theta}(Vcos\theta)\right] - i\omega\varepsilon P = 0; \tag{3.63c}$$

where $\varepsilon = \left(\dfrac{2\Omega a}{gh}\right)^2$ is called the Lamb parameter. Solving for U and V gives

$$U = \left[\frac{\omega s}{cos\theta}P + sin\theta\frac{dP}{d\theta}\right](\omega^2 - sin^2\theta)^{-1},$$

$$V = i\left[s\ tan\theta P + \omega\frac{dP}{d\theta}\right](\omega^2 - sin^2\theta)^{-1},$$

and substituting in Equation 3.63c we obtain the Laplace tidal equation:

$$\frac{d}{d\mu}\left[\frac{(1-\mu^2)}{(\omega^2-\mu^2)}\frac{dP}{d\mu}\right] - \frac{1}{\omega^2-\mu^2}\left[\frac{-s}{\omega}\frac{(\omega^2+\mu^2)}{(\omega^2-\mu^2)} + \frac{s^2}{1-\mu^2}\right]P + \varepsilon P = 0 \tag{3.64}$$

where $\mu \equiv sin\theta$. This equation has been well studied; we summarize a number of properties of its solutions (cf. Longuet-Higgins, 1968; Lindzen, 1970).

(a) For latitudes $\mu < \omega$ Equation 3.64 is hyperbolic; for $\mu > \omega$ it is elliptic.
(b) The singularities at $\omega = \pm\mu$ are apparent; solutions are bounded there.
(c) Equation 3.64 with the boundary conditions at $\mu = \pm 1$ forms an eigenfunction–eigenvalue problem with ε (or equivalently h), the eigenvalue.
(d) For a given s the eigenfunctions P are orthogonal on the sphere.
(e) The eigenvalues h_n are real.

In general there are no closed-form solutions to Equation 3.64. Solutions are generally found by writing P as a sum of associated Legendre polynomials (cf. Longuet-Higgins, 1968). The resulting functions, called Hough functions, after Hough (1898), who first found such solutions, have been extensively tabulated by Longuet-Higgins (1968). A very useful summary is given in Moura (1976). Many of the properties of planetary waves on spheres can be obtained by considering the equations on a beta-plane. Much of this discussion follows Lindzen (1967) and Philander (1978).

The basic idea in the β-plane approximation is to expand the trigonometric functions that appear in Equation 3.63 about a reference latitude θ_0

$$x = a\lambda\cos\theta_0; \quad y = a(\theta - \theta_0) \quad \text{and} \quad f = f_0 + \beta y \qquad (3.65)$$

where

$$f_0 = 2\Omega\sin\theta_0 \quad \text{and} \quad \beta \equiv \frac{df}{dy}\Big|_{\theta=\theta_0} = \frac{2\Omega}{a}\cos\theta_0.$$

Thus, the only effect of the sphericity of the Earth that is retained is the (now linear) variation of f. Equations 3.61 then become (with $f = f_0 + \beta y$)

$$u_t - fv + p_x = 0, \qquad (3.66a)$$

$$v_t + fu + p_y = 0, \qquad (3.66b)$$

$$p_t + gh(u_x + v_y) = 0. \qquad (3.66c)$$

When $f_0 = 0$, we are on an equatorial beta-plane.

4

Boundary layers on both sides of the tropical ocean surface

Sea-surface temperature (SST) is crucial for atmosphere–ocean interactions. Since the sea surface is the common interface between the atmosphere and ocean, either system can change the SST through exchanges of heat and momentum between the two. In general, the regions on both sides of the sea surface are turbulent. In order to understand those turbulent fluxes of heat and momentum capable of changing the SST, we have to be able to characterize the boundary layers on both sides of the interface. The SST can change directly because of changes in boundary-layer mixing, and, in addition, changes in boundary-layer mixing can change the fluxes themselves. For example, an atmospheric boundary layer growing into a vertically sheared wind (say the wind increases upward) will entrain more momentum into the boundary layer and increase the stress at the surface. This, in turn, can change the evaporation and sensible heating into the atmosphere and, consequently, the heat flux into the ocean.

The nature of the boundary layers in the tropical atmosphere and ocean is very different and we will have to treat each boundary layer in a manner that respects this difference. The tropical marine atmospheric boundary layer is driven from below by buoyancy forcing at the ocean surface and is unstable. The tropical ocean boundary layer is also heated at the surface, so warm water overlies cold, creating stable conditions. Mixing has to be forced by wind stresses at the surface.

The tropical atmospheric boundary layer over the ocean is convectively mixed by the buoyancy generated by sensible heat from the surface and by the light weight of water vapor evaporated from the ocean surface. The mixing elements are convective plumes which have large vertical velocities and are dominated by vertical, rather than horizontal, velocities. By contrast, the ocean is stirred mechanically, predominantly by wind stresses at the sea surface. The mixing elements are eddies whose velocities tend to be more homogeneous in the vertical and horizontal dimension. The ocean mixes down according to the working by the wind stress and stops when the mixed layer can no longer entrain heavier laminar fluid upward. The atmospheric boundary layer mixes upward according to buoyancy working from the surface and reaches equilibrium when no additional lighter laminar fluid can be entrained downward into the turbulent boundary layer.

We discuss the atmosphere and ocean separately because of this essential difference in character. We begin by introducing some essential basic concepts and relations common to both the atmosphere and the ocean that are inherent in the geometry of well-mixed turbulent layers. The key concept is entrainment.

4.1 Mixing, inversions and entrainment: general concepts

If we consider a stable profile of potential temperature in the atmosphere (heavy fluid below light fluid) and instantaneously mix the density profile to height h_a with no addition of heat, the profile will exhibit discontinuities (Figure 4.1) simply as a result of the mixing. Note that in this example there is no *entrainment* of fluid from the region above $z = h_a$, and that that the fluid in each region is unchanged.

Exercise 4.1: If the initial temperature profile is given by $T = T_s + \Gamma z$, how big is the discontinuity at h_a after mixing?

Exercise 4.2: Suppose a fluid with an initial profile $\rho(z)$ is mixed uniformly without entrainment or addition of heat to a height h_a. Show that if the initial profile is stable; i.e. $\rho(z)$ decreases with height, then the potential energy of the final profile is greater than that of the initial profile. Note the implication that energy must have been added in the process of mixing. Note too, that if the initial profile below h_a were everywhere unstable (density always increases with height) then the mixing would decrease the potential energy.

The discontinuity (or near discontinuity) of the profile is a characteristic feature of the interface between a well-mixed boundary layer and a stably stratified fluid. For the atmosphere, a discontinuous drop in density at a given height corresponds to a discontinuous rise in temperature (as shown), so that this is usually referred to as an

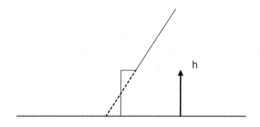

Figure 4.1. The effect of adiabatically mixing an initial, stable potential temperature profile to height h_a in the atmosphere.

inversion. The normal mechanism of mixing is in some way turbulent, so that the density discontinuity marks the interface between a turbulent fluid and a stably stratified laminar fluid. It is a characteristic feature of turbulence that it tends to grow invasively into the laminar fluid and it does this by the mechanism of entrainment: the turbulent elements tend to draw some laminar fluid into themselves and thereby spread the turbulence. To do this, of course, the turbulence must have enough energy to do the work of drawing the lighter laminar fluid downward and mixing it into the denser turbulent fluid below h_a.

We define the entrainment velocity w_e as the rate per unit area at which a volume of laminar fluid passes through the interfacial layer at h_a. It is proper to call it a velocity in that a velocity is a flux (the volume of fluid per unit area crossing the interface). For the atmosphere, with the turbulent layer below, as in Figure 4.1, w_e must be negative, while for the ocean, with the turbulent layer above, it must be positive. If as usual, the ambient vertical velocity dz/dt is denoted by w, then

$$\frac{dh_a}{dt} = w - w_e. \tag{4.1}$$

Equation 4.1 says that if there is no entrainment ($w_e = 0$) then the boundary-layer top at $z = h_a$ moves vertically with the rest of the fluid at the rate w. If the entrainment rate ($-w_e$) exceeds the ambient rate of downward motion $-w$, then the depth of the turbulent fluid will increase. If the opposite is true ($-w > -w_e$) then the depth of the turbulent layer will decrease. This could be the case even while entrainment of fluid from above ($w_e < 0$) persists. The depth of the turbulent layer remains constant in the special case $-w = -w_e$, where the rate that fluid is added by entrainment is exactly balanced by the rate at which it is lost due to ambient divergence. Similar considerations apply to the ocean, except that the entrainment of laminar fluid is upward, since the well-mixed turbulent layer is driven downward from the surface.

In general, in order to examine the changes of density due to boundary-layer mixing, we need an explanation of four separate factors: the height of the interface as it changes; the magnitude of the discontinuity at the interface; the rate of entrainment of mass through the interface; and the energy source for the work needed to entrain the laminar fluid through the interface. This energy source also maintains the turbulence. Equation 4.1 relates interface height and entrainment rate, and, as for example in the first exercise above, the assumption of a uniformly mixed turbulent layer allows the discontinuity to be determined. The difficult issues are the entrainment rate and its relation to energy sources. In the typically stable tropical ocean mixed layer the energy is supplied by the wind, while in the typically unstable tropical atmosphere it is supplied by buoyancy fluxes at the surface. We now turn to the atmospheric surface layer.

4.2 The atmospheric marine boundary layer

4.2.1 Definitions

a. Turbulent fluxes and Reynolds averaging

The vertical flux of a quantity s, i.e the amount of s per unit area crossing a horizontal interface is $\overline{(ws)}$, where the overbar represents horizontal averaging and w is the vertical velocity, i.e. the volume per unit area crossing the interface. In the presence of turbulence, each quantity can be divided into an area-averaged part and a part that is due to the small-scale variations characteristic of turbulence: $w = \bar{w} + w'$ and $s = \bar{s} + s'$, where the horizontal average of the primed quantities vanishes. Then the Reynolds average is

$$\overline{ws} = \overline{w}\,\overline{s} + \overline{w's'}.$$

At the surface, the mean vertical velocity vanishes and the only possible flux is the turbulent flux. In particular, the flux of moisture from the surface is

$$F_q = \overline{\rho(w'q')},$$

where q is the non-dimensional mixing ratio of the mass of water vapor to the mass of air in a unit volume of air and

$$\boldsymbol{\tau} = -\overline{\rho(w'\mathbf{u}')}$$

is the momentum flux (stress) at the interface. An upward moisture flux from the surface would have upward turbulent elements, on average carrying larger amounts of moisture than the downward turbulent elements. Similarly, a positive wind stress would have the downward elements carrying more westerly wind than the upward elements, thereby delivering westerly momentum to the surface.

Other fluxes, buoyancy, heat, dry and most static energy (Chapter 5) will be similarly defined.

b. Friction velocity

For a given stress at the ocean surface $\boldsymbol{\tau}$, the friction velocity u_* is defined as:

$$u_*^2 \equiv \frac{|\boldsymbol{\tau}|}{\rho} \tag{4.2}$$

so that u_* defines a characteristic velocity characteristic of the stress in a medium of density ρ.

Since the stress at the air–sea interface is continuous from the atmosphere to the ocean,

$$|\tau| = (\rho u_*^2)_{\text{air}} = (\rho u_*^2)_{\text{water}},$$

$$\frac{(u_*^2)_{\text{air}}}{(u_*^2)_{\text{water}}} = \frac{\rho_{\text{water}}}{\rho_{\text{air}}} = 1000,$$

therefore

$$(u_*)_{\text{air}} \sim 33(u_*)_{\text{water}}.$$

Typically for $|\tau| = 1$ dyne/cm^2 (or 0.1 Newton/m^2)

$$(u_*)_{\text{water}} = 1 \ \text{cm/s}$$

and

$$(u_*)_{\text{air}} = 33 \ \text{cm/s}$$

c. Monin–Obukhov length

We can define a length:

$$L \equiv \frac{(u_*^3)}{k\overline{(b'w')}_s} \tag{4.3}$$

which measures the vertical distance above the surface in which the mechanical production of turbulence is comparable to the buoyant production of turbulence.

The term $\overline{(b'w')}_s$ is the work done by buoyancy (against gravity) near the surface (subscript s). The buoyancy is measured by θ_v, the virtual potential temperature (☼ see Section 5.1 – Thermodynamic quantities), and $(u_*^3) \propto \tau \cdot \mathbf{u}$ is the work done by mechanical stirring by the wind stress working on the surface of the ocean, and $k = 0.4$ is the von Karman constant. In terms of more familiar quantities,

$$\overline{(b'w')}_s = \frac{g}{\theta_v}\overline{(\theta_v'w')}_s = g\left(\frac{1}{\theta_v}\overline{(\theta'w')}_s + \delta\overline{(q'w')}_s\right)$$

where q is the water-vapor mixing ratio and $\delta = 0.61$ – see Section 5.1. The buoyancy of air is increased by heating it or by adding relatively light water vapor to it.

The Monin–Obukhov length is a length over which the mechanical and buoyancy effects are comparable. Thus, for a mixed layer of depth h: if $h \gg L$, the layer is convectively driven, and if $h < L$, the layer is mechanically driven.

Exercise 4.3: If there are 4 mm/d of evaporation and the Bowen ratio (the ratio of sensible to latent heating) is 0.1, what is the Monin–Obukhov length (in meters)? Take $|\tau| = 1 \ dync/\text{cm}^2$.

d. Characteristic turbulent magnitudes

The moisture flux from the surface is $\rho\overline{(w'q')}_s$ where q is the mixing ratio of moisture. The latent heating from the surface is then $\rho L\overline{(w'q')}_s$ where L is the latent heat of condensation (not to be confused with the Monin–Obukhov length – the context will indicate which is meant). Near the surface, the eddies are mechanically driven so that the vertical and horizontal velocity values are about the same. Therefore, we can define characteristic scales near the surface in analogy to Equation 4.2:

$$\overline{(w'\theta')}_s = ku_*\theta_* \tag{4.4a}$$

and

$$\overline{(w'q')}_s = ku_*q_*. \tag{4.4b}$$

4.2.2 The surface layer

For a fluid near any rigid boundary, in the absence of any heat and moisture fluxes, the velocity follows a classic logarithmic profile ("the law of the wall"):

$$u(z) = \frac{u_*}{k}\ln\frac{z}{z_0} \tag{4.5}$$

where z_0 is the "roughness length." While there have been discussions of the dependence of z_0 on roughness and various other things (e.g. see Kraus and Businger, 1994), we will soon see that the roughness length is related to the neutral drag coefficient. The roughness length is of the order of 0.02 cm over the ocean.

In the presence of upward fluxes from the surface, the profiles are assumed to have universal forms which depend on these fluxes ("Monin–Obukhov similarity theory"). Within a single Monin–Obukhov length of the surface, the mean profiles of temperature, wind and moisture are not well mixed. They take the forms:

$$\frac{d\bar{\theta}}{dz} = \frac{\theta_*}{kz}\varphi_h\left(\frac{z}{L}\right), \quad \frac{d\bar{q}}{dz} = \frac{q_*}{kz}\varphi_w\left(\frac{z}{L}\right) \text{ and } \frac{d\bar{u}}{dz} = \frac{u_*}{kz}\varphi_m\left(\frac{z}{L}\right). \tag{4.6}$$

This scaling seems to match the observations (but with some slight variants in the literature) if we take:

$$\varphi_h\left(\frac{z}{L}\right) = .74\left(1 - 9\frac{z}{L}\right)^{-\frac{1}{2}},$$

$$\varphi_m\left(\frac{z}{L}\right) = \left(1 - 16\frac{z}{L}\right)^{-\frac{1}{2}} \text{ and } \varphi_w\left(\frac{z}{L}\right) = \left(1 - 15\frac{z}{L}\right)^{-\frac{1}{4}}.$$

Note, in particular, that at $z = 0$, $\varphi_m(z/L) = 1$, so that

$$\frac{\partial \bar{u}}{\partial z} = \frac{u_*}{kz}\bigg|_{z=0}$$

and we recover

$$\bar{u}(z \approx 0) = \frac{u_*}{k} \ln \frac{z}{z_0}.$$

Note also that under neutral conditions, i.e. $\overline{b'w'} \to 0$, so that $L \to \infty$ and again,
$\bar{u}(z) = \frac{u_*}{k} \ln \frac{z}{z_0}.$

Define the neutral drag coefficient C_{Dn} as

$$\frac{|\boldsymbol{\tau}|}{\rho} = C_{Dn}|u(z = 10 \text{ m})|^2 = C_{Dn}|u_{10}|^2 = u_*^2$$

in the absence of heat and moisture fluxes from the surface, where the drag coefficient is conventionally defined at $10\,\text{m}$, the nominal height of a ship in olden times. Then

$$u_{10} = \frac{u_*}{k} \ln \frac{10}{z_0} \text{ so that } u_{10}^2 = u_*^2\left[\frac{1}{k^2}\ln^2\frac{10}{z_0}\right]$$

and

$$C_{Dn} = \frac{k^2}{\ln^2 \frac{10}{z_0}}.$$

With $z_0 = 0.02$ cm, $C_{Dn} = 0.0014$ and this is conventionally the value of the neutral drag coefficient for stress.

In the presence of heat and moisture fluxes, one can define a similar relation between stress and winds at $10\,\text{m}$:

$$\frac{|\boldsymbol{\tau}|}{\rho} = C_D|u_{10}|^2, \tag{4.7}$$

but now the drag coefficient C_D depends on the fluxes according to the similarity relations: the source of this variation of drag coefficient with fluxes can be traced to changes of the profiles with fluxes according to the similarity relations. Since the profiles change with fluxes, the conventional 10 m wind used in Equation 4.7 occurs at different points in the vertical profile as the fluxes change, and this changes the effective drag coefficient in Equation 4.7. Although we will not give the results here, the similarity relations Equations 4.6 can be integrated to give explicit profiles for the winds, temperature and moisture near the interface. As $z \gg L$, the profiles approach their mixed-layer values as in Figure 4.2.

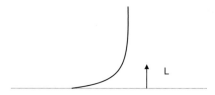

Figure 4.2. Profiles of θ, q or τ above the sea surface.

4.2.3 Fluxes and entrainment in the convectively mixed layer

We will assume that the atmospheric boundary layer is convectively mixed and that the mixed-layer height is large compared to the Monin–Obukhov length.

Let the sensible heat flux at the surface into the atmosphere be $(\overline{\theta' w'})_s$.

The mixed-layer temperature θ_m can be changed by the vertical convergence of this flux and so satisfies, neglecting radiation for the moment,

$$\frac{d\theta_m}{dt} = -\frac{d}{dz}\overline{\theta' w'}.$$

Since the layer is well mixed

$$\frac{d\theta_m}{dz} = 0 \text{ implies } -\frac{d^2}{dz^2}\overline{\theta' w'} = 0$$

so that

$$(\overline{\theta' w'}) = az + b$$

in the interior of the mixed layer. We note that this linear dependence of fluxes in the well-mixed layer is unique to well-mixed layers: it could not have been derived as down-gradient diffusive transfer no matter how high the diffusion coefficient.

The constant b can be evaluated by noting that at $z = 0$, $b = (\overline{\theta' w'})_s$.

To find the other constant, a, we need to know the heat flux at the top of the boundary layer h. If we now allow undulations in the mixed-layer depth so that there can be horizontal advection, the entrainment velocity is $w - \dfrac{dh}{dt}$ where the derivative is now the substantial derivative and, as before, w is the ambient vertical velocity.

There is an inversion at the interface of strength $\Delta\theta = \theta_i - \theta_m$, the difference between the potential temperature of the laminar fluid slightly above the discontinuity θ_i and the mixed-layer potential temperature θ_m. The heat balance at the inversion is

$$\left(w - \frac{dh}{dt}\right)\Delta\theta = \overline{(\theta' w')}_i.$$

In general, the heat flux at the interface will be downward, either because the mixed layer is growing, or because there is a downward environmental velocity, or perhaps because of some combination of the two.

At $z = h$, $ah + b = +\overline{(\theta'w')}_i$ so that

$$\frac{d\theta_m}{dt} = \frac{\overline{(\theta'w')}_s - \overline{(\theta'w')}_i}{h}.$$

To know how convective boundary layers rise, we need an additional equation for the discontinuity, which is simply

$$\frac{d}{dt}\Delta\theta = \frac{d\theta_i}{dt} - \frac{d\theta_m}{dt}$$

$$= +\Gamma w - \left[\frac{\overline{(\theta'w')}_s - \overline{(\theta'w')}_i}{h}\right].$$

If, in the presence of $w \neq 0$, we look for an equilibrium solution $\dfrac{d\Delta\theta}{dt} = \dfrac{dh}{dt} = 0$, then

$$h = \frac{\overline{(\theta'w')}_s - \overline{(\theta'w')}_i}{-\Gamma w}$$

and

$$w\Delta\theta = \overline{(\theta'w')}_i,$$

and we still need a relation for $\overline{(\theta'w')}_i$.

Tennekes (1973), whose method we have followed in the above, recognized that it takes work to bring heat down from above the mixed layer into the mixed layer, since the warmer laminar air is lighter than the cooler turbulent air. The rising plumes gain buoyancy and energy from the surface and lose some to dissipation in the interior of the mixed layer. What is left after dissipation is available to do the work needed to bring the lighter fluid downward into the boundary layer. For most situations, $\overline{(\theta'w')}_i = -0.2\overline{(\theta'w')}_s$ seems to be the right choice (Tennekes, 1973) so the equilibrium solutions are

$$h = \frac{1.2\overline{(\theta'w')}_s}{-\Gamma w} \tag{4.8a}$$

and

$$\Delta\theta = \frac{0.2\overline{(\theta'w')}_s}{-w}. \tag{4.8b}$$

If we know the surface forcing, the temperature gradient Γ into which the mixed layer is rising, and the vertical velocity in the environment, then we can find h and $\Delta\theta$ in the convective mixed layer according to Equations 4.8. The extension to moist boundary layers and the shallow cloud layer is given in Section 5.5.

4.3 The ocean mixed layer

Mixed layers below the tropical ocean surface are stable and wind-stirred from the surface, and further stabilized by downward heat flux through the sea surface: they require a different treatment than the one for convective boundary layers over the tropical sea surface. Furthermore, they see no solid boundary at their upper extent and so the law of the wall need not apply. The models for mixed layers may be divided into two broad classes: 1. Bulk mixed-layer models, and 2. Mixing parameterizations.

4.3.1 Bulk mixed-layer models

Bulk mixed-layer models of the ocean assume a structure like that shown in Figure 4.3. The boundary layer extends to a depth $z = -h$ and, since the layer is assumed well mixed, surface quantities can be identified with boundary-layer quantities. In particular, the temperature of the boundary layer *is* the sea-surface temperature. There are slight corrections to this statement because the molecular "skin" layer near the surface can support a small temperature difference, especially in calm conditions, but for the sake of clarity, we will ignore this.

In the mixed layer, i.e. for $z > -h$, all variables are assumed well mixed:

$$\frac{\partial T}{\partial z} = \frac{\partial \boldsymbol{u}}{\partial z} = \frac{\partial S}{\partial z} = 0.$$

We will assume that all the fluxes of heat, momentum etc., entering the ocean through the surface go into the mixed layer – none get below, with the possible exception of penetrating solar radiation (blue–green light). We will also assume that the properties of the ocean below the entrainment zone, $z = -[h + \delta]$ are unchanged

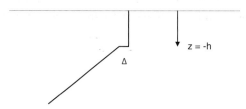

Figure 4.3. Schematic diagram of temperature in a mixed layer extending to depth $-h$ below the ocean surface. Density would have an analogous structure but decreasing with increasing distance upward from the bottom.

by the physics of the ocean mixed layer. So far this is much like the treatment of the atmosphere in Section 4.2.

Note that while it is generally true that $\frac{\partial T}{\partial z}, \frac{\partial \rho}{\partial z}, \frac{\partial S}{\partial z} = 0$ in the ocean mixed layer, it is less true for $\frac{\partial \mathbf{u}}{\partial z}$, which tends not to be so thoroughly well mixed. We also know that the other assumptions are only approximately true: there is some mixing of heat and other quantities out of the base of the mixed layer.

We will first describe bulk-layer formulations; examples include (a) Kraus and Turner (1967), Denman (1973) and Niiler and Kraus (1977); (b) Pollard *et al.* (1973), Price (1979) and Price *et al.* (1986); (c) a constant-depth layer. There are different parameterizations for ocean mixed-layer physics (we return to them below), but regardless of which we use, we already specify a great deal when we fix on the bulk-model structure shown in Figure 4.3. Often in model-physics parameterizations, the model structure is more important than the actual physics used: simply specifying that the quantities are well mixed in the boundary layer can be a major improvement over alternate structures. The simplest choice (c) sometimes works surprisingly well and is usually better than no mixed layer at all.

Suppose for the moment that one or another of these parameterizations provides a way to determine the entrainment velocity w_e, the volume flux per unit area through the base of the ocean mixed layer:

$$\frac{\partial h}{\partial t} + \nabla \cdot (h\mathbf{u}_m) = w_e. \tag{4.9}$$

Exercise 4.4: Show that Equation 4.9, which may be derived in the same manner as shallow-water equation 3.29b is equivalent to Equation 4.1.

Note that for the ocean, the environmental velocity can usually be neglected and that at the base of the mixed layer, the entrainment velocity is upward.

A generic equation for a quantity c (e.g. $c \equiv T, S$) is:

$$\frac{\partial}{\partial t} c + \mathbf{u} \cdot \nabla c + w \frac{\partial c}{\partial z} = -\frac{\partial F}{\partial z} + q_c \tag{4.10}$$

where F is a diffusive flux and q_c a source or sink term and u is the horizontal velocity. Combining Equations 4.9 and 4.10 and integrating yields:

$$\frac{\partial}{\partial t}(hc_m) + \nabla \cdot (h\mathbf{u}c_m) - w_e c_e = F_{TOP} - F_{BOT} + \int_{-h}^{o} q_c dz \tag{4.11}$$

or

$$\frac{\partial}{\partial t} c_m + \mathbf{u} \cdot \nabla c_m + \frac{w_e}{h}[c_m - c_e] = \frac{\Delta F}{h} + \frac{1}{h} \int_{-h}^{o} q_c dz \tag{4.12}$$

where c_e is the value of c in the ocean interior just below the turbulent boundary layer.

In Equation 4.9 we see that the mixed-layer depth can be steady if the divergence in the surface-layer, $\nabla \cdot (hu) = w_e$. (In the ocean the surface-layer divergence is approximately the same as the Ekman pumping since the geostrophic divergence is small; $\nabla \cdot \mathbf{u}_g = O(\beta/f) \approx 0$.)

4.3.2 Mixed-layer parameterizations

We will obtain the salient properties of the steady ocean mixed layers two different ways: by considering the conditions implied by a critical Richardson number, and by considering the turbulent energetics of the layer.

a. The critical Richardson number

Let u_o be the initial ocean mixed-layer velocity, u_m its velocity after deepening, and u_e the velocity of the fluid initially just below the ocean mixed layer. (Here we treat the velocity as a scalar – it refers to either of the horizontal components.) We assume momentum is conserved as the mixed layer deepens from h to $h + \Delta h$.

$$u_m(h + \Delta h) = u_0 h + u_e \Delta h = u_0(h + \Delta h) - [u_0 - u_e]\Delta h$$

so that

$$u_m = u_0 - \frac{\Delta u_0 \Delta h}{h + \Delta h} \tag{4.13}$$

where $\Delta u_0 = u_0 - u_e$. Let $KE = $ kinetic energy; then the change of KE as a result of deepening is:

$$
\begin{aligned}
\frac{\Delta KE}{\rho_0} &= \frac{1}{2}[u_m^2 - u_e^2]\Delta h + \frac{1}{2}[u_m^2 - u_0^2]h \\
&= +\frac{1}{2}[u_m^2 - u_0^2][h + \Delta h] + \frac{1}{2}[u_0^2 - u_e^2]\Delta h \\
&= -\frac{1}{2}[u_m + u_0]\Delta u_0 \Delta h + \frac{1}{2}(u_0 - u_e)\Delta u_0 \Delta h \\
&= -\frac{1}{2}\Delta u_0 \Delta h[u_m - u_e] = -\frac{1}{2}\Delta u_0 \Delta h[u_m - u_0 + \Delta u_0] \\
&= -\frac{1}{2}\Delta u_0^2 \Delta h \frac{h}{h + \Delta h} < 0.
\end{aligned}
\tag{4.14}
$$

Heat and salt are conserved in the deepening; therefore buoyancy is conserved:

$$\rho_m(h + \Delta h) = \rho_0 h + \rho_D \Delta h, \tag{4.15}$$

and the notation for ρ subscripts is the same as that for u. The potential energy is

$$PE = \int g\rho z dz$$

so that PE is increased by pushing warm water *down*.

$$\Delta PE = g\left(\rho_m \frac{(h+\Delta h)^2}{2} - \left\{\rho_0 \frac{h^2}{2} + \rho_d \Delta h\left(h + \frac{\Delta h}{2}\right)\right\}\right)$$

$$\approx \frac{1}{2}g(\rho_0 - \rho_D)h\Delta h > 0. \tag{4.16}$$

Therefore

$$\frac{\Delta PE}{-\Delta KE} = \frac{g(\Delta\rho_0)h}{\rho_0(\Delta u_0)^2}.$$

The Richardson number R_i is defined as

$$R_i = \frac{\dfrac{g}{\rho}\dfrac{\partial\rho}{\partial z}}{\left(\dfrac{\partial u}{\partial z}\right)^2} = \frac{N^2}{\left(\dfrac{\partial u}{\partial z}\right)^2}, \tag{4.17}$$

and the analogous bulk Richardson number R_B is:

$$R_B = \frac{g}{\rho_0}\frac{\left(\dfrac{\Delta\rho}{h}\right)}{\left(\dfrac{\Delta u}{h}\right)^2} = \frac{g}{\rho_0}\frac{\Delta\rho h}{(\Delta u)^2} \tag{4.18}$$

so that

$$\left|\frac{\Delta PE}{\Delta KE}\right| = R_B.$$

The Richardson number measures the ratio of the stabilizing effect of density stratification to the destabilizing effect of vertical shear. For large vertical shear the Richardson number is small, and for a large enough vertical shear, the system is unstable.

The mixed-layer depth is determined by: (a) the conservation rules for momentum and buoyancy Equations 4.13 and 4.15; (b) the condition that all the changes go into the ocean mixed layer without leaking to the ocean below; (c) the well-mixed assumption plus the condition that

$$\left|\frac{\Delta PE}{\Delta KE}\right| = R_B = R_c \leq 1$$

where R_c is the critical bulk Richardson number: the mixed layer mixes down until h is deep enough so that $R_B = R_c$. At this point the ratio of the *PE* gained by mixing to the *KE* lost due to mixing is equal to R_c.

In real applications, the critical bulk Richardson number is not precisely defined with different authors using different values: Pollard *et al.* (1973) use $\frac{1}{4} < R_c \leq 1$, while Price (1979) uses 0.7. If the mixed layer could be identified with the shear layer, Kelvin–Helmholtz instability would imply that $R_c = \frac{1}{4}$. But this would violate our well-mixed condition that says that the vertical shear of the currents in the mixed layer vanishes. The form of Equation 4.18 makes clear that the shear envisaged is measured by the difference between the well-mixed value and the value just below the mixed layer. The mixed layer grows until the shear becomes just small enough to stabilize the layer.

The exact value of R_c probably does not matter too much. We can see this by noting that if there is forcing at the surface by a wind stress τ and a buoyancy B over a time Δt:

$$\Delta u = \frac{\tau \Delta t}{h}; \quad \frac{g \Delta \rho_0}{\rho_0} \equiv \Delta b = \frac{B \Delta t}{h}.$$

At the end of this time interval:

$$R_B = \frac{\Delta b h}{(\Delta u)^2} = \frac{B \Delta t h^2}{\tau^2 \Delta t^2}$$

so that

$$h = \tau \left[\frac{R_c \Delta t}{B}\right]^{1/2} \tag{4.19}$$

and the depth depends relatively slowly on R_c, as the square root. We can see clearly from Equation 4.19 that buoyancy (or heating) is stabilizing: more B results in smaller h. We also see from Equation 4.19 that the longer the wind acts, the deeper the mixed layer. We note still further that since $h \propto (R_c \Delta t)^{1/2}$, the smaller R_c is, the longer the forcings must act to give the same h. This is expected from the interpretation that the smaller R_c is, the smaller the fraction of the kinetic energy lost when the layer deepens available to raise its potential energy; i.e. the smaller R_c is, the greater the fraction of energy that is simply lost.

Why does the ocean mixed layer stop growing? Because the surface-layer current u and hence the shear Δu do not keep increasing even if the wind continues unabated. Typically, the Coriolis effect stops the current from increasing, since the Ekman balance $hu \sim \tau/f$ will eventually obtain. On the equator, the pressure-gradient force or other forces prevent the currents from accelerating without limit.

What if heating goes on forever? For each time increment Δt "new" mixed layer forms at the (same) depth h given by Equation 4.19. Heat is continually put into the mixed layer so its temperature keeps increasing. In reality, the heat flux into the ocean depends on the SST, so the heating will decrease as the temperature increases.

b. Turbulent kinetic energy balance

In this approach, originally due to Kraus and Turner (1967), the TKE (turbulent kinetic energy) balance determines the entrainment rate. It uses the *form* of the vertical profile in Figure 4.3. The essential idea is that the net wind-energy that goes into mechanical stirring is used to raise the potential energy of the water column. The net wind-energy generation is taken to be mu_*^3 ($\tau \cdot \mathbf{u}_{sfc} \propto mu_*^3$) with m a constant. Some versions add an additional dissipation proportional to h, i.e. of the form $-\varepsilon h$:

$$\frac{\Delta(PE)}{\Delta t} = mu_*^3 - \varepsilon h. \tag{4.20}$$

Now

$$\frac{d}{dt}PE = \frac{h^2}{2}\frac{db_m}{dt} + hM(w_e)\delta b = mu_*^3 - \varepsilon h \tag{4.21}$$

where the buoyancy $b = g(\rho_0 - \rho)\rho_0$ and $M(x) = x$ for $x > 0$ and $M(x) = 0$ otherwise.

But all the surface buoyancy forcing B goes into the mixed layer

$$h\frac{db_m}{dt} + M(w_e)\Delta b = B,$$

so that

$$\frac{h}{2}B + \frac{h}{2}M(w_e)\Delta b = mu_*^3 - \varepsilon h. \tag{4.22}$$

In the absence of environmental vertical velocities, equilibrium will obtain when the mixed layer stops deepening and, therefore, when the entrainment velocity vanishes: $w_e = 0$. This implies

$$h = \frac{2mu_*^3}{B + 2\varepsilon} \equiv h_\infty \tag{4.23}$$

which is the ocean version of the Monin–Obukhov length: in the absence of dissipation, the tendency of mechanical stirring to deepen the mixed layer is balanced by the stabilizing effect of a positive buoyancy input. We may write Equation 4.22 as

$$hM(w_e)\Delta b = (h_\infty - h)[B + 2\varepsilon],$$

showing that the layer will continue to deepen until it reaches the depth h_∞. If $h > h_\infty$ there is not enough wind energy to hold the mixed-layer depth where it is; the layer must become shallower. Instead of thinking of the mixed layer as getting shallower one may take the point of view that a *new* mixed layer, shallower than the old one, forms.

Now consider an environmental upward velocity (upwelling) w. An equilibrium value of h is obtained when $w_e = w$. As above

$$w_e h \Delta b = (B + 2\varepsilon)(h_\infty - h),$$

and if the mixed-layer density is constant then $w_e \delta B = B$ so

$$hB = (B + 2\varepsilon)(h_\infty - h)$$

or

$$h = \frac{B + 2\varepsilon}{2B + 2\varepsilon} h_\infty \approx \frac{h_\infty}{2}.$$

The surprising result is that, regardless of upwelling strength, the equilibrium depth is half the Monin–Obukhov depth.

4.3.3 Non-bulk models – mixing at all z

This has been a common approach in numerical general-circulation models of the ocean where mixed layers are not well defined due to the limitations of vertical resolution. Some regard must be given to the need for mixing differently in the mixed layer than in the interior of the ocean. This approach cannot successfully simulate a mixed layer for two reasons: first, the lower boundary of the mixed layer can be resolved only to the accuracy of the vertical discretization and, second, the mixing coefficient required to thoroughly mix the boundary layer would have to be infinite. Yet, the practice of numerical modeling is full of compromises and simulating large (but not infinite) mixing coefficients will suffice in most circumstances.

The problem is to parameterize the mixing terms which take the form:

$$\frac{\partial u}{\partial t} = \cdots - \frac{\partial}{\partial z} \overline{(u'w')},$$

$$\frac{\partial v}{\partial t} = \cdots - \frac{\partial}{\partial z}\overline{(v'w')},$$

and

$$\frac{\partial T}{\partial t} = \cdots - \frac{\partial}{\partial z}\overline{(w'T')}.$$

The basic eddy parameterization is to take κ_m as the eddy viscosity (for momentum), and κ_H as the eddy diffusivity (for heat) by writing

$$-\overline{(u'w', v'w')} = \kappa_m\left(\frac{\partial u}{\partial z}, \frac{\partial v}{\partial z}\right)$$

and

$$\overline{-w'T'} = \kappa_H \frac{\partial T}{\partial z}.$$

There are a number of approaches for the representation of κ_m and κ_m

a. The simplest: κ_H, κ_m = constant.

In general, this is too simple and inaccurate, but there is one important application and this is the Ekman layer (Ekman, 1905). The basic idea of the Ekman layer is that there must be a surface stress layer in which momentum must be transferred from the wind to the water. Pressure cannot "boundary layer" near the surface, so the balance is the "Ekman balance":

$$f\boldsymbol{k} \times \mathbf{u} = \frac{\partial}{\partial z}\boldsymbol{\tau}/\rho = \frac{\partial}{\partial z}\left(\frac{\kappa_m}{\rho}\frac{\partial \mathbf{u}}{\partial z}\right) = \frac{\partial}{\partial z}\left(v\frac{\partial \mathbf{u}}{\partial z}\right) = v\frac{\partial^2 \mathbf{u}}{\partial z^2}$$

where we define a dynamic viscosity $v \equiv \kappa_m/\rho$. The last equality follows from the assumption that v = a constant. The obvious depth scale in this simplified problem is the Ekman depth, $h_{Ekman} = \left(\frac{v}{f}\right)^{1/2}$. Integrate through the surface-layer depth

$$= O\left(\frac{v}{f}\right)^{1/2} = O\left(\frac{u_*}{f}\right)$$ to where the stress vanishes. The transports are the Ekman

drifts: $U_E = \frac{\tau^y}{f}$ and $V_E = -\frac{\tau^x}{f}$. These are independent of the particular form of the mixing coefficient v. The vertical velocity out of the bottom of the Ekman layer is called the Ekman pumping: $w_{ek} = -\nabla \times \left(\frac{\boldsymbol{\tau}}{f}\right)$.

b. Richardson number-dependent coefficients

The basic idea here is to make $\kappa_m = \kappa_m(Ri)$; $\kappa_H = \kappa_H(Ri)$, e.g. Munk and Anderson (1948). The most widely used version is due to Pacanowski and Pilander (1981):

$$\kappa_m = \frac{v_0}{(1 + \alpha Ri)^n} + v_{bm}$$

and

$$\kappa_H = \frac{\kappa_m}{(1 + \alpha Ri)} + v_{bH}$$

where:

$$v_0 = 50 \,\mathrm{cm^2 s^{-1}}, \quad n = 2, \alpha = 5, v_{bm} = 1 \,\mathrm{cm^2 s} \text{ and } v_{bH} = 0.1 \,\mathrm{cm^2 s^{-1}}.$$

As compared to the data, the Pacanowski and Philander parameterization gives values that are too high at high Ri and too low at low Ri. It lacks the sharp transition at $Ri = Ri_c \approx 0.2{-}0.3$.

c. Turbulence closure schemes

Here (e.g. Mellor and Yamada, 1982) the coefficients are parameterized by quantities that depend on the level of turbulence:

$$\kappa_m = lqS_M \text{ and } \kappa_H = lqS_H$$

where $\frac{1}{2}q^2$ = turbulent kinetic energy; ℓ is a turbulent length scale; and S_M, S_H are stability functions taken to be functions of q and are small for $Ri > Ri_c \sim 0.23$ and large for $Ri < Ri_c$.

The turbulent kinetic energy, q, is governed by

$$\frac{d}{dt}\left(\frac{q^2}{2}\right) - \frac{\partial}{\partial z}\left[lqS_q\left(\frac{q^2}{2}\right)\right] = P_s + P_2 - \varepsilon;$$

where P_s is shear production; P_w is buoyancy production; and ε is dissipation assumed to take place in turbulent eddies so that $\varepsilon = \dfrac{q^2}{\beta_1 l}$ where β_1 is assumed constant.

d. KPP, a profile parameterization with nonlocal mixing

At present, the most widely used parameterization of boundary-layer mixing, valid for all states of the ocean mixed layer in ocean general-circulation models is the so-called KPP, the K profile parameterization (Large *et al.*, 1994). Based on theory and observational evidence, KPP assumes specific shapes for the vertical profiles of quantities in the mixed layer rather than taking them to be constant. Fluxes of a quantity χ are defined by:

$$\overline{w'\chi'} = -K_\chi \left(\frac{\partial \chi}{\partial z} - \gamma_\chi \right),$$

where the profile of the diffusivity is given by

$$K_\chi = h w_\chi(\sigma) G(\sigma),$$

where h is the boundary-layer depth, $w_\chi(\sigma)$ is a profile of the turbulent-velocity scale, $G(\sigma)$ is a non-dimensional shape function of the non-dimensional depth $\sigma = \frac{z}{h}$. The vertical velocity profile, taken from universal profiles from boundary-layer theory for both stable and convective boundary layers, is approximated by $G(\sigma) = \sigma(1 - \sigma)^2$ and the nonlocal transport term is given by a set of transports that depend on the condition of the boundary layer.

Finally we note again that in the tropical oceans the ocean mixed layer is never convective for long periods of time. The net heat flux is almost always downward into the ocean, thereby stabilizing the layer. Since the net heat flux acts as a negative feedback to SST, anomalous warm conditions have less net heat flux and anomalous cold conditions more net heat flux downward into the ocean. The ocean outside the tropics *does* become convective – in the winter in particular – and any parameterization for the global ocean must account for this possibility.

5

Atmospheric processes

This chapter deals with the basic atmospheric processes involved in coupled atmosphere–ocean interactions over the tropical oceans, and in particular, those processes needed for a description and analysis of ENSO.

In order to begin the discussion, we have to define some basic atmospheric quantities; in particular, the virtual temperature, the dry static energy and the moist static energy. In terms of these quantities, we examine dry adiabatic ascent, i.e. the temperature changes that would exist if a dry parcel were lifted without the addition of heat, and moist adiabatic ascent, i.e. the temperature changes that a saturated moist parcel would have if lifted with the only internal source of heat being the condensation of parcel water vapor and the subsequent rain out of the water from the parcel.

We then use a classic diagnosis of waves in the tropical Pacific ITCZ to illustrate some unusual differences between tropical and midlatitude atmospheric motions. In particular, the horizontal divergences in these tropical waves are large, in contradistinction to the midlatitudes where geostrophy constrains the horizontal divergences (and therefore the vertical velocities) to be small. The reason for these large tropical divergences is that heating of the atmosphere by deep clouds does not produce much temperature change. Rather, cloud heating by deep cumulonimbus clouds produces synoptic vertical velocities whose adiabatic cooling just balances the cloud condensation heating. The vertical velocity is a measure of divergence and we can therefore say that the essence of tropical atmospheric dynamics is that regions of deep cumulonimbus heating drive divergent circulations rather than changing environmental temperatures. Thermally driven circulations include Walker circulations, Hadley circulations and teleconnections to higher latitudes.

We next examine the basic process that determines the state of the tropical atmosphere: the heating of the atmosphere by clouds. We use a set of simple arguments and models based on the small fractional area covered by the active clouds in a horizontally homogeneous environment to indicate that deep clouds do two distinct things.

If precipitation equals evaporation ($P = E$) over a large synoptic region, the deep cumulonimbus clouds are randomly distributed, there is no net synoptic vertical mass flux, and the latent heating of the clouds is realized outside the clouds by subsiding motion compensating the upward motion in the clouds. Since the clouds cover a very small fractional area, the motion almost everywhere, between the clouds, is downward. In regions of $P > E$, regions of thermal forcing, there is upward synoptic motion but we show that the motion almost everywhere (i.e. between the clouds) is unchanged from the $P = E$ case: the interpretation is that synoptically converged air, which would be expected to cool the environment, instead rises *in* the clouds without cooling the environment – deep cumulonimbus clouds are therefore, in some sense, like insulated tubes. That the air almost everywhere subsides and never changes, regardless of whether $P = E$, $P < E$ or $P > E$, is a realization of the previously noted result that thermal forcing does not change temperatures. It also indicates that in regions of thermal forcing, the heat balance is not effected by additional subsidence compensating the heating but rather by the lack of cooling of the environment by the synoptic upward motion. The discussion will make use of the profound connection between large-scale heat and moisture budgets in the tropics, since both arise almost entirely from the condensation in deep cumulonimbus clouds.

We use the heating of deep clouds to describe the basic structure of the tropical atmosphere in the absence of horizontal temperature gradients and indicate that the tropical atmosphere can be considered to be composed of three vertical layers. The first is a near-surface layer where heat and moisture from the surface mix the atmosphere by dry plumes up to the bottom of the cloud layer (the lifting condensation level). The second is a shallow cloud layer where the shallow clouds condense and re-evaporate without precipitating, leading to a moist layer extending to 2 or 3 kilometers above the surface. The third layer is the interior of the atmosphere where the stratification is set by the deep cumulonimbus clouds. The properties of such a model of the tropical atmosphere are robust even if horizontal temperature gradients at the surface are included.

Three examples of thermally driven circulations are given. The zonally averaged Hadley circulation driven by zonally averaged heating is explored in both its linear and nonlinear forms. A popular linear model for thermally driven circulations in the tropics is the Gill model, which is then discussed and found to be difficult to justify as an explanation for surface winds. An augmented linear theory for thermal forcing by isolated heat sources is described that does explain the conditions under which thermal forcing of the atmosphere can drive surface winds. The *pattern* of the surface-wind forcing resembles the Gill model and this, no doubt, is responsible for its considerable popularity.

We proceed to a simple discussion of the basic process that anchors the regions of deep cumulonimbus convection to regions of warm sea-surface temperatures. The basic process is simple and was first described by Lindzen and Nigam: warm water implies low overlying pressure so that surrounding air converges into the vicinity of the warm water. The boundary-layer winds, as given by the Lindzen–Nigam mechanism, also satisfies the Gill equations.

Once justified and understood, the Gill model is a useful heuristic for thermally driven surface winds in the tropics. The modifications needed to use the theory for surface winds in atmospheric-anomaly models concludes the chapter.

5.1 Thermodynamic quantities

5.1.1 Virtual temperature T_v

We first introduce the concept of virtual temperature which allows us to compare the relative buoyancy of dry and moist parcels.

The equation of state of dry air is (where p is the pressure):

$$p_d = \rho_d R_d T,$$

while the similar equation of state of water vapor is (where e_v is the partial pressure of water vapor):

$$e_v = \rho_v R_v T,$$

where ρ_u, ρ_d are the densities of water vapor and dry air and the dry and vapor gas constants are, respectively,

$$R_d = \frac{R}{29}, \quad R_v = \frac{R}{18} \quad \text{where } R = 8.31 \, \frac{J}{mole. \, K} \text{ is the universal gas constant.}$$

Note that water vapor (H_2O atomic weight 18) is *lighter* than air, which has a mean atomic weight of 29, being composed of about 80% nitrogen (N_2) and about 20% oxygen (O_2) with numerous additional small constituents.

Define the vapor-mixing ratio q as $\rho_v \equiv q\rho_d$ so that the pressure of the mixture is

$$p_{mixture} = p = p_d + e_v = T[\rho_v R_v + \rho_d R_d]$$
$$= T\left[q\rho_d \frac{R}{m_v} + \rho_d \frac{R}{m_d}\right].$$

The density of the mixture is $\rho = \rho_d + \rho_v = \rho_d(1 + q)$, so that

$$p_{mix} = T \frac{R}{m_d} \rho_d \left[q \frac{m_d}{m_v} + 1\right]$$
$$= TR_d \frac{\rho}{1 + q}\left[1 + \frac{q}{\varepsilon}\right]$$

$$\approx \rho T R_d \left[1 + \frac{q}{\varepsilon} - q\right] = \rho R_d T[1 + \delta q],$$

where $\delta = \dfrac{1}{\varepsilon} - 1 = 0.61$ and $\varepsilon = \dfrac{m_v}{m_d} = 0.62$ where m is the relevant atomic weight.

We can now define the virtual temperature

$$T_v = T[1 + \delta q], \tag{5.1}$$

so that the *mixture* satisfies an equation of state which resembles the ideal gas law:

$$p = \rho R_d T_v. \tag{5.2}$$

Note that the equation of state for the mixture uses the *dry* gas constant.

5.1.2 Saturated vapor q_s (T)

When a parcel of air has water evaporated into it until it cannot hold any more, then

$$\rho_{vsat} \equiv q_s(T)\rho,$$

where

$$q_s(T) = \frac{\rho_{vsat}}{\rho} = \frac{e_s(T)/R_v T.}{p/R_d T_v}$$

$$\simeq \frac{0.622}{p} e_s(T).$$

By the Clausius–Clapeyron equation ☼

$$q_s(T) = \frac{0.622}{p} \exp\left[c\left(\frac{1}{T_o} - \frac{1}{T}\right)\right] \tag{5.3}$$

where

$$T_o = 273K \text{ and } c = 17.67 T_o.$$

For example, at $p = 1000\,\text{mb}$, $q_s(30\,°\text{C}) = 27\,\text{gm/kg}$ and $q_s(10\,°\text{C}) = 8\,\text{gm/kg}$, while at $p = 500\,\text{mb}$, $q_s(-10\,°\text{C}) = 1.5\,\text{gm/kg}$ and $q_s(20\,°\text{C}) = 2.5\,\text{gm/kg}$.

In general, the saturated mixing ratio gets very small above the lowest 2 or 3 kilometers of the atmosphere, mostly because it is cold.

5.1.3 Dry adiabatic ascent

Under adiabatic conditions (i.e. condensation is not occurring so that no heat is added to the parcel), a parcel of unit mass, if raised, expands, and its temperature decreases. If lowered, the parcel contracts and compresses, so the temperature of the parcel goes up. Potential temperature θ is the actual temperature a parcel would have if brought adiabatically to a standard pressure p_o, usually taken to be 1000 hPa.

$$\theta = T\left(\frac{p_o}{p}\right)^\gamma, \quad \text{where} \quad \gamma = 2/7 = R/c_p \tag{5.4}$$

so that at $p = p_o$, $\theta = T$.

As the pressure changes by raising or lowering a parcel, θ stays the same and the T of the parcel changes according to:

$$T = \theta\left(\frac{p}{p_o}\right)^\gamma.$$

Since we do not add heat, θ is constant so that $\dfrac{d\theta}{dz} = 0$. The temperature gradient of a parcel raised or lowered adiabatically may be found from Equation 5.4 as follows:

$$0 = \frac{1}{\theta}\frac{d\theta}{dz} = \frac{1}{T}\frac{dT}{dz} - \gamma\frac{1}{p}\frac{dp}{dz} = \frac{1}{T}\frac{dT}{dz} - \gamma\frac{1}{p}(-\rho g);$$

$$\frac{1}{T}\frac{dT}{dz} = -\gamma g\rho\frac{1}{\rho RT} = -\gamma g\frac{1}{RT} = -\frac{1}{T}\frac{g}{c_p}.$$

Hence

$$\left(\frac{dT}{dz}\right)_{ad} = -\frac{g}{c_p}, \tag{5.5}$$

which determines the temperature a parcel will have if raised adiabatically. Note that the dry adiabatic lapse rate $-g/c_p$ is $-9.8°$C/km.

Now consider a parcel raised adiabatically in an atmosphere of background stratification

$$\left(\frac{dT}{dz}\right)_{env} = -\Gamma.$$

At height z, the force per unit mass on the parcel is

$$F = -g\frac{\rho(z) - \rho_{env}(z)}{\rho(z)},$$

so that if the parcel arrives at z with density $\rho(z) > \rho_{env}(z)$, then the force will be downward and the parcel will sink back toward its original level. At a level z

$$\rho(z) = \frac{p}{RT}; \quad \rho_{env}(z) = \frac{p}{RT_{env}}$$

so that

$$F \approx -g\,\frac{\dfrac{1}{T(z)} - \dfrac{1}{T_{env}(z)}}{1/T(z)} = -g\,\frac{T_{env}(z) - T(z)}{T_{env}(z)},$$

i.e. if the temperature of parcel is *cooler* than that of the environment, the parcel will sink back toward its original position. Now, the temperature of the environment is

$$T_{env} = T_0 - \Gamma_{env} z$$

where T_0 is the environmental temperature at $z = 0$, and the temperature of a parcel starting at $z = 0$ is:

$$T_{parcel} = T_0 - \frac{g}{c_p} z$$

so that

$$F = -g\,\frac{\dfrac{g}{c_p} - \Gamma_{env}}{T}\,z.$$

And a parcel will be stable (will sink back: $F < 0$) if

$$\frac{g}{c_p} > \Gamma_{env} = -\frac{dT_{env}}{dz},$$

i.e. the environmental lapse rate must be less steep than the adiabatic lapse rate.

Note that $g/c_p = 9.8\,°\mathrm{C/km}$, while in the tropics $\Gamma_{env} \approx 6\,°\mathrm{C/km}$.

Since the acceleration of the parcel $\ddot{z} = F$,

$$\ddot{z} + \frac{g}{T}\left(\frac{g}{c_p} + \left(\frac{dT}{dz}\right)_{env}\right) z = 0.$$

Define

$$N^2 \equiv \frac{g}{T}\left[\frac{g}{c_p} + \left(\frac{dT}{dz}\right)_{env}\right], \tag{5.6}$$

where N is the Brunt–Väisälä frequency. N is the frequency the parcel will have when adiabatically displaced from its neutrally buoyant position. When $N^2 > 0$, the environmental stratification is stable and the parcel will vertically oscillate with frequency N. When $N^2 < 0$, the environmental stratification is unstable and the parcel will move far from its neutral position.

Note that, since

$$\frac{d\theta}{dz} = \frac{\theta}{T}\left[\frac{g}{c_p} + \frac{dT}{dz}\right],$$

$$N^2 = \frac{g}{\theta}\frac{d\theta}{dz}; \tag{5.7}$$

while in the ocean

$$N^2(z) \approx -\frac{g}{\rho(z)}\frac{d\rho(z)}{dz}.$$

A commonly used quantity is the "dry static energy" s:

$$s = c_p T + gz \tag{5.8}$$

so that

$$\frac{ds}{dz} = c_p\frac{dT}{dz} + g$$

and

$$\frac{ds}{dz} = c_p\frac{T}{\theta}\frac{d\theta}{dz} = c_p T\frac{d}{dz}\ln\theta.$$

We see that s is conserved in *dry* (i.e. non-condensing) ascent.

The gross, dry static-energy profile of the tropical atmosphere looks like Figure 5.1.

The dry static energy is constant in the bottom mixed layer, z_m, has a positive slope up to the tropopause z_T and then is very stable in the stratosphere above the tropopause.

If a parcel rises from the mixed layer, it will have a temperature $(s_m - gz)/c_p$ while the environment will have a temperature $(s - gz)/c_p$.

Therefore,

$$T_{parcel} - T_{env} = \frac{s_m - gz}{c_p} - \frac{s - gz}{c_p} = \frac{s_m - s}{c_p} < 0$$

and the parcel will sink back.

5.1.4 Moist adiabatic ascent

When a *moist* parcel is lifted from near the surface, the temperature first falls adiabatically until the parcel saturates (i.e. s stays constant until saturation). After saturation the "moist" static energy h,

$$h = c_p T + gz + Lq \tag{5.9}$$

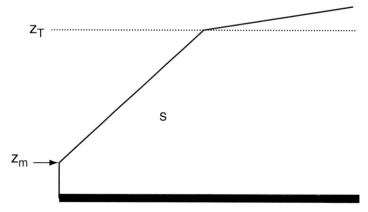

Figure 5.1. Schematic of stable stratification of troposphere: z_m is mixed-layer height, z_T is height of tropopause.

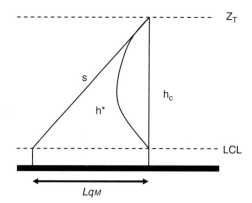

Figure 5.2. Schematic of the moisture structure of the tropical atmosphere. See text for explanation.

stays constant. Note: the liquid is assumed to rain out so that h is constant as the vapor condenses.

Exercise 5.1: What is the relationship between h and $\theta_e = \theta exp\left[\dfrac{Lq}{c_p T}\right]$ where θ_e is equivalent potential temperature?

Exercise 5.2: The buoyancy of a rising parcel is proportional to $T_{parcel} - T_{env}$. Prove that $(T_{parcel} - T_{env} \propto h_{parcel} - h^*_{env})$ where $h^*_{env} = c_p T_{en} + gz + Lq_{sat}(T_{env})$.

Since the environment holds less and less vapor as it gets colder aloft, a typical environmental profile looks like Figure 5.2.

The saturated environmental h^*_{env} first decreases upward and then increases as the environmental value of q^*, the saturated value of the vapor-mixing ratio, decreases with temperature according to the Clausius–Clapeyron relation (Equation 5.3).

In general the lifting condensation level (LCL) is at or near the top of the mixed layer. Parcels rise from within the mixed layer and conserve their value of h – this is the cloud h_c in the diagram. Since the parcel becomes saturated at the lifting condensation level, the cloud h_c is saturated above the lifting condensation level. Therefore, a mixed-layer parcel will rise with constant h_c until it loses its buoyancy, which is proportional to $(h_c - h^*)$. The parcel (cloud) will reach the tropopause at height z_T if

$$\frac{ds}{dz} = \frac{s(z_T) - s_m}{z_T \quad z_M} \approx \frac{Lq_M}{z_T}. \tag{5.10}$$

This means that the water vapor condensing out of the cloud parcel supplies enough bouyancy to the parcel to overcome the dry static stability of the troposphere.

The moist adiabatic lapse rate is given when h^* is constant:

$$\frac{dh^*}{dz} = \frac{d}{dz}(s + Lq^*) = c_p \frac{dT}{dz} + g + L\frac{dq^*}{dz} = 0$$

so that

$$\frac{dT}{dz} = -\left(\frac{L}{c_p}\right)\frac{dq^*}{dz} - \frac{g}{c_p} \tag{5.11}$$

is the moist adiabatic lapse rate. We can get an estimate of the average moist adiabatic lapse rate throughout the troposphere by examining Figure 5.2. An alternate to Equation 5.11 is

$$\frac{ds}{dz} = c_p \frac{dT}{dz} + g = \frac{Lq_M}{z_T - z_{LCL}} \quad \text{so that}$$

$$\frac{dT}{dz} = -\frac{g}{c_p} + \frac{Lq_M}{c_p(z_T - z_{LCL})}.$$

We can estimate the average moist adiabatic lapse rate throughout the troposphere from Equation 5.11 by noting that

$$q^* = q_M \text{ at } z = z_{LCL} \text{ and } q^* \approx 0 \text{ at } z = z_T.$$

Since $L/c_p \approx 2500$, $z_T - z_{LCL} = 16\,\text{km}$ and $q_M = 0.02$, so that the second term is about $+3.3\,\text{K/km}$. The gross moist adiabatic lapse rate in the tropics is therefore about $-6.3\,\text{K/km}$.

Notice that the difference of h_c and h^* is exaggerated in Figure 5.2. The parcel has only a small buoyancy above the LCL and therefore, since the cloud is saturated, the difference between the saturated cloud moist static energy and that of the saturated environmental h^* is, in fact, quite small. Remember also that the environment is *not* saturated – the quantity h^* is useful because $(h_c - h^*)$ measures the buoyancy. Since the environmental lapse rate above the mixed layer is determined by moist parcels raised from the mixed layer, and the buoyancy of these parcels is small, the lapse rate of the free atmosphere is close to moist adiabatic. We will examine a model for this in Section 5.4.

5.1.5 Buoyancy flux from the surface

Since comparing virtual temperatures measures the relative buoyancy of two parcels, the buoyancy flux from the surface is defined as:

$$F \equiv \frac{g}{\theta_v} \overline{w'\theta'}_v,$$

and it is this flux that mixes the boundary layer.

There are two sources of buoyancy – heat flux and water-vapor flux (remember that water vapor is lighter than air so evaporation produces light vapor which produces buoyancy).

Since we are near the surface where $p = p_0$, $\theta_v = T_v$, and since

$$T_v = T(1 + \delta q),$$

$$\overline{w'T_v} \approx \overline{w'T'} + T\overline{\delta q'w'}$$

neglecting the third-order quantity $\overline{w'q'T'}$.

Define the sensible heat flux from the surface

$$S - \rho c_p \overline{T'w'}$$

and the latent heat flux

$$LE = \rho L \overline{q'w'}$$

where $L = 2.5 \times 10^6 J/\mathrm{kgK}$ and $c_p = \frac{7}{2}(287)\ J/\mathrm{kgK}$

Also define the Bowen ratio to be the ratio of sensible to latent heat from the surface, $b = S/LE$, so that

$$\overline{w'T_v} = \frac{S}{\rho c_p} + \frac{\delta T}{\rho L}(LE) = \frac{1}{\rho c_p}S\left[1 + \frac{c_p \delta T}{L}b^{-1}\right] \approx \frac{S}{\rho c_p}[1 + 0.07b^{-1}]. \quad (5.12)$$

We see that for $b \sim 0.07$, the latent and sensible heat contribute equally to the buoyancy flux from the surface. It turns out that this is indeed the order of magnitude of the Bowen ratio in the tropics.

5.2 The diagnosis of Reed and Recker

Reed and Recker (1971), and since then many others, examined 5-day waves in the ITCZ of the western Pacific using data gathered by an enhanced observing system implemented to monitor the thermonuclear bomb tests of the 1960s. Then, instead of averaging, which would have wiped out the structure of the waves since the wavelength was not constant, they averaged all waves into eight separate bins and constructed a composite wave from the results. The composite wave structure is shown in Figure 5.3.

Since these disturbances are *in* the ITCZ, the mean vertical velocity averaged over the entire wave is upward, and this is accomplished by low-level convergence necessarily concomitant with upper-level divergence (Figure 5.4).

The precipitation through the various phases of the wave is shown in Figure 5.5 (left). At the peak of the wave, the rainfall is $>2\,\mathrm{cm/d}$, which corresponds to a heating rate of about 6 K/d. (Note that $1\,\mathrm{mm/d} \Leftrightarrow 29\,\mathrm{W/m^2}$ and $100\,\mathrm{W/m^2}$ heats an atmospheric column 0.8K/d – see Appendix 1.)

We would expect the moisture budget to be

$$P = E + \frac{1}{g}\int_{p_0}^{300mb} \nabla \cdot \overline{(q\mathbf{u})}dp + \text{storage}$$

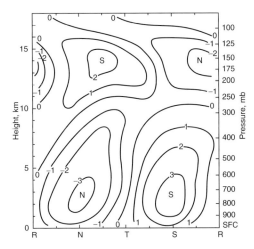

Figure 5.3. The wave composited by meridional wind speed (m/s). The trough and ridge, T and R, have near-zero meridional winds and N and S have meridional winds from the north and south, respectively. (From Reed and Recker, 1971.)

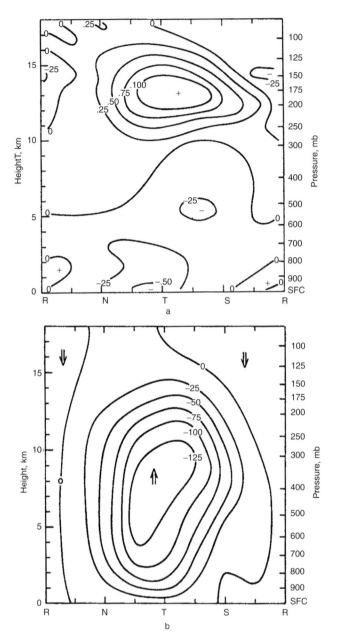

Figure 5.4. Wave composite of horizontal velocity divergence (Upper panel), in units of 10^{-5} per s) and vertical velocity (Lower panel) in units of 10^{-5} mb/s. (From Reed and Recker, 1971.)

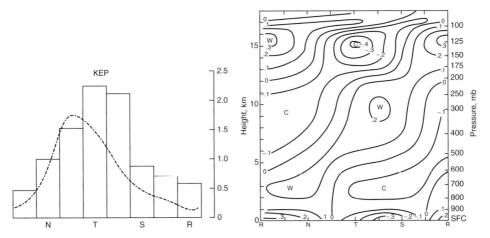

Figure 5.5. Left panel: Precipitation (cm/d) in the various parts of the composite wave (the dot-dashed line gives an estimate based on the moisture budget). Right panel: Temperature structure (degrees K) through the wave. (From Reed and Recker, 1971.)

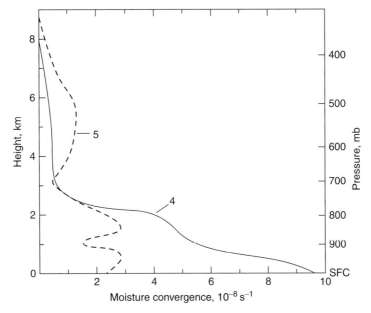

Figure 5.6. Moisture convergence at trough (category 4) and next bin to the east of the trough (category 5) in units of 10^{-8}/s. (From Reed and Recker, 1971.)

and, while there are observational problems, the budget is roughly satisfied. As Figure 5.6 shows, most of the convergence of moisture into the trough part of the wave occurs at low levels, below 2 km.

What is truly extraordinary in this diagnosis is shown in Figure 5.5 (right). Although the heating rates are of the order of 6°/d in the trough part of the wave, the temperature response is no more than 0.1 °/d, *two orders of magnitude smaller.* Thus, see that despite large amounts of precipitation, $\frac{\partial \bar{s}}{\partial t} \approx 0$, so that the basic heat balance is

$$\rho \bar{w} \frac{d\bar{s}}{dz} = Q + Q_R. \qquad (5.13)$$

It is hard to overemphasize the importance of Equation 5.13. It basically says that the net heating, $(Q+Q_R)$, rather than changing temperatures, produces vertical velocities and therefore drives circulations. It is a crucial fact about the tropics that all its various circulations are thermally driven. This is a basic difference from the dynamics of midlatitude circulations, where quasi-geostrophy implies that the divergent circulations are small (see Section 3.4) where the ratio of divergence to vorticity is of the order of the Rossby number. In the tropics, the divergence is of the order of the vorticity and thermally driven divergent circulations are the essence of the tropics.

5.3 How clouds heat

This section is based on the classic paper of Yanai *et al.* (1973). They derived a set of relations that allowed them to use the data set previously mentioned in Section 5.2 to diagnose the properties of clouds in the tropics.

   The essence of the argument is that the active cumulus clouds that heat the tropics cover a very small fractional area of the tropics, something of the order of 2–3%. (This only includes the active core of the cumulonimbus clouds – the cirrus outflow, which is dynamically inactive, but radiatively active, may cover a much larger fractional area.) Thus, while great amounts of latent heat are produced in the atmosphere, the fractional area covered by deep clouds is so small that it is simply not possible for the condensed heat to effectively diffuse from the clouds into the environment. It is possible, however, for the deep cumulonimbus clouds to have large amounts of moisture converged into their base and, because the vertical velocities within the clouds are so large, to rain many times the local evaporation. As will also be shown below, we have to distinguish between the heating that balances the mean evaporation (here taken as spatially uniform, for the sake of simplicity) and the heating that occurs when, in the presence of mean evaporation, there is localized precipitation that exceeds the evaporation. In this latter case, the cumulative effects of deep clouds produce heating because the air going up into the clouds, which would normally produce large-scale cooling, does *not* produce cooling. In this sense the air going up into the clouds has large-scale synoptic effects but the clouds themselves act like insulated pipes – the clouds heat by the synoptically rising air not cooling. In this view, the mean evaporation sets the net number of deep cumulonimbus clouds in the tropics, estimated to be of the order of 2000 deep cumulonimbus clouds around the tropics (Malkus, 1962). It is these cumulonimbus clouds that are responsible for establishing and maintaining the mean state of the tropical atmosphere (see Section 5.4). Local precipitation greater than evaporation corresponds to bunching of deep cumulonimbus clouds. Local precipitation less than evaporation corresponds to scattering and thinning of deep cumulonimbus clouds. This point of view, first pointed out to us by R. S. Lindzen, will be explained in more detail below.

We start by considering the budgets of both heat and moisture. Since all the heat realized in the atmosphere was produced by convergence of low-level moisture, considering both budgets is always a good scientific strategy in the tropics.

Define the quantities Q_1 and Q_2:

$$Q_1 = \rho \frac{D\bar{s}}{Dt} = \rho \frac{\partial \bar{s}}{\partial t} + \rho \bar{\mathbf{u}} \cdot \nabla \bar{s} + \rho \bar{w} \frac{\partial \bar{s}}{\partial z},$$

where $\bar{s}$ is the environmental dry energy per unit mass and the averaging is defined by:

$$\begin{cases} \bar{a} \equiv \text{average } a \text{ over all areas – cloud and non-cloud} \\ \tilde{a} \equiv \text{average } a \text{ over non-cloud area only} \\ a_c \equiv \text{average } a \text{ over cloud area only.} \end{cases}$$

$$Q_2 = -\rho \frac{D\bar{q}}{Dt} = -\rho \left(\frac{\partial \bar{q}}{\partial t} + \bar{\mathbf{u}} \cdot \nabla \bar{q} + \bar{w} \frac{\partial \bar{q}}{\partial z} \right).$$

Q_1 is the apparent heat source and Q_2 is the apparent moisture sink. The dynamic equations governing heat and moisture are then:

$$Q_1 = Q_R + L(c - e) - \frac{\partial}{\partial z} \rho \overline{s'w'} \tag{5.14}$$

and

$$Q_2 = (c - e) + \frac{\partial}{\partial z} \rho \overline{q'w'}, \tag{5.15}$$

where Q_R is the radiative heating rate (per unit volume), c is the condensation rate of water vapor (per unit volume) due to the clouds, e is the evaporation rate of cloud liquid water, and the last terms in Equations 5.14 and 5.15 are the vertical convergence of the heat and moisture fluxes due to the *ensemble* of clouds. s', q' and w' are the anomalies of dry static energy, moisture, and vertical velocity, respectively, at the level z, and the overbar is the horizontal average at level z.

We have assumed, on the right-hand sides of Equations 5.14 and 5.15 that the horizontal convergence of flux due to clouds vanishes, i.e. that clouds do not converge net moisture or heat by their *horizontal* motion, i.e. that the same number of clouds of the ensemble stay within the averaging area.

We would like to simplify the right-hand side of Equations 5.14 and 5.15 so that measurements of the *observable* quantities Q_1, Q_2 and Q_R can tell us something about the properties of cloud ensembles.

Assume that we have a horizontal area large enough so that many clouds are included. The fractional area σ covered by the clouds is assumed to be small. This will be the **Basic assumption**: $\sigma \ll 1$.

Then the average over both cloud and non-cloud areas gives:

$$\bar{s} = \sigma s_c + \tilde{s}(1 - \sigma),$$

where s_c is dry static energy in the cloud ensemble and $\tilde{s}$ is the dry static energy between the clouds. Then

$$\bar{s} = \tilde{s} + \sigma(s_c - \tilde{s}).$$

Since observations show that the temperature difference between the cloud and non-cloud regions is small (this is equivalent to the clouds being almost neutrally buoyant as noted in the previous section)

$$s_c - \tilde{s} \ll \tilde{s}.$$

Therefore

$$\bar{s} \approx \tilde{s} + O(\sigma).$$

Similarly

$$\bar{q} \approx \tilde{q} + O(\sigma).$$

We see that the static enthalpy and the moisture averaged over both cloud and non-cloud regions are given by the values *between* the clouds since the fractional area covered by the clouds is so small.

When we look at the vertical mass flux averaged over between-cloud and non-cloud areas, the situation is quite different:

$$\overline{\rho w} = \sigma(\rho w)_c + (1 - \sigma)(\widetilde{\rho w}),$$

but the vertical velocities in clouds are so large that

$$(\rho w)_c \sim O\left(\frac{1}{\sigma}\right)(\widetilde{\rho w})$$

and we cannot make any simplifications. Hence we can approximate s and q by ignoring cloud effects, but not so for mass flux:

$$\bar{M} = M_c + \widetilde{M},$$

where $\bar{M} = \overline{\rho w}$ is the net synoptic mass flux in a region (given by the total convergence of air entering the region from the sides), M_c is the mass flux *in* the clouds and $\widetilde{M}$ = mass flux *between* the clouds. In general, only $\bar{M}$ is directly observable from synoptic observations.

Three cases will help to fix ideas:

(a) $\bar{M} = 0$, so that there is no net mass flux into the region from the sides, so that the averaged vertical velocity is zero. (Later, taking the region to be the *entire* tropics, we will consider that $\bar{M} = 0$.) There can still be clouds and precipitation as long as the mass flux between the clouds balances the vertical mass flux in the clouds. In this case there is no net large-scale mass convergence and therefore there is no large-scale moisture convergence. One way this can happen is that there is a random distribution of deep cumulonimbus clouds whose precipitation simply balances evaporation. In this case $\widetilde{M} = -M_c$.

There is an intimate relationship between the moisture budget and the low-level mass budget. If we converge no moisture into the region, then $P = E$. There *must* be precipitation balancing evaporation and there must be the correct number of deep cumulonimbus clouds to make this happen. If we make a model of the moisture structure so that almost all the water vapor q_m is well mixed and confined to within 2 km of the surface, then flux balance into the atmosphere above 2 km requires

$$P = E = M_c q_m.$$

This is the tropical water budget. It follows that $M_c = E/q_m$, implying that between the clouds, $\widetilde{M} = -E/q_m$ almost everywhere (i.e. in approximately 98% of the area between the active deep clouds). Outside of the clouds, which is almost the entire area, the motion is downward. Therefore, the steady heat budget of the atmosphere must be

$$LP = \int_0^{Z_T} Q_R \, dz = LE = -Lq_m\widetilde{M};$$

the radiative heat loss of the free atmosphere is balanced locally by the heating due to downward motion. We have, for the purposes of illustration, neglected the specific radiative effects of the deep cumulonimbus clouds and, in particular, their cirrus out-flows, which are known to be important. It is important to remember for this example that the synoptically measured vertical velocity is zero, yet, almost everywhere, the motion is downward – the net upward mass flux *in* the clouds just balances the slow downward motion between the clouds.

(b) $\widetilde{M} = 0$ implies $\bar{M} = M_c$. All the mass goes up into the clouds and is exported laterally. There is *no* descending motion and nothing to balance radiative cooling of the clear air between the clouds, i.e. almost everywhere. This is a most unrealistic case. The measured synoptic mass flux is upward but, almost everywhere (between the clouds) there is no upward or downward motion.

(c) $M_c \neq 0, \quad \tilde{M} \neq 0, \quad \overline{M} \neq 0$: there are now no simplifications. We have to consider convergence in the region, partitioned into upward vertical mass flux in the clouds and environmental mass flux outside the clouds.

We return to the problem of evaluating the vertical convergence of the heat and moisture fluxes in Equations 5.14 and 5.15. We first evaluate the flux terms carried by the cloud ensemble.

$$\rho \overline{s'w'} \approx M_c(s_c - \bar{s}) + O(\sigma)$$

and

$$\rho \overline{q'w'} \approx M_c(q_c - \bar{q}) + O(\sigma).$$

In order to further evaluate the vertical derivatives in Equations 5.14 and 5.15, we need a cloud *model*.

To illustrate, let us assume that the cloud ensemble can be characterized by a single $M_c(z)$. The clouds comprising the ensemble will entrain environmental air and detrain cloud air, and $M_c(z)$ will be the mass flux left in the ensemble due to all the clouds that have not detrained below level z. We further assume that a cloud detrains (i.e. stops rising) only where it has lost buoyancy, i.e. at $s_c = \bar{s}$.

Call the entrainment rate ε and the detrainment rate d. The cloud model is then the set of budget equations for the cloud ensemble:

$$\frac{dM_c}{dz} = \varepsilon - d \qquad\qquad \text{Mass,}$$

$$\frac{d}{dz} M_c s_c = \varepsilon \bar{s} - d s_c + Lc \qquad \text{Heat,}$$

$$\frac{d}{dz} M_c q_c = \varepsilon \bar{q} - d q_c - c \qquad \text{Moisture.}$$

This is our cloud model. We can now use the cloud model to evaluate the vertical derivative of cloud flux:

$$\frac{\partial}{\partial z} \rho \overline{s'w'} = \frac{\partial}{\partial z} M_c(s_c - \bar{s}) = \frac{\partial}{\partial z} M_c s_c - \bar{s} \frac{\partial M_c}{\partial z} - M_c \frac{\partial \bar{s}}{\partial z}$$

$$= \varepsilon \bar{s} - d s_c + Lc - \bar{s}(\varepsilon - d) - M_c \frac{\partial \bar{s}}{\partial z}$$

$$= d(s_c - \bar{s}) + Lc - M_c \frac{\partial \bar{s}}{\partial z}.$$

But clouds detrain where $s_c - \bar{s} = 0$ so that

$$-\frac{\partial}{\partial z}\rho\overline{s'w'} = M_c\frac{\partial\bar{s}}{\partial z} - Lc. \tag{5.16}$$

Similarly

$$\frac{\partial}{\partial z}\rho\overline{q'w'} = \frac{\partial}{\partial z}M_c(q_c - \bar{q}) = \frac{\partial}{\partial z}M_cq_c - \bar{q}\frac{\partial M_c}{\partial z} - M_c\frac{\partial\bar{q}}{\partial z}$$

$$= \varepsilon\bar{q} - dq_c - c - \bar{q}(\varepsilon - d) - M_c\frac{\partial\bar{q}}{\partial z}$$

$$= d(\bar{q} - q_c) - c - M_c\frac{\partial\bar{q}}{\partial z};$$

$$\frac{\partial}{\partial z}\rho\overline{q'w'} = -M_c\frac{d\bar{q}}{dz} - c + d(\bar{q} - q_c). \tag{5.17}$$

Using Equations 5.16 and 5.17 in 5.14 and 5.15 gives the final expression for how the cloud ensemble heats and moisturizes:

$$Q_1 = \rho\frac{D\bar{s}}{Dt} = M_c\frac{d\bar{s}}{dz} - Le + Q_R, \tag{5.18}$$

$$-Q_2 = +\rho\frac{D\bar{q}}{Dt} = M_c\frac{d\bar{q}}{dz} + e + d(q_c - \bar{q}). \tag{5.19}$$

The first two terms on the right-hand side of Equation 5.18 describe how the cloud ensemble heats the atmosphere. The first term is a direct heating term where positive mass flux in the clouds acts on the stable atmospheric stratification. The second term is the cooling due to evaporation of liquid that has detrained from the clouds.

The three terms on the right-hand side of Equation 5.19 describe how the cloud ensemble moisturizes the atmosphere. Since $\frac{d\bar{q}}{dz} < 0$, the first term describes drying due to upward motion in the clouds. The second term is the moisturizing due to evaporation of liquid water from detraining clouds, and the third is direct injection of moisture to the environment by detraining clouds.

Yanai *et al.* (1973) were able to use the observed values of Q_1 and Q_2 and some reasonable assumptions on the other terms to diagnose the mass fluxes and other properties of the cloud ensemble using data from the synoptic network set up in the western Pacific to monitor the aforementioned nuclear tests.

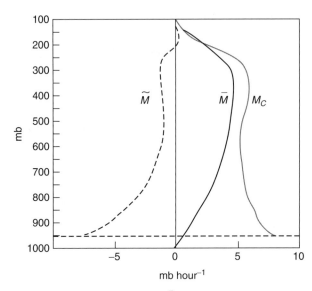

Figure 5.7. The synoptic mean mass flux $\bar{M}$, the cloud mass flux M_c and the mass flux between the clouds $\tilde{M}$ as diagnosed from observations. (From Yanai *et al.*, 1973.)

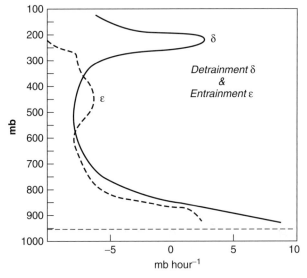

Figure 5.8. The entrainment and detrainment rates as diagnosed from observations. (From Yanai *et al.*, 1973.)

Figure 5.7 shows the synoptically observed $\bar{M}$, the diagnosed cloud mass flux M_c and the inferred mass flux between the clouds $\tilde{M}$. Figure 5.8 shows the entrainment and detrainment in the cloud ensemble. Clearly from Figure 5.8 the mass flux has a large amount of low-level detrainment (below 700 hPa) and a large amount of detrainment at the top of the troposphere. The simplest interpretation of these results are that the distribution of clouds is essentially bi-modal, with deep and shallow clouds. The shallow clouds are mostly below 2 or 3 kilometers above the surface and the deep clouds essentially extend to the tropopause. This interpretation gains added weight from considerations of the heat and moisture balances of the atmosphere.

Figure 5.9 indicates that the major part of the heating is accomplished by the cloud mass flux term. We see from Figure 5.10 that the evaporative cooling (coming from

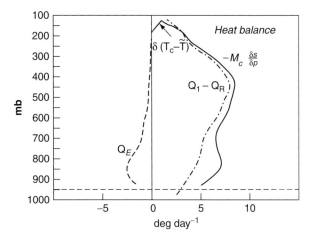

Figure 5.9. The heat balance as diagnosed from observations. (From Yanai *et al.*, 1973.)

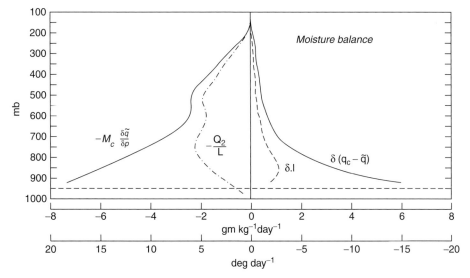

Figure 5.10. The moisture balance as diagnosed from observations. (From Yanai *et al.*, 1973.)

detrained liquid water) is only large in the lower 2 km, and from Figure 5.8 that most of the detrainment of liquid and vapor also occurs only at lower levels, while the drying of the atmosphere occurs throughout the atmosphere, due to the M_c of deep clouds. The action of the shallow clouds, therefore, seems mostly confined to the lower levels of the atmosphere and mostly seems to involve moisturizing these levels.

This leads us to suggest a major simplification, namely that the tropical atmosphere may be considered to be deep precipitating cumulonimbus clouds interacting with a moist boundary layer consisting of shallow clouds. The major action of the deep clouds is to precipitate and the major action of the shallow clouds is to moisturize the moist boundary layer. The deepest clouds can only reach the tropopause if they do not entrain very much, and the shallow clouds can only effectively moisturize the boundary layer if they do not precipitate. Because deep clouds are assumed to rain out all their moisture, they have none left upon reaching their detrainment level near the tropopause. Because they have become so cold on detrainment near the tropopause, they have no moisture left to detrain. The equations for deep non-entraining clouds then become:

$$Q_1 = M_c \frac{d\bar{s}}{dz} + Q_R,$$

$$Q_2 = -M_c \frac{d\bar{q}}{dz}.$$

Let us go back to our model of moisture confined to a boundary layer of depth about 2 km.

$$\rho \frac{d\bar{s}}{dt} \equiv \rho \frac{\partial \bar{s}}{\partial t} + \rho \bar{\mathbf{u}} \cdot \nabla \bar{s} + \overline{M} \frac{d\bar{s}}{dz} = Q_1 = M_c \frac{d\bar{s}}{dz} + Q_R.$$

Assume we can neglect $\nabla \bar{s}$ since the horizontal temperature gradients are relatively small. Then

$$\rho \frac{\partial \bar{s}}{\partial t} = (M_c - \overline{M}) \frac{d\bar{s}}{dz} + Q_R,$$

$$\rho \frac{\partial \bar{s}}{\partial t} = -(\tilde{M}) \frac{d\bar{s}}{dz} + Q_R. \qquad (5.20)$$

Equation 5.20 indicates that all the temperature changes are due to subsidence between the clouds and radiation. As we have seen, the temperatures hardly change in the tropical atmosphere so that the same between-the-cloud subsidence always balances the radiative cooling between the clouds.

Now let us return to the general case (c): $\overline{M} \neq 0$ and consider the moisture budget: $P = E +$ moisture converged into a tropical domain A by mean motions. Since all the moisture to speak of is in the boundary layer:

$$P = E + \frac{1}{A} \int_0^{zm} dz \int q_m \nabla \cdot \rho \mathbf{u} dA;$$

or

$$P = E + \overline{\rho w_m q_m}.$$

Since we have assumed that all the water going up into high clouds is precipitated out, the precipitation is equal to the moisture flux in clouds going up above the boundary layer:

$$P = M_c q_m.$$

Therefore

$$M_c = \frac{P}{q_m} = \frac{E}{q_m} + \overline{\rho w_m}.$$

And, since

$$\tilde{M} = \overline{\rho w_m} - M_c,$$

$$\tilde{M} = -\frac{E}{q_m} \tag{5.21}$$

independent of the mean synoptic mass flux $\bar{M} = \overline{\rho w_m}$ and, therefore, independent of the convergence that leads to precipitation. The downward motion between the clouds (i.e. the downward motion almost everywhere) is the same regardless of whether or not there is precipitation, or equivalently, whether or not there is an M_c.

Now consider the heat budget:

$$\overline{\rho w_m} \frac{d\bar{s}}{dz} = M_c \frac{d\bar{s}}{dz} + Q_R$$

$$= \left[\frac{E}{q_m} + \overline{\rho w_m} \right] \frac{d\bar{s}}{dz} + Q_R;$$

$$-\frac{E}{q_m} \frac{d\bar{s}}{dz} = Q_R.$$

Hence

$$\tilde{M} \frac{d\bar{s}}{dz} = Q_R.$$

We see that the radiative cooling of the atmosphere is balanced by the subsidence between the clouds which, as we have seen, is independent of the precipitation. Note that this downward motion will also dry the air.

Alternately, we may write

$$-\frac{E}{q_m}\frac{d\bar{s}}{dz} = Q_R$$

and find the *total* heat budget

$$\int_0^{z_T} Q_R dz = -\frac{E}{q_m}\int_0^{z_T}\frac{d\bar{s}}{dz}dz = -\frac{E}{q_m}\left(\bar{s}(z_T) - s_m\right)$$

$$= -\frac{E}{q_m}\left(Lq_m\right) = -LE;$$

so that radiative cooling is balanced by the latent heat of evaporation even when there is precipitation present. (Recall from Figure 5.2 that $Lq_m = \bar{s}(z_T) - s_m$.) As a corollary to Equation 5.21 and the heat budget, we notice that temperature does not change in the presence of local regions of precipitation or in regions where precipitation is absent. This provides a rationale for the result in the previous section that the temperature in regions of large thermal heating remains unchanged.

The net result of these considerations is that it is enlightening to think of deep cumulonimbus clouds simply as conduits for upward mass fluxes. The total number of such clouds is given by the evaporation, and this number of clouds may be considered randomly distributed when $P = E$ with no synoptic mass flux: $\bar{M} = 0$. The mean heating of the atmosphere is given by the downward mass flux between the clouds $\widetilde{M}$ and if this downward mass flux can be said to compensate anything, it is the radiative cooling of the atmosphere: $-\widetilde{M}\frac{d\bar{s}}{dz} = Q_R$. That $\widetilde{M}$ does not depend on whether or not $P>E$ or $P<E$ means that the deep cumulonimbus clouds can be considered simply to bunch together ($P>E$) or disperse apart ($P<E$). When there is upward synoptic mass flux, ($\bar{M}>0$), the normally expected added compensatory adiabatic cooling does not occur. In fact since $\widetilde{M} = \bar{M} - M_c$ stays constant, any additional mass flux $\bar{M}$ goes up into the clouds as additional M_c. When $\bar{M}<0$, there are fewer clouds in the region and more of the mass flux between the clouds is synoptic mass synoptic flux. The downward motion almost everywhere stays constant regardless of synoptic scale convergence or divergence, although the interpretation will be different for each. This means that the old idea of compensating subsidence must be modified: subsidence *does* compensate the mass flux in the clouds due to evaporation. But when deep clouds congregate, i.e. when there is local upward synoptic vertical velocity, the between-the-cloud subsidence does not change, so that there is no *additional* subsidence compensating the low-level convergence.

Consider the circulation in the two cases $P > E$ and $P < E$ and let X be a positive quantity:

$$
\begin{array}{ll}
P > E & P < E \\
P = M_c q_m = E + X & P = M_c q_m = E - X \\
M_c = \dfrac{E}{q_m} + \dfrac{X}{q_m} & M_c = \dfrac{E}{q_m} - \dfrac{X}{q_m} \\
\bar{M} = \dfrac{X}{q_m} \text{ and } \widetilde{M} = -\dfrac{E}{q_m} & \overline{M} = -\dfrac{X}{q_m} \text{ and } \widetilde{M} = -\dfrac{E}{q_m}
\end{array}
$$

We see that regardless of the magnitude of X, the downward motion $\widetilde{M}$ in both cases is the same everywhere because the heat goes up in the clouds, i.e. the part of the heating due to net moisture convergence is balanced by adiabatic cooling of the mean motion. It is almost as if the clouds were insulated pipes. In convergent regions, many insulated pipes gather together and $P > E$, yet the between-the-cloud motion is $\widetilde{M} = -\dfrac{E}{q_m}$. In divergent regions, the insulated clouds disperse and thin out and the synoptic mass flux $\bar{M}$ is downward and $P < E$, yet it is still true that $\widetilde{M} = -\dfrac{E}{q_m}$. In both cases, the vertical motion, almost everywhere, is the same and is given by $-\dfrac{E}{q_m}$. Since the downward motion almost everywhere (i.e. the between-the-cloud downward motion) does not change, the temperature does not change even in regions of heavy precipitation where the latent heating is large.

The key assumption in making this grand interpretation was horizontal homogeneity – that E is the same everywhere and that horizontal temperature variations are small. When there are horizontal temperature variations, such as the East–West surface-temperature gradient in the tropical Pacific, much of the preceding survives as atmospheric interior motion, with the interior being connected to the surface through a boundary layer. The observed temperature structure above the boundary layer is very nearly identical in the eastern and western tropical Pacific Ocean.

5.4 A model for the vertical structure of the tropical atmosphere

This section will indicate that the typical schematic structure of the tropical atmosphere (Figure 5.11) is a robust and ubiquitous feature that can be understood in terms of the roles of convection in carrying vapor, liquid and heat. It will further

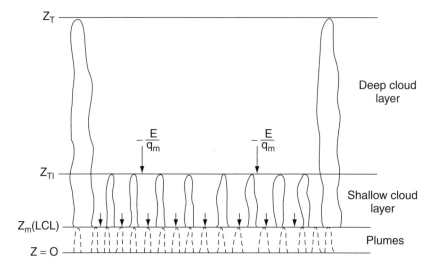

Figure 5.11. Schematic of the vertical structure of the tropical atmosphere. (From Sarachik, 1978.)

show that in the absence of ocean dynamics the temperature of the surface of the ocean is determined by a one-dimensional radiative–convective equilibrium in the atmosphere. Deviations away from this radiative–convective equilibrium temperature are due to ocean dynamics forced by heat and momentum fluxes at the ocean surface.

Assume a tropical atmosphere on top of an ocean with no mean horizontal transports (of anything) in either the atmosphere or the ocean: for the purposes of this discussion, we will assume complete horizontal homogeneity.

The resulting atmospheric state is in moist radiative–convective equilibrium and its properties can be calculated by one-dimensional considerations. Vertical diffusion will guarantee that the ocean will be at a constant temperature throughout its depth, which will therefore be the temperature at the surface. Everything (the vertical structure of the atmosphere, the SST, the evaporation rate etc.) can be calculated solely in terms of the solar constant. The only assumptions will be on the radiative properties of the system, and some assumptions on the wind variance. In the spirit of this book, the presentation will be simplified. More complex radiative–convective calculations can be found in the literature.

The structure of the resulting equilibrium tropical atmosphere can be described in terms of the convective elements. The bottom boundary layer is a mixed layer whose mixing is due partly to the sensible heat from the surface and partly from the evaporation from the surface. The convective elements mixing the heat and vapor are invisible plumes. The mixed layer rises until it reaches the lifting condensation layer. Deep and shallow clouds break out in such a way as to keep the top of the mixed layer just below the lifting condensation level. The shallow clouds (trade cumulus) are small and highly entraining so they detrain within the shallow cloud layer without precipitating. The deep cumulonimbus clouds rise to the tropopause and precipitate heavily, thereby determining the lapse rate of the free atmosphere to be moist adiabatic. The system is in dynamic equilibrium: the shallow cloud layer is held to about 2 km by the downward motion in the environment.

We might consider each of the layers illustrated in Figure 5.11 as a layer characteristic of the nature of the convective elements and their relationship to the forms of water. The mixed layer is characterized by invisible plumes and is partly driven by the light weight of water vapor without condensation. The shallow cloud layer is driven by the condensation and subsequent re-evaporation of water leaving, as a net effect, the mixing and moisturizing of the shallow cloud layer. The deep cumulonimbus clouds condense water with the water falling as precipitation. The realization of the heat of condensation in the cumulonimbus clouds occurs in the downward motion outside the clouds, which balances the radiative cooling of the free atmosphere and sets the lapse rate of the free atmosphere at the moist adiabatic lapse rate.

We assume horizontal homogeneity and, in equilibrium, because there is no convergence or divergence,

$$P = E = M_c q_m,$$

where M_c is the mass flux in the deep cumulonimbus clouds and q_m is the mixed-layer moisture. With no net low-level convergence, $\overline{\rho w} = 0$, and we see that $M_c = +\dfrac{E}{q_m}$ and $\tilde{M} = -\dfrac{E}{q_m}$. The downward motion, almost everywhere, is given by $\tilde{M}$.

We assume we know the net radiation at the surface R_{net}: in a one-dimensional model, it is best not to assume too much about the properties of clouds, so that radiative transfer though the clouds is finessed by assuming the net radiation at the surface is known. Because the ocean has no net transports, there is no heat flux into the ocean so that at the surface

$$R_{net} = LE + S,$$

where S is the sensible heat transfer from the sea surface into the atmosphere.

We will take the radiative cooling in the free atmosphere to be given by Rayleigh cooling

$$Q_R = \rho \frac{T(z) - T_e(z)}{\tau},$$

with a fixed radiative-cooling rate τ taken to be about 15 days and $T_e(z)$ is the radiative–dry convective equilibrium temperature structure, i.e. the temperature structure that would exist if there were radiation and dry convection but no moist processes.

In the free atmosphere, the heat balance is

$$M_c \frac{d\bar{s}}{dz} = + \frac{E}{q} \frac{d\bar{s}}{dz} = \frac{\rho}{\tau}(T - T_e).$$

The lapse rate is determined by moist convection due to the deep cumulonimbus clouds. The atmosphere is assumed dry above the trade inversion (TI) since the deep clouds are non-entraining and we assume that precipitation out of the rising clouds reaches the surface and does not re-evaporate on the way down.

The mean lapse rate of the free atmosphere is

$$\frac{s(z_T) - s_m}{z_T - z_m} = \frac{Lq_m}{z_T - z_m} \approx \frac{Lq_m}{z_T},$$

so that

$$\frac{LE}{z_T} \approx \frac{\rho}{\tau}(T(z) - T_e(z))$$

or

$$LE = c_p \int_0^{z_T} \frac{\rho}{\tau}(T - T_e(z))dz. \tag{5.22}$$

We will use this later to calculate z_T. (Once we know z_T and the temperature at the tropopause, we know s_m.)

We relate s_m and q_m to their surface values $c_p T_s$ and $q_{sat}(T_s)$ by the similarity relations;

$$\frac{dT}{dz} = \frac{\theta_*}{kz}f(z/L_*) \qquad \frac{dq}{dz} = \frac{q_*}{kz}g(z/L_*),$$

where $\overline{w'T'} \approx u_*\theta_* = S$ and $E = \overline{q'w'} \approx u_*q_*$ and u_* is specified. (There is no mean wind but there is wind *variance*.) L_* denotes the Monin–Obukhov length.

If we integrate the moisture and temperature equations we have $T_s - T_m = $ some function of $(\theta_*, z/L_*)$ and $q_{sat}(T_s) - q_m = $ some other function of $(q_*, z/L_*)$. The exact expressions are given in Sarachik (1978).

So once the E and S are known, the relationship between surface- and mixed-layer quantities are known.

The mixed-layer temperature is calculated from the heat balance

$$LE = c_p \int_0^{z_T} \frac{\rho}{\tau}(T - T_e(z))dz.$$

The radiative-equilibrium temperature profile is assumed to be a profile given for radiative–dry convective equilibrium. We know that such a profile exists because radiative equilibrium calculations alone (e.g. Manabe and Möller, 1961) give a statically unstable troposphere which is then unstable to dry convection. Dry convection will then mix the troposphere to a depth z_1 so that

$$T_e = T_o - \Gamma_a z \quad z < z_1$$

and $T = T_{strat}$ for $z > z_1$ where

$$\Gamma_a = \frac{g}{c_p}.$$

The profile we are calculating is

$$T = T_m - \Gamma_t z \qquad z < z_T$$

where

$$-\Gamma_t = \frac{Lq_m}{c_p z_t} - \Gamma_a$$

and

$$T = T_{strat} \qquad z > z_T,$$

i.e. we are assuming an isothermal stratosphere at its radiative equilibrium value.

Exercise 5.3: Do the integration in Equation 5.22 and show that:

$$z_T = \frac{Lq_m}{2g} \pm \frac{1}{2}\sqrt{\left(\frac{Lq_m}{g}\right)^2 + z_1^2 + 2\frac{LE}{g}\frac{\tau}{\rho}}.$$

(Keep $\frac{\tau}{\rho}$ constant.) Note that this result says that the effect of evaporation is clearly to raise the tropopause over its non-moist value z_1.

Once we have z_T, we know Γ_t. Then we know $T_m = T_{strat} - \Gamma_t(z_1 - z_m)$.

We know z_m implicitly in terms of q_m and T_m because it is the lifting condensation level.

We also know that the mass flux into the trade cumulus clouds is the mass flux necessary to hold the mixed layer to the lifting condensation level, i.e.

$$M_{TC}(z_m) = 1.2\frac{\overline{(\theta'_v w')}_s}{\Gamma z_m} \qquad \overline{(\theta'_v w')}_s = \frac{LE}{\rho c_p}S[b + 0.07]$$

and

$$\Delta\theta_v = \frac{0.2\overline{(\theta_v' w')}}{-M_{TC}(z_m)}.$$

The partition of sensible and latent is gotten by noting that all the sensible heat is used in balancing radiation in the mixed layer.

$$\frac{S}{LE} = \frac{\int_o^{z_m} \rho \left(\frac{T_m - T_e}{\tau}\right) dz}{\int_{z_m}^{z_T} \rho \left(\frac{T - T_e}{\tau}\right) dz}$$

or

$$S \approx \int_o^{z_m} \rho \frac{(T_m - T_e)}{\tau}.$$

(Note that we may need a small correction due to entrainment at the interface.)

The height and jump across the *TI* layer is obtained by considering the *TI* layer as a convective boundary layer but the closure relation involves $\overline{\theta_e' w'}$ since moist convection mixed the layer. It is held down by the mass flux compensating the deep cumulonimbus clouds.

There are now enough relations to determine everything in terms of the specified R_{net}. The results are almost totally insensitive to u_*.

We find by doing the calculation (Sarachik, 1978, 1985) that, for the known solar constant, and the observed (radiative equilibrium) temperature of the tropical stratosphere, the resulting surface temperature is $27\,°C$, the evaporation rate is 5.8 mm/s, the trade inversion height is 2 km and the tropopause height is of the order of 15 km. These values are close to those of the western Pacific, where the heat flux into the ocean is certainly small, so that this should be the region of the ocean in which the ocean dynamics participate least in the sea-surface temperature.

The basic result of this section is that the mean state of the tropical atmosphere can be simply understood in terms of deep cumulonimbus clouds interacting with a boundary layer composed of shallow clouds. Since, as we saw in the previous section, the bunching or dispersal of deep cumulonimbus clouds into regions of $P>E$ and $P<E$, respectively, changes neither the mass flux between the clouds nor the mean temperature: the structure of this mean state is therefore robust under these various conditions.

5.5 Theories of thermal forcing of the atmosphere

In this section, we will give three examples of thermal forcing of the atmosphere. The first is the zonally averaged Hadley circulation, the second is the Gill model so commonly used in models and interpretations of the tropical atmosphere, and the third is the linear theory of forcing by an isolated heat source. The two-dimensional (zonally averaged) Hadley circulation is one of the most basic thermally forced circulations in the tropics, and the papers by Schneider and Lindzen (1977) and Schneider (1977) give credence to the important role of the Hadley circulation in the general circulation of the Earth, in particular in the maintenance of the midlatitude jet. (Note that the simple thermally forced theory must be completed by the addition of midlatitude eddies.) The simple Gill model has been a standard tool used by both modelers and diagnosticians for describing the atmospheric response to thermal forcing in the atmosphere, yet it raises serious problems of interpretation. We will solve the full linear problem of the forcing of the tropical atmosphere by an isolated thermal source and use the full solution to decide on the applicability of the simplified Gill model. The crucial application of the Gill model is determining the surface winds forced by thermal sources determined by SST anomalies. In particular, as we saw in Chapter 2, the evolution of ENSO indicates that the surface winds are westerly to the west of a warm anomaly and this property of the Gill model has been one of the sources of its popularity. A critical evaluation of this property will be given in Section 5.7.

5.5.1 *The zonally averaged Hadley circulation*

The following exposition is based of the work of Schneider and Lindzen (1977) and Schneider (1977). This work changed the paradigm for the role of the Hadley circulation in the Earth's general circulation. It shows that the full nonlinear (but two-dimensional) Hadley cell has dynamics constrained by angular-momentum conservation producing *stronger* than observed subtropical jets. The role of the eddies is then to move, broaden and weaken the jet. This contrasts with the earlier observational work of Victor Starr and collaborators on the physics of "negative viscosity" – the idea that the eddies associated with the atmospheric jets flux angular momentum into the jet to strengthen them. The interaction of jets and eddies is currently an area of active research.

The basic idea is that we are looking for steady two-dimensional motions in the yz-plane that are independent of x. The circulations are assumed to be driven by cloud and radiative heating.

$$\rho[u_t + uu_x + vu_y + wu_z - \beta yv] = -p_x + F^{(x)},$$

$$\rho[v_t + uv_x + vv_y + wu_z + \beta yu] = -p_y,$$

$$u_x + v_y + w_z = 0,$$

$$\frac{\partial p}{\partial z} = -g\rho = -g\frac{p}{RT},$$

$$\frac{dT}{dt} + wN^2 = \frac{Q}{c_p},$$

The term vT_y is shown a posteriori to be small and is neglected.

The total heating is $Q = Q_{rad} + Q_{cloud}$ where $Q_{cloud} = M_c N^2$ is specified (since we are looking for the response to thermal forcing) and taken to be *symmetric* with respect to the equator, and the radiative cooling of the atmosphere is taken to be of Rayleigh cooling form:

$$Q_{rad} = c_p \rho \frac{T_e(y,z) - T}{\tau};$$

where T_e is the radiative-equilibrium temperature. There are also diffusive processes in the surface boundary layer.

The momentum transport due to clouds is given by a simple parameterization introduced by Schneider and Lindzen (1976)

$$F^{(x)} = \frac{d}{dz}[M_c(u - u(z_c))],$$

which basically corresponds to the cloud picking up momentum from the surface and subsequently detraining it into the ambient atmosphere. The divergence of this transport, represents "cumulus friction." The result of this process is to reduce the upper-level flow by injecting slower moving surface air directly into the upper troposphere.

The frictional forcing, $F_x = -h\dfrac{d\tau^x}{dz}$ and the eddy friction is taken as $\tau^x = -h\left(\mu \dfrac{du}{dz}\right)$ where the eddy viscosity μ is significant in the surface boundary layer *only.*

The u boundary condition at the top is stress free, $\mu\dfrac{du}{dz} = 0$, and at the bottom is $\mu\dfrac{du}{dz} = c_D u|v|$.

The thermal boundary condition at the top $\dfrac{dT}{dz} = 0$, i.e. $T = const$ (constant temperature stratosphere consistent with a stratosphere that is extremely stable and is close to radiative equilibrium) and at the bottom $T = T_0(y)$; a specified surface temperature taken to be symmetric around the equator.

a. The linear Hadley circulation

The steady, linear u momentum equation becomes

$$-\beta y v = F^{(x)}$$

and we see that, in the linear case, friction is absolutely essential to get a meridional circulation. The v momentum equation reduces to geostrophy.

Since the solution is steady, integrating the u momentum equation from the surface to the stratosphere gives:

$$\int_{bot}^{top} v \, dz = 0 \Rightarrow \int_{bot}^{top} F^{(x)} dz = 0 \Rightarrow \int_{bot}^{top} \frac{d}{dz}\left(\mu \frac{du}{dz}\right) dz = 0$$

and since the stress at the top vanishes, the stress at the surface must vanish. Since the surface stress is given by the drag formulation, $u = 0$, *so the linear model cannot generate zonal surface winds.*

We list the results of the calculations:

(a) In the absence of cumulus friction and heating, there is no interior meridional circulation, only a low-level cell (where boundary-layer eddy friction is not zero), the temperature is almost in radiative equilibrium, and the zonal winds are in thermal wind balance with the radiative-equilibrium temperature. (Case I – Figure 5.12.)

(b) With broad imposed heating (symmetric about the equator) but no cumulus friction, there can still be no interior meridional circulation but because there is internal heating, we get a stronger meridional temperature gradient maintained by heating (i.e. the interior is raised to temperatures far above radiative equilibrium) and a very strong westerly jet in thermal wind balance with the temperature. (Case III – Figure 5.13.) There is no momentum conservation in this linear case so it is pointless to ask where the momentum in this strong jet came from.

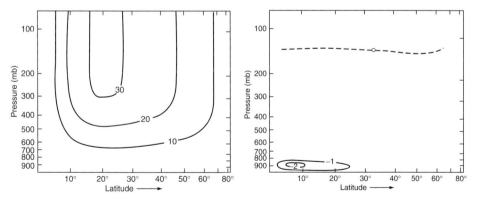

Fig 5.12. Case I. Left: Zonal winds in cm/sec. Right: Meridional circulation contour interval 10^{10} kg/sec. (From Schneider and Lindzen, 1977.)

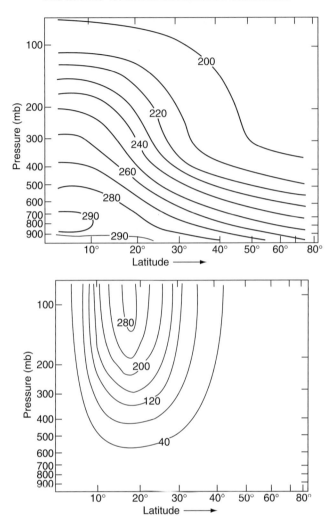

Figure 5.13. Case III. Top: Zonally averaged meridional temperature distribution. Bottom: Zonal velocity (m/sec). (From Schneider and Lindzen, 1977.)

(c) Putting in cumulus friction drives a meridional circulation that has the horizontal scale of the forcing. The temperature patterns become flattened within the latitude of the cell's influence and sharpened just at the northern boundary of the cell. (Case IV – Figure 5.14.)

We can see why this is so from the u momentum equation:

$$-\beta y v = M_c \frac{du}{dz} \propto M_c \frac{1}{\beta y}\frac{dT}{dy}$$

by the thermal wind relation so that

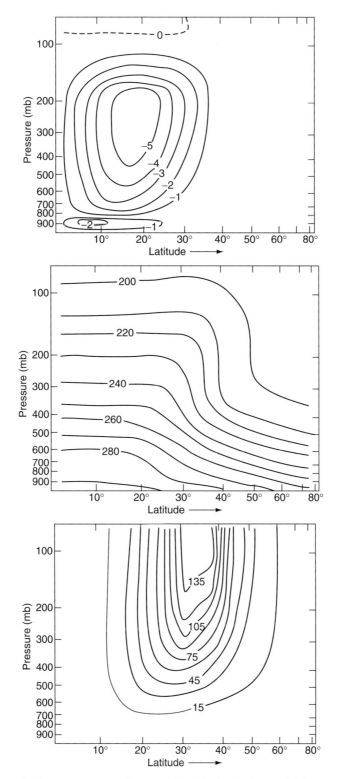

Figure 5.14. Case IV. Top: The meridional circulation. Middle: Meridional distribution of temperature. Bottom: The zonal wind. (From Schneider and Lindzen, 1977.)

$$\frac{dT}{dy} \quad \propto \quad \frac{y^2 v}{M_c}.$$

Since the forcing is symmetric, v is antisymmetric, so v must go at least as near the equator as y. Therefore

$$\frac{dT}{dy} \quad \propto \quad \frac{y^3}{M_c} \Rightarrow T \sim \frac{y^4}{M_c},$$

which is a very slow y dependence. The temperature gradient is therefore flattened under the influence of the meridional circulation.

(d) Using "observed heating" and cumulus friction gives a "reasonable looking" Hadley circulation with jets of the right order of magnitude. (Case VII – Figure 5.15.)

   While the linear solution for the thermally driven Hadley circulation in Figure 5.15 looks reasonable, remember that linearity has constrained the solution to have no zonal surface winds.

b. The nonlinear Hadley circulation

The fully nonlinear u momentum equation is

$$\frac{D_2}{D_t} u - \beta y v = F^{(x)},$$

where the two-dimensional advective derivative is:

$$\frac{D_2}{Dt} = v \frac{d}{dy} + w \frac{d}{dz}.$$

Since

$$v = v \frac{\partial y}{\partial y} + w \frac{\partial y}{\partial z} = \frac{D_2 y}{Dt},$$

$$\frac{D_2}{Dt} \left(u - \frac{1}{2} \beta y^2 \right) = F^{(x)}.$$

The quantity $\hat{u} = u - \frac{1}{2} \beta y^2$ is the angular momentum and is conserved on parcels of the Hadley circulation if friction can be ignored.

   Note that

$$\frac{\partial \hat{u}}{\partial y} = u_y - \beta y = -[\zeta + f] = \text{absolute vorticity},$$

so when angular momentum is constant on a horizontal branch of the "inviscid" Hadley circulation, the absolute vorticity vanishes.

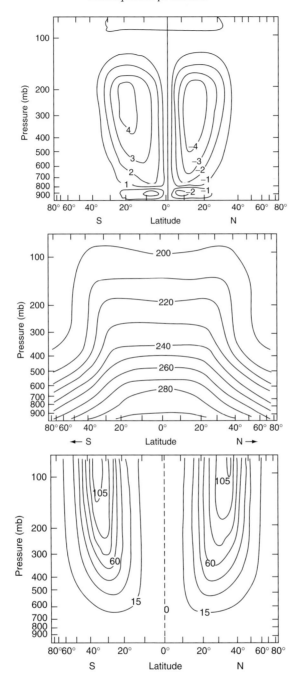

Figure 5.15. Top: Meridional circulation. Middle: Temperature field. Bottom: Zonal winds, Case VII. (From Schneider and Lindzen, 1977.)

For two-dimensional circulations, and two-dimensional circulations only, there is an extremely useful theorem, Hide's theorem, that provides a vital constraint on the tropical zonal velocity (Hide, 1969). The content of Hide's theorem is that there cannot be steady westerlies on the equator.

An extension of Hide's theorem (Schneider, 1977) is that $\hat{u}$ cannot be maximum or minimum in the interior of a fluid or at a stress-free upper boundary. This assumes diffusion is down gradient in angular momentum. (Any diffusion that depends only on z *is* down gradient in angular momentum.)

The extremum must be at the lower boundary. If it's a minimum, the atmosphere will always deliver angular momentum to the minimum, which will speed up the Earth eventually, since u gets converted to surface stress.

If it is a maximum, angular momentum gets delivered to the atmosphere by the Earth. The maximum must be zero **(Exercise 5.4**: Why?) and only obtains at the surface where $u < 0$.

Therefore, the content of the theorem is:

$$\hat{u} \leq 0 \quad \text{everywhere.}$$

We see that on the equator we cannot have steady westerlies, either at the surface or aloft. We emphasize that this only holds for two-dimensional circulations: for a fully three-dimensional circulation, stationary or transient eddies can converge westerly momentum onto the equator and possibly produce steady westerlies.

In the fully nonlinear case, the results are shown in Figure 5.16. The midlatitude jet is much stronger than observed and this is purely due to angular-momentum conservation. The meridional streamfunction is about the right strength and the interior and the boundary layer flows seem separate. There is a weak midlatitude circulation carrying heat equatorward, a Ferrell cell, even in the absence of eddies. The surface winds are easterly in the tropics and westerly in midlatitudes. The meridional temperature gradient in the influence zone of the meridional cell is remarkably flat and there is an indication of an inverse gradient near the tropopause.

The argument for the flatness of the temperature is different than for the linear case. In the linear case it was necessary for (cumulus) friction to enter the dynamics at leading order. If, with the more complete nonlinear dynamics, we assume that the northward branch of the circulation aloft is relatively inviscid, then

$$\hat{u} \cong 0, \quad u = \frac{1}{2}\beta y^2$$

so

$$\frac{\partial u}{\partial z} \approx \frac{\frac{1}{2}\beta y^2}{Z},$$

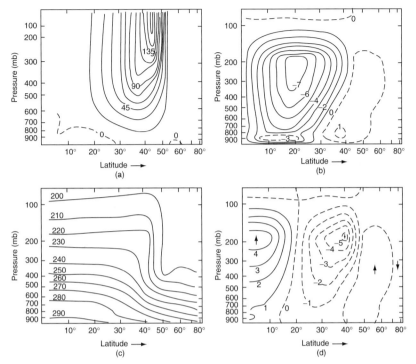

Figure 5.16. (a) Zonal winds, (b) Meridional streamfunction (units 10^{13}g/s), (c) Temperature field, and (d) Vertical velocity (mm/s). (From Schneider, 1977.)

where Z is tropopause height. By the thermal wind relation,

$$T_y \sim y\frac{\partial u}{\partial z} \sim \frac{\frac{1}{2}\beta^2 y^3}{gZ}$$

and this implies $T \sim y^4$ and we recover the flat temperature gradients, but now due to angular-momentum conservation by the meridional circulation.

The inverse temperature gradient near the tropopause can be understood as follows. The parameterization for cumulus friction takes the near-surface zonal velocity and dumps it into the top of the troposphere. This creates a vertical shear that decreases with height so that, by the thermal wind relation, the meridional gradient of temperature must have the temperature increasing northward.

What determines size of the Hadley cell? Call the meridional extent y_1. The heat equation is

$$\overline{\rho w}N^2 = \frac{Q_c}{c_p} - \frac{\rho}{\tau}(T - T_e).$$

Integrating over the horizontal extent of the cell gives $\int_0^{y_1} \overline{\rho w} dy = 0$ since there cannot be any net vertical mass flux. We see that, since N^2 is constant in the Hadley cell,

$$\int_0^{y_1} Q_c dy = c_p \int_0^{y_1} \frac{\rho}{\tau} (T - T_e) dy.$$

Now we know, with $T_0 \equiv T(y = 0)$

$$T = T_0 - \frac{y^4}{8gZ} \beta^2.$$

At $y = y_1$, $T = T_e$, so $T_e = T_0 - y_1^4 \frac{\beta^2}{8gZ}$ and $T - T_e = (y_1^4 - y^4) \frac{\beta^2}{8gZ}$. Hence

$$c_p \int_0^{y_1} (T - T_e) dy = \frac{\tau}{\rho} \int_0^{y_1} Q_c dy = c_p \int_0^{y_1} (y_1^4 - y^4) \frac{\beta^2}{8gZ} dy = \frac{1}{10} \frac{c_p \beta^2 y_1^5}{gZ}.$$

Now integrating both sides over z and noting that $\int_0^z \int_0^{y_1} Q_c dy dz = LEy_1$ since the total heat released comes from the latent heat of precipitation which, integrated over the Hadley cell, is simply the evaporation (assuming no moisture transport out of the Hadley cell, i.e. neglecting eddies). Therefore,

$$\frac{y_1^4}{10} \frac{\beta^2 \rho c_p}{g\tau} = LE$$

or

$$y_1^4 = 10(LE) \frac{g\tau}{\beta^2 \rho c_p}.$$

We saw that the cumulus clouds mix the troposphere and are responsible for the observed lapse rate. The tropopause height can be calculated, as before, to be

$$Z \sim 2 \frac{\tau}{\rho} \frac{E}{q_m}$$

so

$$y_1^4 = 5Lq_m \frac{gZ}{c_p \beta^2}.$$

Since

$$\frac{ds}{dz} = \frac{Lq_m}{Z} = \frac{c_p}{g}N^2,$$

$$y_1^4 = 5\frac{N^2 Z^2}{\beta^2} = 5L_{eq}^4,$$

where $L_{eq} = \left[\frac{NZ}{\beta}\right]^{1/2}$.

So

$$y_1 = 5^{1/4}L_{eq}$$

and we see that the meridional extent of the Hadley cell is simply the equatorial radius of deformation characteristic of a fluid with stratification N^2 and equivalent depth equal to the depth of the troposphere. A clear implication of this result is that as the surface radiation increases (either due to solar radiance or due to increases of greenhouse gases), the tropopause rises and the jet moves northward.

5.5.2 The Gill model

The basic simplifying assumption of the Gill (1980) model is that the heating and response is confined to a single vertical mode. Take

$$Q \propto sin\frac{\pi Z}{D}$$

so the vertical wavelength is $2D$ and the vertical wavenumber is π/D. If the atmosphere is isothermal then

$$\frac{\pi^2}{D^2} = \frac{\kappa}{H_s H} - \frac{1}{4H_s^2},$$

where H_s is the scale height and H is the equivalent depth (we will derive the relationship between vertical wavelength and equivalent depth in the next section). If instead the atmosphere has a constant N^2, the gravity wave speed $c = ND/\pi$ so $\sqrt{gH} = c = ND/\pi$. In either case, $H \sim 1\,\mathrm{km}$ (Gill takes it to be $400\,\mathrm{m}$ so $c = 60\,\mathrm{m/s}$). The linearized equations for the amplitudes of the modes (u,v,p), which are functions of (x,y,t) are then

$$\frac{\partial u}{\partial t} - \beta y v = -\frac{\partial p}{\partial x} - \varepsilon u,$$

$$\frac{\partial v}{\partial t} + \beta y u = -\frac{\partial p}{\partial y} - \varepsilon v,$$

$$\frac{\partial p}{\partial t} + \frac{\partial u}{\partial x} + \frac{\partial v}{\partial y} = -Q - \varepsilon_T p,$$

which are the shallow-water equations for a single vertical mode of equivalent depth H, where we have scaled velocities by c, lengths by $\sqrt{\frac{c}{\beta}}$ and time by $\frac{1}{\sqrt{\beta c}}$, ε is the Rayleigh damping in the momentum equations and ε_T is Newtonian cooling (thermal damping) in the pressure equation.

The vertical structure of the heating and of the vertical velocity is $sin\frac{\pi z}{D}$; since $u, v \sim \frac{\partial w}{\partial z}$ the vertical structure of the u and v fields is $cos\frac{\pi z}{D}$.

Consider steady forcing Q independent of t; the equations for the steady response are:

$$\varepsilon u - \beta y v = -\frac{\partial p}{\partial x}, \tag{5.22a}$$

$$\varepsilon v + \beta y u = -\frac{\partial p}{\partial y}, \tag{5.22b}$$

$$\varepsilon_T p + \frac{\partial u}{\partial x} + \frac{\partial v}{\partial y} = -Q, \tag{5.22c}$$

$$w = -\left(\frac{\partial u}{\partial x} + \frac{\partial v}{\partial y}\right) = \varepsilon_T p + Q. \tag{5.22d}$$

The forcing here is thermal (i.e. it forces the divergence directly). Note that these equations are formally similar to the shallow-water system introduced in Chapter 3, but with Rayleigh damping replacing the time dependence. Since it is a model for the atmosphere, here the forcing is thermal, in contrast to the wind stress driving the ocean as in Chapter 3 and Chapter 6.

In the example of Gill, $\varepsilon_T \equiv \varepsilon$ and the forcing is chosen to be

$$Q(x, y) = F(x)e^{-\frac{1}{2}y^2}, \tag{5.23a}$$

so that it projects only onto the first Rossby and Kelvin mode (see Chapter 6). The zonal dependence is taken to be:

$$F(x) = \cos kx \text{ for } |x| < L \text{ and } = 0 \text{ for} |x| > L, \tag{5.23b}$$

where $k = \frac{\pi}{2L}$.

For $c = 60\,\text{m/s}$ $L_{eq} = \sqrt{\frac{c}{\beta}} = 1700\,\text{km}$ and $T_{eq} = \frac{1}{\sqrt{\beta c}} = 0.3\,\text{d}$.

The solution for $L = 2$ (i.e. 3400 km) and $\varepsilon = 0.1$ (i.e. the damping time is 3 d) is shown in Figure 5.17.

We may note the following points:

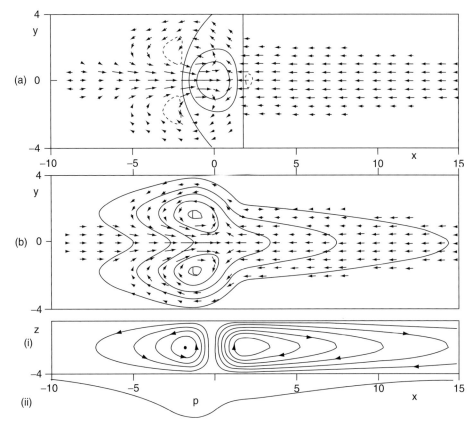

Figure 5.17. Solution of Equations 5.22 with forcing given by Equation 5.23. (a) The contours are vertical velocity while the arrows show the velocity field in the lower layer. (b) Contours show pressure field (everywhere negative) while the arrows are repeated from (a). (i) Meridionally integrated streamfunction – Walker circulation. (ii) Meridionally integrated surface pressure. (From Gill, 1980.)

(a) The surface winds are westerly to the west of the heat source and easterly to the east. The low-level winds converge into the location of the heat source.

(b) The response to the east of the forcing has the meridional form of a Kelvin mode, $u, p \propto e^{-y^2/2}$ and $v = 0$, and to the west of the forcing it has the form of a Rossby mode, $u \sim (1 - y^2)e^{-y^2/2}$.

(c) The zonal extent is the distance the relevant wave will propagate before decaying in time $\sim \varepsilon^{-1}$ (in the long-wave approximation the phase and group velocities are identical). To the east of the heat source centered at $x = 0$ the response is $\exp\left[-\dfrac{x\varepsilon}{c}\right]$ and to the west $\exp\left[\dfrac{-3x\varepsilon}{c}\right]$.

(d) At low levels, the convergence into the heating region is all provided by the u component of the velocity: the vorticity equation is

$$\varepsilon u_y + \varepsilon \beta y p - \beta v = -\beta y Q$$

so for small ε

$$v \approx y Q,$$

positive north of the equator, negative to the south, thereby indicating meridional divergence.

(e) Because of point (c), the damping must be taken as large enough to constrain the effects of the forcing to a fraction of the Earth's circumference so

$$\frac{c}{\varepsilon} < 10\,000\,\text{km} \quad \Rightarrow \quad \frac{1}{\varepsilon} \leq \frac{1}{1.5}\,\text{d}$$

It is very difficult to see where such strong damping comes from in the free atmosphere. Furthermore, the interpretation of the Gill model is that the thermal forcing Q is forcing the low-level winds, which seems unlikely in reality. The heating in the Gill model extends to the ground which by Equation 5.22c is the direct cause for low-level convergence. Yet, as we know, the condensation heating in the real atmosphere does not start until the lifting condensation level, which is at or above 600 m in the tropics, so that there is no Q forcing in the planetary boundary layer near the surface. Further, while a forcing of a given form (e.g. $\sin(\pi z/D)$) will tend to excite a response with that form, unless the form has the structure of a vertical mode, energy will leak away into other structures. Now the tropical atmosphere does not have a single vertical mode because it does not have *any* vertical modes: the tropopause in no way acts as a rigid lid that can sustain standing modes, as we will see in Section 5.5.3. This leaves two issues with the Gill model: how can upper-level heating force surface winds and what is the correct vertical structure in an atmosphere that does not support free modes?

The Gill model takes the damping for temperature and momentum to be the same. This is the simplest assumption, and a great analytic convenience in the time-dependent case, where it allows the results of the free (wave) solutions to be applied to the forced case. However, it is unnecessary for the steady-state equations. Keeping Equation 5.22a, and 5.22b as before and multiplying Equation 5.22c by $\varepsilon/\varepsilon_T$ returns the equations to a form with apparently equal damping on the wind and pressure, but the equivalent depth is now $(\varepsilon/\varepsilon_T)H$ and (as originally scaled by Yamagata and Philander, 1985) the new equatorial length scale is $(c/\beta)^{1/2}(\varepsilon/\varepsilon_T)^{1/4}$ and the new Kelvin decay scale is $c/(\varepsilon\varepsilon_T)^{1/2}$. When $\varepsilon_T = \varepsilon$, this reduces to the Gill results above. When the thermal damping (Newtonian cooling) is reduced, both the meridional scale and the zonal damping scale increase, but when the momentum

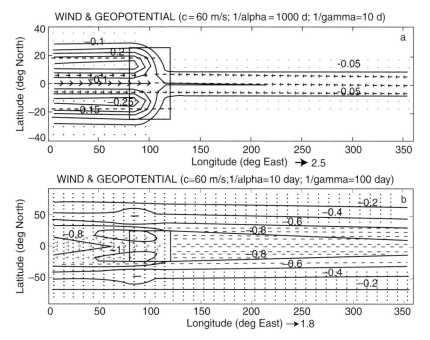

Figure 5.18. Upper panel: Newtonian damping dominant with 100 times the damping of Rayleigh damping. Lower panel: Rayleigh damping dominant with ten times the damping of Newtonian damping. Note the difference in scale of meridional axes between the two panels. (From Wu *et al.*, 2001.)

damping (Rayleigh) friction is reduced then the meridional scale decreases but the zonal damping scale increases.

Realistic values for the damping times range from a few days in the boundary layer to perhaps a month in the free atmosphere, so in a realistic model these stretching effects are not large. Figure 5.18 shows the modified geopotential and winds in a realization of thermal forcing in a stratified atmosphere (as in the next section) with Rayleigh and Newtonian damping having widely different values. The Newtonian damping dominant case is confined to the heating region while the Rayleigh damping dominant case extends far poleward meridionally.

The Gill model has been used extensively in interpreting atmospheric responses to heat sources. For example, the work of Wallace *et al.* (1998) in Figure 5.19 shows responses to ENSO cold-tongue anomalies that resemble Figure 5.17. But, despite this success, there are problems of interpretation when the Gill model is applied to the internal forcing of the atmosphere by cumulus heating. But then what *is* the correct theory of the forcing of surface winds by upper-level thermal forcing? And why does the Gill model seem to work, at least in the lower layers, as a model for the surface winds? We turn to these problems in the next two sections.

a   rainfall and trosposheric temperature

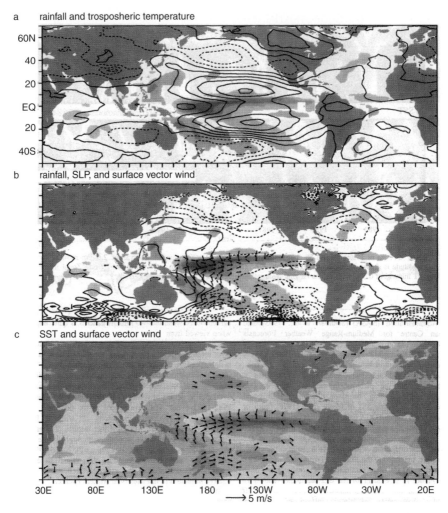

b   rainfall, SLP, and surface vector wind

c   SST and surface vector wind

Figure 5.19. Regression of (a) rainfall (shaded) and vertically averaged tropospheric temperature, (b) rainfall (shaded), sea-level pressure (contours) and surface winds (arrows), and (c) sea-surface temperature (shaded) and surface winds (arrows) all on a cold-tongue index representing temperatures around the equator in the eastern Pacific. (From Wallace *et al.*, 1998.)

5.5.3 *Linear theory of thermal forcing by an isolated heat source*
on an equatorial beta-plane

This section treats the general linear theory of the atmospheric response to an isolated thermal forcing. In particular, we will see that, since the atmosphere has no top, vertical propagation of energy away from the thermal source is a very general feature of thermal forcing near the equator and this distinguishes the problem from the oceanic case. Second, we will consider the processes that *prevent* the vertical propagation of energy away from the heat source and which confine the response vertically to the heating region. We follow Lindzen (1967) in setting up the problem.

Consider the general set of linear equations for thermally forced motion on a β-plane in an unbounded atmosphere, i.e. we will not assume $w=0$ at the top of the atmosphere.

For convenience, take the atmosphere to be isothermal (nothing important depends on this assumption – we could work more generally in log p coordinates).

The basic state is given by

$$p_o = \rho_o R T_o \quad \text{where} \quad \rho_o = \rho_o(z) \; p_o = p_o(z) \quad \text{but at} \quad T_o = const,$$

$$\frac{\partial p_o}{\partial z} = -g\rho_o = \frac{\partial \rho_o}{\partial z} R T_o \Rightarrow p_o(z) = p_o(0)e^{-z/H},$$

$$\rho_o(z) = \rho_o(0)e^{-z/H}.$$

The dynamical equations are

$$\frac{\partial u}{\partial t} - (f_o + \beta y)v = -\frac{1}{\rho_o}\frac{\partial p}{\partial x}, \tag{5.24a}$$

$$\frac{\partial v}{\partial t} + (f_o + \beta y)u = -\frac{1}{\rho_o}\frac{\partial p}{\partial y}, \tag{5.24b}$$

$$\frac{\partial p}{\partial z} = -g\rho \quad \text{Hydrostatic,} \tag{5.24c}$$

$$\frac{\partial \rho}{\partial t} + w\frac{\partial \rho_o}{\partial z} + \rho_o\left(\frac{\partial u}{\partial x} + \frac{\partial v}{\partial y} + \frac{\partial w}{\partial z}\right) = 0 \quad \text{Continuity.} \tag{5.24d}$$

The heat equation is $dQ = c_v dT + p dv = c_p dT - v dp; \quad v = 1/\rho$ and the ideal gas law is $p = \rho RT; \quad R = c_p - c_v, \quad \gamma = \frac{c_p}{c_v}, \kappa = \frac{R}{c_p} = 1 - \frac{1}{\gamma}.$

Define the heating rate Q:

$$\rho Q = \rho c_p \frac{dT}{dt} - \frac{dp}{dt}$$

$$= \rho c_p \frac{d}{dt}\left[\frac{p}{\rho R}\right] - \frac{dp}{dt} = \rho c_p\left[\frac{1}{\rho R}\frac{dp}{dt} - \frac{p}{\rho^2 R}\frac{d\rho}{dt}\right] - \frac{dp}{dt},$$

$$= \left(\frac{c_p}{R} - 1\right)\frac{dp}{dt} - \frac{c_p\,p}{R\,\rho}\frac{d\rho}{dt},$$

$$= \frac{1}{\gamma - 1}\frac{dp}{dt} - c_p T\frac{d\rho}{dt} = \frac{1}{\gamma - 1}\frac{dp}{dt} - \frac{\gamma}{\gamma - 1}RT\frac{d\rho}{dt};$$

$$\rho(\gamma - 1)Q = \frac{dp}{dt} - \gamma RT\frac{d\rho}{dt}.$$

Linearizing and using $RT_o = gH$, the heat equation becomes

$$\frac{\partial p}{\partial t} + w\frac{dp_o}{dz} = \gamma gH\left(\frac{\partial \rho}{\partial t} + w\frac{d\rho_o}{dz}\right) + (\gamma - 1)\rho_o Q. \qquad (5.24e)$$

Now look for Fourier components that go as $e^{i(kx+\omega t)}$ and transform to variables

$$v \Rightarrow \rho_o^{1/2}v = \exp\left[-\frac{z}{2H}\right]v; \quad u \Rightarrow \rho_o^{1/2}u, \quad w \Rightarrow \rho_o^{1/2}w,$$

$$p \Rightarrow \rho_o^{1/2}p = \exp[z/2H]p \quad \rho \Rightarrow \rho_o^{1/2}\rho.$$

The equations become

$$i\omega u - (f_o + \beta y)v = -ikp, \qquad (5.25a)$$

$$i\omega v + (f_o + \beta y)u = -\frac{\partial p}{\partial y}, \qquad (5.25b)$$

$$\frac{\partial p}{\partial z} - \frac{1}{2H}p = -g\rho, \qquad (5.25c)$$

$$i\omega\rho + \frac{\partial w}{\partial z} - \frac{1}{2H}w + iku + \frac{\partial v}{\partial y} = 0, \qquad (5.25d)$$

$$i\omega p = i\omega\gamma gH\rho + g(1-\gamma)w + (\gamma-1)\rho_o^{1/2}Q. \qquad (5.25e)$$

We can express everything in terms of p from Equations 5.25a and 5.25b where $(f = f_o + \beta y)$:

$$(f^2 - \omega^2)u = \omega kp - f\frac{\partial p}{\partial y}, \qquad (5.26a)$$

$$(f^2 - \omega^2)v = ikfp - i\omega\frac{\partial p}{\partial y}. \qquad (5.26b)$$

Using Equation 5.25c in Equation 5.25e to eliminate ρ gives

$$g(1-\gamma)w = i\omega\left(1 - \frac{\gamma}{2}\right)p + i\omega\gamma H\frac{\partial p}{\partial z} + (1-\gamma)\rho_o^{1/2}Q. \qquad (5.27)$$

Equations 5.26 and 5.27 express u,v and w in terms of p. Now use them in the divergence equation 5.27d to get an equation in p,u,v:

$$-\frac{i\omega H}{k}\left[\frac{\partial^2 p}{\partial z^2} - \frac{p}{4H^2}\right] + g\left(iku + \frac{\partial v}{\partial y}\right) = -\left(\frac{\partial}{\partial z} - \frac{1}{2H}\right)\left[\rho_o^{1/2}Q\right]. \qquad (5.28)$$

Note that *if* we were looking at a forced problem in the ocean, we would separate the problem by solving

$$-\frac{H}{\kappa}\left[\frac{\partial^2}{\partial z^2} - \frac{1}{4H}\right]p = \frac{p}{h}$$

or

$$\frac{\partial^2 p}{\partial z^2} + \left[\frac{\kappa}{Hh} - \frac{1}{4H^2}\right]p = 0 \tag{5.29}$$

so that

$$i\omega p + gh\left(iku + \frac{\partial v}{\partial y}\right) = \text{forcing} \tag{5.30}$$

and Equations 5.25a, 5.25b and 5.30 would form a set of shallow-water equations using the equivalent depths given by the solutions to Equation 5.29.

The boundary conditions for the problem are obtained by putting $w = 0$ at $z = 0$ and assuming finiteness or outgoing energy flux as an upper boundary condition.

Exercise 5.5: From Equations 5.27 and 5.26b show that $w = 0$ implies the boundary condition that:

$$\frac{\partial p}{\partial z} - \frac{\kappa}{H}\left(1 - \frac{1}{2\kappa}\right)p = 0 \quad \text{at} \quad z = 0,$$

where $\kappa = 1 - \dfrac{1}{\gamma}$.

In the atmosphere in general, Equation 5.29 with the above boundary condition has no solutions, except one that travels with the speed of sound: the so-called Lamb wave with $h = \gamma H = \gamma\dfrac{RT}{g}$, i.e. with $c_s^2 = gh = \gamma RT$ or $c_s = \sqrt{\gamma RT}$ the soundwave speed. Except for this mode (sometimes called "external"): the atmosphere has no solutions to Equation 5.29. Forgive us for shouting, but the result is important and generally widely misunderstood, i.e.:

THE ATMOSPHERE HAS NO DISCRETE FREE-VERTICAL MODES.

Now that we are calm again, we will mention one possible, if arcane, pseudo-lid for the atmosphere. Although the atmosphere surely has no lid, it does have a rich structure of winds in the stratosphere and it is possible that upward-propagating waves regard these as a lid. We have in mind the consequence of a "critical layer," a region where the local wind velocity matches the speed of the wave. In such a case the layer may either absorb the wave energy, or reflect it. Typically, these layers are absorbing, which means that the wave energy does not propagate up far beyond the critical layer. It is, in this sense, a lid, but it does not do what is needed for standing modes: reflect the energy downward with little loss. As more and more of the wave energy is absorbed, the critical layer might get to be nonlinear and saturated, and begin to reflect instead of absorb. Then it would be the lid we need. However, it appears that the critical layer just propagates downward as more waves are absorbed and never has a chance to become too reflecting. (We thank K. K. Tung for insight into this issue.)

So, unlike the ocean, the atmosphere has no lid and therefore standing vertical modes do *not* occur. We must therefore proceed differently than in the ocean (an ocean version of this is given in Philander [1978]; it is useful for vertically propagating modes).

Following Lindzen (1967) we can rewrite Equation 5.28 in terms of v only.

$$\frac{H}{\kappa} L_z[f^2 v] + \left(\frac{\partial}{\partial y} - \frac{k}{\omega}f\right)\left(\frac{\partial}{\partial z} - \frac{1}{2H}\right)(\rho_o^{1/2}Q) + gM_y[v] = 0 \qquad (5.31)$$

where

$$L_z \equiv \frac{\partial^2}{\partial z^2} + \left(\frac{\kappa g}{H}\frac{k^2}{\omega^2} - \frac{1}{4H^2}\right)$$

and

$$M_y \equiv \frac{\partial^2}{\partial y^2} + \frac{k}{\omega}\left(\beta - \frac{k}{\omega}f^2\right);$$

recalling that $f = f(y) = f_o + \beta y$.

We separate Equation 5.33 by expanding in eigenfunctions of the *horizontal-structure equation*:

$$M_y[\Psi_{k,\omega,n}] = \left(\frac{1}{gh_{k,\omega,n}} - \frac{k^2}{\omega^2}\right)(f^2 - \omega^2)\Psi_{k,\omega,n}, \qquad (5.32)$$

where the separation parameters $h_{\omega,k,n}$ are determined from the horizontal equation: with the boundary conditions that $v = 0$ at the North and South Poles, $y = y_n, y_s$; then $h_{\omega,k,n}$ are the equivalent depths. Note that they depend on frequency and spatial wavenumbers. The problem of solving the linear response to thermal forcing in the atmosphere is then solved as follows.

The thermal forcing appearing in Equation 5.31 is expanded in a set of horizontal eigenfunctions of Equation 5.32:

$$\left(\frac{\partial}{\partial y} - \frac{k}{\omega}f\right)\left(\frac{\partial}{\partial z} - \frac{1}{2H}\right)(\rho_o^{1/2}Q)_{k,\omega} = (f^2 - \omega^2)\sum_n S_{\omega,k,n}(z)I_{\omega,k,n}(y).$$

In terms of this forcing, the vertical-structure equation

$$\frac{d^2 V_{k,\omega,n}}{dz^2} + \left(\frac{\kappa}{Hh_n} - \frac{1}{4H^2}\right)V_{k,\omega,n} = -\frac{\kappa}{H}S_{k,\omega,n} \qquad (5.33)$$

is solved, using the equivalent depths found when solving the horizontal-eigenvalue Equation 5.32. The boundary conditions are

$$\frac{\partial p}{\partial z} - \frac{\kappa}{H}\left(1 - \frac{1}{2\kappa}\right)p = 0 \; at \; z = 0 \text{ and outgoing radiation conditions at } z = \infty.$$

Since the atmosphere has only the (trivial) Lamb mode as a free solution to Equation 5.33, or equivalently Equation 5.29, in general the response to Equation 5.33 will be a set of *forced* modes that will either propagate with vertical wavenumber m, where

$$m^2 = \frac{\kappa}{Hh_n} - \frac{1}{4H^2};$$

if h_m is small and positive, or will be trapped with decay scale $|m|$ if h_n is negative or large and positive ($h_{\omega,k,n} > 4\kappa H$).

It may be surprising that an equivalent depth could be negative, but remember that this term was introduced because Equation 5.33 looks like Equation 5.29. Again, we emphasize, there is but one free mode in the atmosphere, a sound wave, so that we must solve the *forced* Equation 5.33 with eigenvalues given by the horizontal Equation 5.32. This contrasts sharply with the oceanic case where we expand the forcing in the eigenfunctions of the vertical equation (with *positive* equivalent depths only) and then solve a forced horizontal equation for each equivalent depth separately.

We proceed to look at the solutions to the horizontal-structure Equation 5.32 on (a) a midlatitude β-plane $f = f_0 + \beta y$ and (b) an equatorial β-plane $f = \beta y$.

(a) Midlatitude β-plane

$$\frac{d^2\Psi}{dy^2} + \left\{\frac{1}{gh}(\omega^2 - f_0^2) + k\left(\frac{\beta}{\omega} - k\right)\right\}\Psi = 0 \qquad (5.34)$$

where we have made the usual midlatitude β-plane assumption that $\beta y \ll f_0$ and $f^2 \approx f_0^2$ a constant, then the solutions to Equation 5.34 are clearly sines and cosines and the eigenvalue h will be determined by the boundary conditions, which we take to be $\Psi = 0$ at $y = \pm d$.

The solutions are then $\Psi_m = \cos\dfrac{n\pi}{2d}y$ and $\dfrac{1}{gh}(\omega^2 - f_0^2) + k\left(\dfrac{\beta}{\omega} - k\right) = \dfrac{n^2\pi^2}{4d^2}$

or

$$gh = \frac{\omega^2 - f^2}{\dfrac{n^2\pi^2}{4d^2} + k^2 - \dfrac{\beta k}{\omega}}. \tag{5.35}$$

For large enough ω, the equivalent depths are positive and the forced response propagates vertically. There are negative equivalent depths for small enough ω (in particular $\omega^2 < f_0^2$) and a suitable range of k.

This result was first used by Lindzen (1966) to show that the solar wavenumber 1 semi-diurnal tide propagated vertically, while the diurnal tide was trapped. Since the excitation for both was mostly ozone heating in the stratosphere, only the semi-diurnal tide reached the ground.

(b) Equatorial β-plane

$$\frac{d^2\Psi}{dy^2} + \left\{\frac{k}{\omega}\beta - k^2 + \frac{\omega^2}{gh} - \frac{\beta^2}{gh}y^2\right\}\Psi = 0. \tag{5.36}$$

Introduce a length scale L such that $L^4 = \dfrac{gh}{\beta^2}$ and a dimensionless variable $\eta = y/L$

$$\frac{d^2\Psi(y)}{d\eta^2} + \left\{\left(\frac{k}{\omega}\beta - k^2 + \frac{\omega^2}{gh}\right)L^2 - \eta^2\right\}\Psi(\eta) = 0.$$

With boundary conditions $\Psi = 0$ at Y_N, Y_s the solutions are parabolic-cylinder functions. On an infinite equatorial beta-plane the boundary conditions are $\Psi \to 0$ as $\eta \to \pm\infty$ and the solutions are Hermite functions (✢ Appendix 2)

$$\Psi_n(\eta) = e^{-y^2/2}H_n(\eta) \tag{5.37}$$

and the eigenvalues are given by

$$\left(\frac{k}{\omega}\beta - k^2 + \frac{\omega^2}{gh}\right)\frac{\sqrt{gh}}{\beta} = 2n + 1.$$

Solving this eigenvalue equation for the equivalent depth gives

$$
\sqrt{gh} = \frac{(2n+1)\beta}{\frac{k}{\omega}\beta - k^2} \left[1 \pm \sqrt{1 - \frac{4\left(\frac{k}{\omega}\beta - k^2\right)\omega^2}{\beta^2(2n+1)^2}} \right].
$$

For long waves in x at low frequency we may neglect $k \ll \frac{\beta}{\omega}$ and $\omega \ll \frac{\beta}{k}$:

$$
\sqrt{gh} = (2n+1)\frac{\omega}{k}\left[1 \pm \left(1 - \frac{2k\omega}{\beta(2n+1)^2} \right) \right]. \qquad (5.38)
$$

Although we have skimped on subscripts, it should be remembered that h is really a function of k, ω and n, so that both the y and the η in Equation 5.37 depends on n and, therefore, the horizontal scale changes with n.

The $+$ sign generally does not correspond to solutions of the equations on a sphere so

$$
\sqrt{gh} = \frac{2\omega^2}{\beta(2n+1)},
$$

independent of k. The vertical wavenumber is

$$
m \approx \left(\frac{\kappa}{Hh_n} \right)^{1/2} = \left(\frac{\kappa g}{H} \right)^{1/2} \frac{\beta(2n+1)}{2\omega^2}
$$

and the vertical group velocity is negative with amplitude

$$
\frac{d\omega}{dm} = C_v = \frac{\omega^3}{\beta(2n+1)}\sqrt{\frac{H}{\kappa g}}.
$$

Suppose we introduce a damping time τ into the problem. The distance the forced response reaches in time τ is

$$
C_v\tau = \frac{\omega^3\tau}{\beta(2n+1)}\sqrt{\frac{H}{\kappa g}}.
$$

To fix ideas, let $\omega = \frac{1}{30}\Omega$ (~ 30 d period) then $\sqrt{gh} = \frac{1}{2(2n+1)}$ m/s so the longest h corresponds to $\sqrt{gh} = \frac{1}{2}$ m/s $\Rightarrow h = \frac{1}{40}$ m $\Rightarrow L_v = 150$ m and

$$
C_v = 2 \times 10^{-5}\,\text{m/s} = 2\text{m/d};
$$

thus a damping time of something less than a month, a reasonable value for radiative damping in the atmosphere, would mask the short wavelengths of the response and confine the response to the region of the forcing. This argument would seem to imply that the response to thermal forcing could not travel from the bottom of the cloud layer to the surface, so that thermal forcing due to deep cumulonimbus clouds whose cloud base is 600 m from the surface could not drive surface winds at the surface. This turns out not to be true.

The solution to this conundrum, and others posed by the Lindzen (1967) paper, is that the solutions on an equatorial β-plane, as presented, do not form a complete set. There are additional solutions to Equation 5.36 with negative equivalent depth that are part of the continuous spectrum of the solution, as pointed out by Wu *et al.* (1999).

At low frequencies, the negative equivalent-depth solutions of Equation 5.36 satisfy

$$\frac{d^2\Psi}{dy^2} + \left\{\frac{k}{\omega}\beta + \frac{\beta^2}{g\hat{h}}y^2\right\}\Psi = 0, \tag{5.39}$$

where $\hat{h} = -h$ is positive. The solutions to Equation 5.39 are Weber parabolic-cylinder functions and are needed to form a complete set of horizontal solutions. It is these extra negative equivalent-depth modes that, when included, allow a wind response at the surface and everywhere below the forcing level when Newtonian cooling is the damping term and included in the thermal Equation 5.25e. For a Newtonian cooling with damping time τ the derivation of Equation 5.33 goes through as before and becomes:

$$\frac{d^2 V_{k,\omega,n}}{dz^2} + \left(\frac{\kappa\omega}{H(\omega - i\tau^{-1})h_n} - \frac{1}{4H^2}\right) V_{k,\omega,n} = -\frac{\kappa}{H}S_{k,\omega,n}. \tag{5.40}$$

For low frequencies (i.e. periods that are long compared with the damping time), the first term in brackets is small and the equation becomes:

$$\frac{d^2 V_{k,\omega,n}}{dz^2} - \frac{V_{k,\omega,n}}{4H^2} = -\frac{\kappa}{H}S_{k,\omega,n},$$

which states that the response decays exponentially in the vertical with an exponential decay scale of $H/2$. When the mass scaling of v by $\rho_0^{1/2}$ introduced at the beginning is included, it is seen that the winds are uniform below the forcing. Thermal sources in the free atmosphere *can* force surface winds, but only in the presence of Newtonian cooling.

Wu *et al.* (2000b) approached the problem in a manner formally similar to the method used for the ocean. That is, they first solved the free vertical-structure equation for equivalent depths and vertical eigenmodes:

$$\frac{d^2 V_m}{dz^2} + \lambda_m V_m = 0$$

where

$$\lambda_m = \left(\frac{N^2}{gh_m} - \frac{1}{4H^2} \right).$$

This equation, has, in addition to the discrete Lamb mode, a continuous spectrum of eigenmodes (Wu *et al.*, 2000b):

$$V_m(z) = \left(\cos \; mz - \frac{1}{2Hm} \sin \; mz \right) \frac{2Hm}{(4H^2m^2 + 1)^{1/2}}$$

where *m* is *any* positive real number. This is a profound difference from the ocean case: because there is no lid to quantize m_n, the spectrum of eigenvalues is continuous, not discrete (the discrete Lamb mode excepted).

The major modification arises because, unlike in the idealized Gill model, the equivalent depths (and therefore the vertical wavelengths) are forced by a heating from a spectral *continuum*, which in the absence of damping would propagate vertically, unlike the single-standing mode in the Gill model. For example, for thermal forcings with vertical structures shown in Figure 5.20 (arbitrarily labeled CP and MC), located at 93°E and having a Gaussian structure of about 20° meridionally, the decomposition in the vertical can be done in terms of the complete spectrum of vertical modes.

The spectral density of the projections of the forcings in Figure 5.20 are shown in Figure 5.21 and it is seen that although the spectrum is continuous, there are vertical wavelengths that are favored. We would therefore expect the forcing to be dominated by wavelengths of 13 km and 25 km in the CP case and 14 km in the MC case. (Note that these scales are imposed by the *forcing* and not by the structure of the atmosphere, as would have been true if there were free modes.) For damping parameters taken to be equal in momentum and heat, so that the horizontal structure should look like the classical Gill model of Figure 5.17, if a single vertical mode of about 15 km were involved, we can plot the actual three-dimensional structure: the MC and CP responses are shown in Figures 5.22 and 5.23.

The horizontal plan view of the velocity field looks very much like the Gill results.

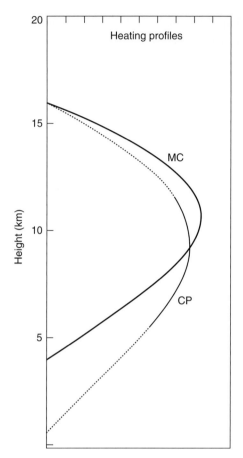

Figure 5.20. Heating profiles referred to in the text. (Redrawn after Wu *et al.*, 2000b.)

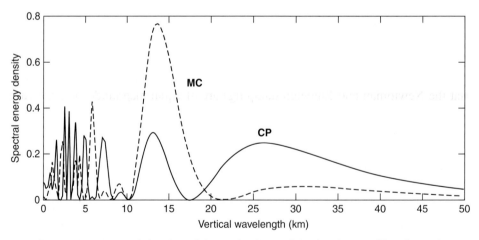

Figure 5.21. Spectral density of the projections of the heating profiles shown in Figure 5.20. (From Wu *et al.*, 2000b.)

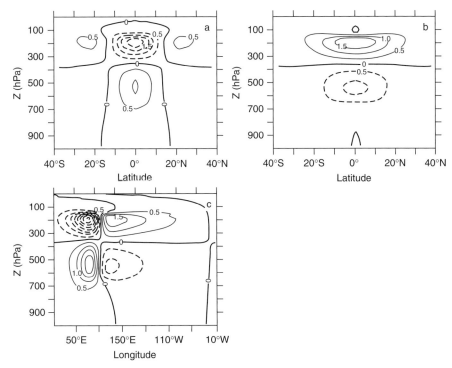

Figure 5.22. Zonal velocities at different locations for the MC case. (a) At 53.4°E (to the west of the heating), (b) at 132°E (to the east of the heating), and (c) at the equator. The contour interval is 0.5 m s^{-1}. Westerlies are represented by the solid lines and easterlies are represented by dashed lines. (From Wu *et al.*, 2000b.)

Clearly, while the velocity field looks Gill-like for an individual level, it does not look identical for all levels and, therefore, departs from the Gill paradigm. Further, the zero line of the zonal velocities does not coincide in the vertical with the level of maximum heating as it does for the Gill model. The damping in this case is such that the Newtonian and Rayleigh dampings are of equal magnitude. The CP case generates surface winds because it starts nearer to the ground. The MC case does not (compare Figures 5.22 and 5.23).

A conclusion that emerges forcefully from the work of Wu *et al.* (2000b) is that some thermal-damping mechanism appears to be essential if thermal forcing in the free troposphere is to generate low-level winds. In contrast, Rayleigh friction alone confines the wind response to the vertical extent of the forcing. This difference is evident in the wind profiles in Figure 5.25. Strong radiative damping tends to eliminate temperature perturbations below the forced region, so that the pressure changes at the bottom of the forcing region extend to the surface, creating

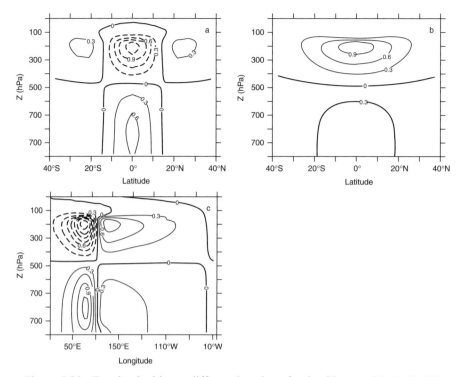

Figure 5.23. Zonal velocities at different locations for the CP case. (a) At 53.4°E (to the west of the heating), (b) at 132°E (to the east of the heating), and (c) at the equator. The contour interval is $0.3 \, \mathrm{m \, s^{-1}}$. Westerlies are represented by the solid lines and easterlies are represented by dashed lines. (From Wu *et al.*, 2000b.)

a vertically uniform momentum forcing and hence a vertically uniform wind change.

Wu *et al.* also found that the thermally forced convergences of moisture are not sufficient to account for the assumed strength of the thermal sources. It must be some other mechanism that produces the convergence to maintain the precipitation. We turn to the Lindzen–Nigam boundary-layer model.

5.6 The processes that anchor regions of persistent precipitation to SST

We saw in Chapter 2 that the regions of persistent precipitation occur over the warmest water. Neelin (1989) showed that the Gill equations could be transformed into a form that resembled boundary layer equations forced by SST anomalies. In order to see why this should be true, we present a simplified version of arguments originally given by Lindzen and Nigam (1987), Neelin

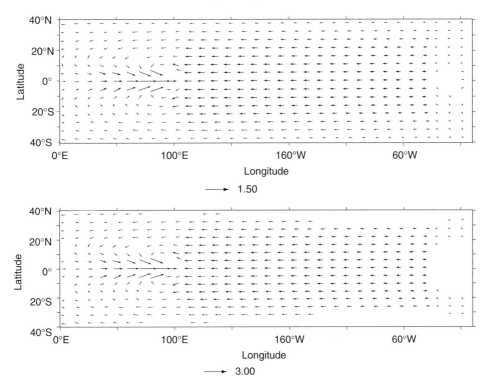

Fig 5.24. Upper: The horizontal velocities at 810 hPa for the CP case. Lower: The horizontal velocities at 560 hPa for the MC case. (From Wu *et al.*, 2000b.)

(1989), and Battisti *et al.* (1999). The basic idea is that warm (cold) SST hydrostatically induces low (high) pressure over the SST perturbations and the pressure variations subsequently induce low-level convergence (divergence). Low-level convergence of moisture then produces precipitation and the regions of persistent precipitation above the warm SST then drive circulations in a manner described in Section 5.5.3.

We consider the low-level flow below a well-mixed atmospheric-boundary layer extending to an undisturbed height H_b of about 2 or 3 kilometers – we can think of this boundary layer as extending to the trade inversion. We define the vertically density-averaged horizontal velocity $\mathbf{U}$:

$$\mathbf{U} = \frac{1}{H_b \rho_0} \int_0^{H_b} \rho \mathbf{u} \, dz,$$

where ρ_0 is the mean density in the layer.

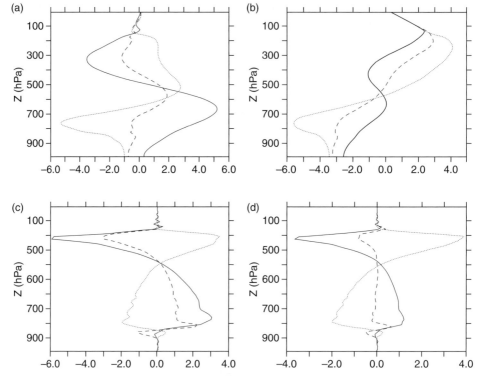

Figure 5.25. Vertical profiles of zonal velocity (a) at the equator, Newtonian cooling alone, (b) at 5.4°N, Newtonian cooling alone, (c) at the equator, Rayleigh friction alone, (d) at 5.4°N, Rayleigh friction alone. Solid lines are at 53.4°E, dashed lines are at 92.4°E, dotted lines are at 132.2°E. The thermal forcing is a Gaussian pattern centered at (93°E, 0°) with a longitudinal scale of 20° and a meridional scale of 11°. In the vertical it is a half sine wave extending from 840 hPa to 160 hPa. For more details see Wu *et al.* (2000b).

This vertically averaged horizontal velocity (U,V) satisfies a linearized equation similar to Equations 5.22a and 5.22b:

$$\varepsilon U - \beta y V = -P_x, \tag{5.41a}$$

and

$$\varepsilon V + \beta y U = -P_y, \tag{5.41b}$$

where

$$P = \frac{1}{H_b \rho_0} \int_0^{H_b} p \, dz$$

is the vertically mass-averaged pressure gradient and ε is a drag coefficient on the horizontal wind. Pressure perturbations are hydrostatic and due to two distinct sources:

(a) a mean perturbation to the boundary-layer density, ρ', which means that

$$\frac{dp'}{dz} = -\rho'g \Rightarrow p'(z) = -\rho'g(H_b - z) \Rightarrow \bar{p}' = \frac{1}{2}\rho'gH_b$$

where $\bar{p}'$ is the average perturbation pressure in the layer;

(b) a change h' in the boundary-layer height, which means a pressure change throughout the layer of $\Delta\rho gh'$, where $\Delta\rho = \rho_0 - \rho_a$ and ρ_a is the density just above the boundary layer.

Hence, the total layer average pressure perturbation is

$$P = \frac{1}{2}gH_b\frac{\rho'}{\rho_0} + g\frac{\Delta\rho}{\rho_0}h', \tag{5.42}$$

where the first term is due to changes of density within the boundary layer and the second is due to changes of the height of the boundary layer.

The changes of density are approximately given by

$$\frac{\rho'}{\rho_0} = -\frac{\theta'}{\theta_0}; \quad \frac{\Delta\rho}{\rho_0} = -\frac{\Delta\theta}{\theta_0},$$

where $\Delta\theta$ is the potential-temperature jump between the boundary layer and the free atmosphere immediately above. It is here assumed that the boundary layer sees variations of SST which then extend throughout the layer.

If we now assume that perturbations in the boundary-layer height are due to changes in total convergence within the boundary layer (i.e. we assume that changes in entrainment rate may be neglected), then

$$\varepsilon_T h' = -H_b \nabla \cdot U, \tag{5.43}$$

where ε_T is the boundary-layer relaxation time.

Combining Equations 5.42 and 5.43 gives

$$\varepsilon_D P + gH_b\left[\frac{\varepsilon}{\varepsilon_T}\frac{\Delta\theta}{\theta_o}\right]\nabla \cdot U = -\varepsilon\frac{gH_b}{2}\left(\frac{\theta'}{\theta_o}\right). \tag{5.44}$$

Note that Equations 5.41 and 5.44 have the form of the one-and-a-half-layer model of Section 3.5.3, but with the time dependence replaced by a drag term to yield steady solutions. The "equivalent depth" here is

$$H_b \left[\frac{\varepsilon}{\varepsilon_T} \frac{\Delta\theta}{\theta_o} \right]$$

and the system is driven by a thermal-forcing term proportional to the potential-temperature perturbation, θ'. Since θ' is assumed constant throughout the well-mixed boundary layer, this directly connects the sea-surface temperature perturbations to the mean convergence in the lower layer. We note that Equations 5.42 and 5.45 have the precise form of the Gill model (see Section 5.5.2) with forcing proportional to the temperature perturbation.

The original work by Lindzen and Nigam showed that the theory correctly gave the low-level pressure perturbations as a function of SST perturbations but only gave the realistic induced low-level convergence given by Equation 5.44 when the relaxation coefficient ε_T was very large, implying a very rapid relaxation of the boundary layer (of the order of minutes). However, Lindzen and Nigam mistakenly omitted the term $\Delta\theta/\theta_o$ in Equation 5.44; since it is $0(10^{-2})$ it allows ε_T to be two orders of magnitude larger – something of the order of a day, not a few minutes, which is a more plausible boundary-layer relaxation time.

5.7 Surface winds for simple atmospheric models

We have noted that Wu *et al.* (1999) found that upper-level forcing can force surface winds but cannot force enough convergence to maintain upper-level heat sources. The convergence must be maintained by boundary-layer processes. If we are to get the surface winds right, both processes must be present. We have seen that both can be treated by the Gill equations, although this admits only the simplest vertical structure for momentum and thermal damping. One cannot even take values in the boundary layer that differ from those in the free atmosphere.

Chiang *et al.* (2001) investigated whether the combination of upper-level heating and SST-induced boundary-layer convergence in a linear model could indeed simulate observed surface-wind fields. (The heating was inferred from precipitation data.) The full forcing (both boundary layer and elevated heat source) gave the results in Figure 5.26 (Upper) where the contours are the SST anomalies. This compares favorably with the observational analysis in Figure 5.19. The easterly anomalies to the east of the heating (which bears the same relationship to the SST as Figure 5.19) are much reduced, primarily by including the Andean topography which tends to reflect and cancel the atmospheric Kelvin mode.

They also separated the upper-level heating (loosely, the "Gill mechanism") from the boundary-layer SST forcing (the "Lindzen–Nigam" mechanism). The

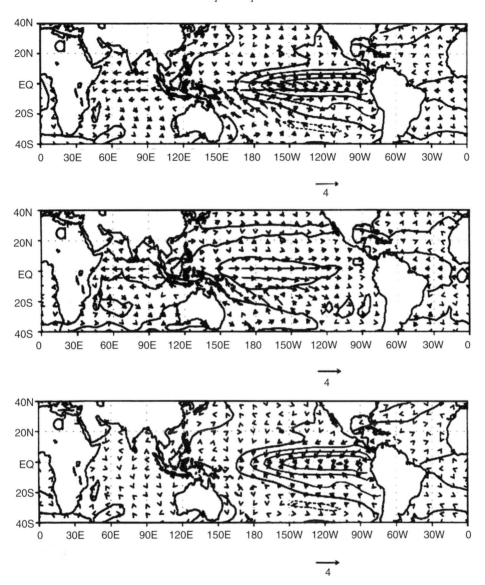

Figure 5.26. Upper: The surface-wind field forced by both boundary-layer SST variations and upper-level heating. Middle: The surface-wind field forced by upper-level heating only. Lower: The surface-wind field forced by boundary-layer SST variations only. (From Chiang *et al.*, 2001.)

elevated thermal forcing by itself gave the results in Figure 5.26 (Middle), while the boundary-layer SST forcing alone gave Figure 5.26 (Lower). They conclude that it is upper-level heating that drives the surface zonal winds, which are the stronger component of the wind field, but that the meridional component is largely attributable to the SST boundary forcing. The latter is the primary contributor to surface-layer convergence, so this work again leads to the conclusion that it is the SST influence that creates local moisture convergence to drive convective heating. The upper-level heating does produce convergence in some places, such as the region of the South Pacific convergence zone (SPCZ), but in the eastern tropical Pacific the boundary-layer effect is dominant. As with Wu *et al.*, Chiang *et al.* (2001) found that the surface influence of upper-level thermal forcing is very sensitive to the vertical distribution of the heating. Observations show that one vertical-heating profile does not account for all tropical heating (or even all tropical Pacific heating), and since the model does not allow for this it may be underestimating the impact of upper-level thermal forcing on wind convergence in some regions.

Chiang *et al.* (2001) confirm the need to have strong non-adiabatic effects in order for upper-level heating to influence the surface, but depart from the simple uniform Newtonian cooling used by Wu *et al.* (1999) and use a very strong damping in the boundary layer of up to 1/(0.5 d). What justifies such a large value when the radiative-relaxation time is known to be in the order of a week or more? They argue that it is a consequence not of radiative relaxation but of turbulent mixing in the boundary layer.

Why should the Newtonian cooling term be so large (damping time of the order of a day) when we might expect the radiative-relaxation rate to be small (in the order of 2 weeks)? The temperature changes *in* the boundary layer by sensible heating is given by:

$$\frac{dT_a}{dt} = \frac{S}{\rho c_p H} = \frac{c_s |\mathbf{u}|}{H}(T_s - T_a),$$

where the sensible heating S is given by its drag-coefficient parameterization, with T_a some boundary-layer temperature, and H the boundary-layer height. Clearly the effective damping rate is $\dfrac{c_s |\mathbf{u}|}{H}$. Taking H of order 500 m, c_s of order 10^{-3} and velocity of order 5 m/s gives an effective damping of 1/day or a damping time of 1 day. It might be said that the investigations we have recounted here of how tropospheric thermal forcing influences the surface-layer flow have led us to believe that the remaining uncertainties are largely tied up in issues of boundary-layer physics. For example, we have yet to examine the possible impact of entrainment through the inversion layer of free atmosphere momentum.

At a minimum we can conclude that the *location* of the thermal sources is determined by convergences given by SST gradients within the boundary layer by the Lindzen–Nigam mechanism – the upper-level thermal sources cannot force enough convergence to do this themselves. (It should be noted that this comment is scale-dependent – if the horizontal scale of the thermal source is small enough then it can – individual clouds or cloud clusters *can* force their own convergence.) This eliminates an entire class of mechanisms, the conditional instability of the second kind (CISK) mechanisms, that have been on the table for years. Since both the boundary-layer convergence and the upper-level forcing are given by Gill-like patterns, it makes much sense that a simplified atmospheric model might assume a Gill form for the surface winds (as shown in Section 5.6). A more complex model, of course, must obtain the Gill form as a consequence of its internal dynamics. We might note in conclusion that these considerations hold for deep convection and do not necessarily hold for shallow precipitating convection (Wu, 2003), which we have assumed never occurs. The remainder of this section will, on the basis of what has been learned in the previous two sections, deal with a reasonable parameterization of surface winds for use in simple models.

In applying the previous considerations to the type of intermediate model we will introduce in Chapter 7, in particular the Zebiak–Cane model, we have to adapt the Gill model, which is used for both surface winds forced by upper-level thermal forcing and by surface-temperature gradients in the boundary layer. The atmospheric model in this coupled model assumes the climatology is specified and only the anomalies are calculated. Two modifications, both introduced by Zebiak, are needed before the atmospheric model can be used to determine the surface winds.

The first (Zebiak, 1982), is to allow for arbitrary distributions of surface-temperature anomalies. First Fourier transform in x so that each Fourier component is forced by the Fourier component of the heating, solve for each Fourier component separately, and then re-synthesize the field so that the low-level winds for arbitrary distributions of heating could be obtained. In this first model Zebiak took $Q \propto T$ so the model he was solving was more like the Lindzen–Nigam equations for the boundary layer than a model for upper-level heating.

The second modification (Zebiak, 1986) was to consistently calculate the *anomaly* of convergence in response to SST anomalies.

For an SST anomaly, first calculate the anomaly of evaporation in terms of the SST anomaly T' which then acts as a "seed" heating to the atmosphere:

$$Q_s = (\alpha T') \exp[(\overline{T} - 30\,^\circ\text{C})/16.7\,^\circ\text{C}],$$

where α is a constant. The total heating, $Q_s + Q^n$ is used to drive the Gill equations and calculate a new anomalous heating which at the nth step has the form

$$Q^n = \beta[M(\overline{c} + c^n) - M(\overline{c})] \qquad (5.45)$$

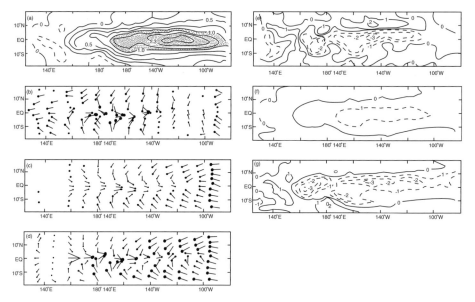

Figure 5.27. (a) Observed SST anomaly in composite warm phase of ENSO for December of year (0) (see Chapter 2). (b) Observed surface-wind field in composite. (c) Model surface winds with no feedbacks. (d) Model surface winds with feedbacks. (e) Observed divergence in composite. (f) Model divergence without feedback. (g) Model divergence with feedbacks. (From Zebiak, 1986.)

where

$$M(a) = a \quad \text{if} \quad a > 0 \quad \text{and} = 0 \quad \text{if} \quad a \leq 0.$$

$Q^0 = 0$, β is a constant, and c^n is the anomalous convergence heating while $\bar{c}$ is the climatological convergence. Normally, this iteration process continues until the heating converges. Recall that the atmosphere equations are steady state. Advancing to a new time means changing the forcing boundary conditions $T'(x, y)$.

$M(\bar{c} + c^n)$ is proportional to the total precipitation (hence must be positive) and $M(\bar{c})$ is the climatological precipitation (which is also positive). The anomaly can be positive or negative of course.

The form of Equation 5.45 introduces an interesting nonlinearity. If the anomalous divergence c^n is so large that the first term becomes zero, then the anomalous heating is simply the negative of the climatological heating. If it is even larger, the anomaly is still the negative of the climatological heating and no more: there can be a negative rainfall anomaly, but there cannot be negative rainfall.

Figure 5.27 shows a comparison of the modeled low-level (i.e. surface) wind fields and surface convergence for two characteristic months in the "composite"

El Niño. The no-feedback case is where the iteration given by Equation 5.45 is not performed. The model of the atmospheric surface winds shown in Figure 5.27 is the one used in the Zebiak–Cane coupled atmosphere–ocean model to be treated in the next chapter.

We have come, by a very circuitous route, to a model for surface winds, which is much like a modified Gill model.

6

Ocean processes

This chapter deals with the ocean processes needed to understand the interaction of the ocean with the atmosphere in the tropics. Since the interaction between the atmosphere and the ocean occurs entirely at the surface, through the interchange of heat, water and momentum fluxes, the key quantity the ocean provides is the sea-surface temperature (SST). The chapter opens with a discussion of the upper-ocean processes that change SST and rapidly focuses in on the effects of upwelling on the upper-ocean stratification, as indexed by the depth of the tropical thermocline. Upwelling is easily calculated from frictional processes near the surface but the change in the depth of the thermocline is a subtle process that responds mostly to wind stresses at the surface.

The time-dependent response of the thermocline to wind stresses is treated in a number of simplified contexts. As an introductory example, the non-divergent (barotropic) case is worked out in some detail. The steady interior solution in this case is the Sverdrup solution, where the curl of the wind stress balances the meridional mass transport. How the Sverdrup relation is set up in a time-dependent manner is demonstrated and the role of both viscous and inviscid western-boundary layers in balancing the vorticity constantly put in by the curl of the wind stress is considered. The signaling properties (carried by signal fronts having properties of waves) that tell parts of the basin whether forcing is either present or absent is emphasized, and the common misunderstanding of the sense in which waves are present is discussed in some detail. This barotropic example has all the major features, albeit in a simplified context, that characterize equatorial adjustment of the thermocline to surface forcing.

Using this barotropic example as a conceptual model of adjustment, we successively proceed to introduce the complications of the midlatitude beta-plane and the equatorial beta-plane. We introduce the basic simplifications of wind stresses that are spatially constant but limited in zonal extent in a basin laterally bounded by meridians. The signal fronts that carry the information that the forcing is limited in spatial extent have many of the properties of waves on an equatorial beta-plane. If the forcing were infinite in extent and impulsively applied, the low-frequency response would be resonant at zero frequency, thereby growing linearly with time. Signal fronts, traveling with the dispersion properties and meridional structure of equatorial waves, deliver the message that the forcing is limited in extent and the linear time dependence is modified behind these signal fronts.

The presence of meridional boundaries causes reflections of these signal fronts: on an equatorial beta-plane, western boundaries concentrate the signals into an equatorially confined Kelvin signal and eastern boundaries spread the signal meridionally as a series of Rossby signals. Adjustment to a suddenly applied forcing therefore proceeds from the equator outward to higher latitudes: the equatorial region is adjusted basically upon the passage of a single Kelvin and low-order Rossby signal, while progressively higher latitudes adjust more slowly by similar processes.

The details of how to calculate adjustment are given in some detail: first the expansion of the forcing in parabolic-cylinder functions; next the steady solution to which the solution adjusts; next the unbounded response; next the signals that indicate that the forcing region is limited; and finally the reflections of signals at the boundaries. The adjustment to constant winds applied suddenly in a basin is limned out in some detail.

Finally, since ENSO has aspects that are event-like and aspects that are quasi-periodic in time (with periods of 3 to 7 years), the critical properties of periodically forced thermocline motions are outlined and the essential differences between thermocline adjustment and periodically forced thermocline motions are explained.

6.1 The processes that change SST

We will take the upper ocean to be well mixed at all times, i.e. it will be assumed that there will always be a mixed layer of depth h at the top of the ocean so that the temperature T_m of the well-mixed layer *is* the sea-surface temperature (SST). The heat budget integrated over the mixed layer is:

$$c_p \left[\frac{\partial}{\partial t} + u \frac{\partial}{\partial x} + v \frac{\partial}{\partial y} \right] (hT_m) = Q_s - w_e c_p \Delta T, \tag{6.1}$$

where u and v are horizontal velocities averaged over the mixed-layer depth h, Q_s is the heat flux into the surface of the ocean at the top of the mixed layer, ΔT is the discontinuity at the bottom of the mixed layer (as described in Chapter 4) and w_e is the entrainment velocity at the bottom of the mixed layer:

$$w_e = \frac{\partial h}{\partial t} + w. \tag{6.2}$$

As described in Chapter 4, the entrainment velocity is the volume flux crossing the (possibly moving) interface at the bottom of the mixed layer per unit time and is defined only when it is positive. The heat flux at the ocean surface is given by:

$$Q_s = R_{net} + LE + S, \tag{6.3}$$

where R_{net} is the net radiative flux at the ocean surface, LE is the latent heat flux due to evaporation E into the atmosphere at the ocean surface, and S is the sensible heat flux from the ocean surface into the atmosphere (we, as before, define all heat fluxes as positive upward). An interpretation of Equation 6.3 is that the heat flux into the ocean is the amount of net (downward) radiation that is left unbalanced by the sum of the latent heat of evaporation of water and the sensible heating.

The heat flux at the bottom of the mixed layer arises only because cooler water enters the mixed layer from below. The effect of the rest of the ocean below the mixed layer is seen in the heat budget mainly in the term $\Delta T = T_m - T_{sub}$ where T_{sub} is the temperature of the ocean just below the mixed layer. As we will see, T_{sub} depends primarily on where the thermocline is: T_{sub} is larger when the thermocline is closer to the bottom of the mixed layer (i.e. shallower) and smaller when the thermocline is farther away (i.e. deeper). The location of the thermocline relative to the bottom of the mixed layer is therefore a crucial part of the ocean's role in changing SST. The thermocline changes on the timescales of interest mostly in response to tropical winds (there are longer-term ocean effects involving the slower, deeper parts of the ocean reached by the thermohaline circulation that are outside the scope of this book). The process of thermocline response to the winds is called adjustment and is the major topic of the rest of this chapter.

6.2 The barotropic adjustment problem

We first will use a simple barotropic analog to introduce the problem of equatorial adjustment. The problem has many of the elements of the equatorial problem as well as being important in its own right.

We consider the adjustment of non-divergent motions to the imposition of wind stresses on a midlatitude β-plane: i.e. $f = f_0 + \beta y$. In the spirit of the midlatitude β-plane, f is taken as a constant $f = f_0$ unless differentiated, $\frac{df}{dy} = \beta$.

We will start with the forced *non-divergent* shallow-water equations:

$$u_t - fv = -p_x + F - ru, \tag{6.4a}$$

$$v_t + fu = -p_y + G - rv, \qquad (6.4b)$$

$$u_x + v_y = 0, \qquad (6.4c)$$

where F and G are the horizontal components of the wind stress and r is a Rayleigh drag coefficient. Because the divergence is taken to be zero, we can define a streamfunction ψ such that:

$$u = -\psi_y, \quad v = \psi_x.$$

In terms of the streamfunction the vorticity is:

$$\zeta = v_x - u_y = \psi_{xx} + \psi_{yy} = \nabla^2 \psi,$$

so that the vorticity equation becomes:

$$\nabla^2 \psi_t + \beta \psi_x = C - r\nabla^2 \psi, \qquad (6.5)$$

where $C = G_x - F_y$ is the curl of the wind stress. Equation 6.5 says that an imposed wind-stress curl C can increase the local vorticity, can move the parcel meridionally in the gradient of planetary vorticity β or can dissipate the local vorticity.

6.2.1 Free planetary waves

We can construct any forced solutions in terms of the solutions to the free frictionless equation:

$$\nabla^2 \psi_t + \beta \psi_x = 0, \qquad (6.6)$$

which has planc-wave solutions of the form

$$\psi = \psi_0 \exp[i(kx + my - \omega t)],$$

with

$$\omega = -\frac{\beta k}{k^2 + m^2}. \qquad (6.7)$$

Equation 6.7 is the dispersion formula for inviscid divergenceless planetary waves on a midlatitude β-plane (Rossby waves) and is plotted in Figure 6.1. Note that we choose m to have some specified North–South dimension: for example, in the Atlantic, we can choose a scale characteristic of the transition from easterlies in the subtropics to westerlies in the midlatitudes so that $m = \dfrac{2\pi}{L_{NS}}$ where L_{NS} is of the order 1500 km.

The zonal-phase velocity of these waves is:

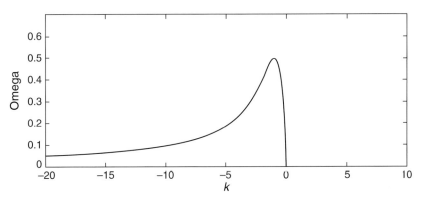

Figure 6.1. The dispersion relation Equation 6.13. Omega (ω) in units of β, and $m = 1$.

$$c = \frac{\omega}{k} = \frac{-\beta}{k^2 + m^2} \qquad (6.8)$$

which, since it is the slope of the line from the origin to points on the curve in Figure 6.1, is westward with a value of $-\beta/m^2$ for $k=0$ and decreases monotonically to zero as $k \to -\infty$. The zonal-group velocity is:

$$c_g = \frac{\partial \omega}{\partial k} = \beta \frac{k^2 - m^2}{(k^2 + m^2)^2} \qquad (6.9)$$

and is the slope of the curve at each point on the curve of Figure 6.1. For long waves (i.e. near $k=0$), the waves are nearly non-dispersive and

$$c = c_g = -\frac{\beta}{m^2}.$$

For short waves (i.e. $k \to -\infty$), the phase velocity is small and westward while the group velocity is small and eastward:

$$c = -\frac{\beta}{k^2} \text{ and } c_g = +\frac{\beta}{k^2}.$$

We turn now to the problem of the forced solutions to Equation 6.5 in the presence of boundaries at the east and west, roughly representing the problem of the barotropic solution to the wind-forced ocean response in a bounded basin. (A totally bounded basin would require consideration of possible boundary layers at the northern and southern boundaries, as in, for example, Cane [1979]. We ignore this complication here.)

6.2.2 The steady response

We take boundaries at $x=0$ and $x = X_E$ with boundary conditions $u=0$ at $x = 0, X_E$. Since $u = -\psi_y$, the boundary condition becomes $\psi = 0$ at $x = 0, X_E$.

The steady inviscid solution to Equation 6.5 is simply

$$\beta\psi_x = \beta v = C, \tag{6.10}$$

which simply says that, in order to satisfy the vorticity equation, the vorticity put in by the wind-stress curl C induces a meridional velocity that changes vorticity by moving in the gradient of planetary vorticity β. Equation 6.10 representing the interior flow away from boundaries is called the Sverdrup relation.

The solution valid in the interior, Equation 6.10, does not satisfy the boundary conditions. But since Equation 6.10 is first order in x, it can by itself satisfy only a single boundary condition. To see this, choose the wind-stress curl C to be independent of x. Then the interior solution is

$$\psi_{\text{int}} = \frac{1}{\beta}Cx + const$$

and the constant can be chosen to satisfy $\psi_{\text{int}} = 0$ either at $x = 0$ or $x = X_E$. To see which of the boundary conditions is to be chosen, we have to understand how the other boundary condition can be satisfied.

The total interior meridional mass flux is:

$$\int_0^{X_E} v\,dx = \frac{C}{\beta}X_E,$$

and the only way this interior mass flux can be returned is by a frictional boundary layer at the east or at the west. The boundary-layer equation, i.e the equation for the term that must be added to ψ_{int} in order to satisfy the boundary condition $\psi = 0$ is a reduced version of Equation 6.5:

$$\beta\psi_x = -r\psi_{xx}. \tag{6.11}$$

We drop the term ψ_{yy} because, in the spirit of boundary layers, we anticipate that the zonal scale in the boundary layer at the east or the west will be much smaller than the meridional scale – this should be checked a posteriori. Equation 6.11 has solutions proportional to $\exp\left[-\dfrac{x}{l_0}\right]$ where $l_0 = \dfrac{r}{\beta}$ is the boundary-layer width which must be small compared to X_E. Since the boundary layer decays eastward, we must choose the boundary layer at the western boundary. We therefore impose $\psi_{\text{int}} = 0$ at the eastern boundary X_E so that the interior solution becomes:

$$\psi_{\text{int}} = \frac{C}{\beta}(x - X_E)$$

and adding the boundary-layer solution to the interior solution to satisfy $\psi = 0$ at $x = 0$ yields the final solution:

$$\psi = \frac{C}{\beta}(x - X_E) + \frac{CX_E}{\beta} \exp\left[-\frac{x}{l_0}\right]. \tag{6.12}$$

The meridional velocity becomes:

$$v = \psi_x = \frac{C}{\beta} - \frac{CX_E}{\beta} \exp\left[-\frac{x}{l_0}\right]. \tag{6.13}$$

The first term is the interior flow and the second, in the opposite direction and of order X_E/l_0 larger, is the boundary-layer flow along the western boundary of the ocean.

Exercise 6.1: Show that the total meridional mass flux integrated across the basin is zero, i.e. that the western boundary layer returns all the interior mass flux.

Could we have told in advance (i.e. without solving the equations) on which side of the ocean basin the boundary-layer flow would be? There are two distinct ways we could have known, both of them illuminating.

The first is to note that vorticity of the correct sign can only be dissipated on the western boundary. If we add vorticity C per unit time by the wind-stress curl, this vorticity input must be dissipated at this same rate if the circulation is to be steady. Since the dissipation in the ocean interior is negligibly small in our solution, all the dissipation must take place in the boundary layer. For definiteness, take C as negative, or anticyclonic (as in the midlatitude Atlantic with easterlies in the subtropics and westerlies in the midlatitudes). Then the interior meridional flow will be negative (Equation 6.10) so the boundary flow must be positive. The possible configurations for the meridional velocity are given in Figure 6.2.

The rate of vorticity dissipation is $r\zeta$ and must also be negative (anticyclonic) since the input rate C is negative; thus the boundary-layer vorticity must be negative. Clearly from Figure 6.2 this can only occur if the boundary layer occurs at the western boundary.

x = 0   x = X$_E$

Figure 6.2. Schematic of possible boundary-layer meridional velocities at the eastern and western boundaries of the basin.

The second way to tell which side of the ocean the boundary layer has to be on is due to a nifty argument given by Pedlosky (1965) that makes essential use of the dispersion relation Equation 6.7 (diagrammed in Figure 6.1). In the ocean interior the large-scale, low-frequency (note that steady is zero frequency and therefore certainly low frequency) wind forcing excites long waves. Since these have westward group velocity, they travel westward to the western boundary where they are reflected as short waves with eastward group velocity. The distance the short waves travel in one dissipation time r^{-1} is $\dfrac{c_g}{r} = \dfrac{\beta}{rk^2}$, which by definition is the boundary-layer width l_0. The short waves are therefore of size $k \sim \dfrac{1}{l_0}$ and $\dfrac{c_g}{r} = \dfrac{\beta}{rk^2} = \dfrac{\beta l_0^2}{r} = l_0$ so that, as before, $l_0 = \dfrac{r}{\beta}$.

Pedlosky also noted that the same argument can be used to estimate the width of an inertial boundary layer. If the eastward propagating short waves are, instead of being dissipated, trapped by a westward zonal velocity of magnitude $|U|$, then $c_g = \dfrac{\beta}{k^2} = |U|$, so that $l = \sqrt{\dfrac{|U|}{\beta}}$ which is the inertial boundary-layer width.

Finally let us close this subsection on steady responses by asking whether or not a steady (linear) solution could exist in the absence of friction. A steady *interior* solution to Equation 6.5 could exist if, in the boundary layer,

$$\nabla^2 \psi_t^B = -\beta \psi_x^B. \tag{6.14}$$

Again neglecting the ψ_{yy} term, the approximate solution to Equation 6.14 is

$$\psi^B = -\frac{\psi_{\text{int}}(y)}{\beta} J_0(2\sqrt{\beta x t}), \tag{6.15}$$

which satisfies the condition that $\psi = \psi_{\text{int}} + \psi^B = 0$ at $x = 0$. J_0 is a Bessel function of zero order. The Bessel function $J_0(x)$ has a maximum at zero argument and decreases uniformly to its zero at $x = 2.2$, and wiggles with decreasing amplitude after that. Therefore the solution, Equation 6.15, has a constantly thinning boundary layer of width $x \sim \dfrac{(1.1)^2}{\beta t}$. The thinning boundary layer corresponds to increasing meridional velocity and to increasing vorticity in the boundary layer. If the thinning were to be stopped by friction with timescale $t \sim \dfrac{1}{r}$, so that, from Equation 6.15, $\dfrac{\beta x}{r} \sim 1$ and $x \approx \dfrac{r}{\beta}$, as obtained before for the frictional boundary-layer width. The thinning boundary layer has constantly increasing vorticity (rather than dissipating the vorticity) at a rate that just balances the vorticity put in by the wind-stress curl.

6.2.3 Adjustment to the steady response

Let us consider how the steady Sverdrup solution in the interior, $\psi_{\text{int}} = \frac{C}{\beta}(x - X_E)$, is approached if the wind-stress curl were suddenly turned on. We consider this inviscid adjustment by looking at the solutions to:

$$\nabla^2 \psi_t + \beta \psi_x = CH(t), \tag{6.16}$$

where the Heaviside function $H(t) = 0$ unless $t \geq 0$ whereupon it has a value of unity. If we take the wind-stress curl $= C \sin(my)$ to be independent of x across the basin, then there can be two long-term solutions to Equation 6.16 corresponding to each of the first two terms in Equation 6.16 balancing the last:

$$\psi = -\frac{C}{m^2} t, \tag{6.17a}$$

or

$$\beta \psi_x = C. \tag{6.17b}$$

The first solution, Equation 6.17a, corresponds to forcing on resonance (i.e. at the origin of Figure 6.1) resulting in secular growth. The second is the steady Sverdrup relation. To see how they are related, we note that for C independent of x, Equation 6.16 has the form of a simple wave equation:

$$-m^2 \psi_t + \beta \psi_x = CH(t),$$

which admits solutions with wave fronts moving westward with velocity $\frac{x}{t} = -\frac{\beta}{m^2}$. (This is only an approximation, since there is a contribution to the wave front from the ψ_{xxt} term in Equation 6.16. The actual solution is the integral of an Airy function which, as the wave front evolves, becomes more and more like a square wave front – the full solution is given in Cane and Sarachik [1976].)

Four examples will be given to fix ideas. The first example is wind-stress curl forcing (of wavenumber 1 and arbitrarily of unit amplitude) independent of x, except that the forcing is everywhere to the east of $x = 0$ on an unbounded plane:

$$-m^2 \psi_t + \beta \psi_x = H(x) H(t) \tag{6.18}$$

with solution:

$$\psi = \frac{1}{\beta} x H(x) - \frac{1}{\beta}\left(x + \frac{\beta}{m^2} t\right) H\left(x + \frac{\beta}{m^2} t\right), \tag{6.19}$$

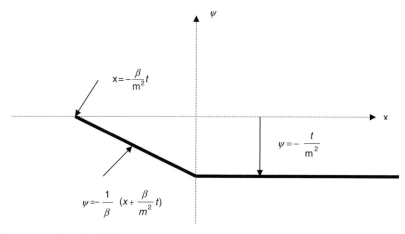

Figure 6.3. Sketch of the response Equation 6.19 of the barotropic streamfunction to a uniform wind-stress curl forcing imposed at $t=0$, confined to the right-half plane, Equation 6.18.

which is illustrated by Figure 6.3.

The nature of the solution can be described as follows: to the east of $x=0$ there is always secular growth; since all signals propagate westward they all originate in the region of forcing and no signal can ever arrive to indicate the absence of forcing. At a point to the west of $x=0$, no response exists until a signal reaches that point. When the signal arrives, the streamfunction begins to grow. The signal constantly propagates westward so that at a point to the west of $x=0$, either the streamfunction is zero (before the signal reaches that point) or the streamfunction is growing linearly with t (after the signal has reached that point). Note that while the signal front propagates with the Rossby wave velocity $-\dfrac{\beta}{m^2}$, the signaling is *not* done by Rossby waves, but rather by packets of waves that do not look at all wave-like: no amount of observation of the streamfunction would ever see waves, only the onset of growth when the signal arrives.

The second example is similar to the first, except that the forcing is everywhere to the west of $x=0$ on an unbounded plane:

$$- m^2\psi_t + \beta\psi_x = H(-x)H(t) \tag{6.20}$$

with solution:

$$\psi = \frac{1}{\beta}xH(-x) - \frac{1}{\beta}\left(x + \frac{\beta}{m^2}t\right)H\left(-\left(x + \frac{\beta}{m^2}t\right)\right), \tag{6.21}$$

which is illustrated in Figure 6.4.

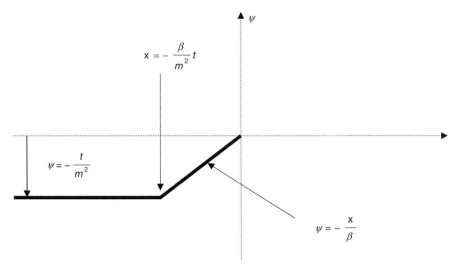

Figure 6.4. Sketch of the response Equation 6.21 of the barotropic streamfunction to a uniform wind-stress curl imposed at $t=0$ and confined to the left-half plane, Equation 6.20.

Here again, far to the west, where the westward propagating signal (indicating that there is no forcing to the east of $x = 0$) has not yet reached, the streamfunction grows linearly with t. At a point to the west of $x=0$, when the signal does reach that point, the secular growth stops, leaving $\psi = -\dfrac{x}{\beta}$ (the Sverdrup relation) in its wake.

Again, all signals propagate with the Rossby wave speed, but the streamfunction either grows with t or is constant with t – no waves are ever evident.

The first example has the message that the streamfunction should start growing when the signal arriving from the east says that there is forcing everywhere to the east – the streamfunction begins to grow when the signal arrives and never stops. The second example has growth until the signal arrives that there is no forcing to the east: then the growth stops.

The third example is a combination of the first two and is simply an impulsively applied wind-stress curl between $x=0$ and $x=X_E$ (Figure 6.5a). The streamfunction between 0 and X_E grows until the message that there is no forcing to the east of $x = X_E$ reaches and leaves the Sverdrup relation behind. At $t = \dfrac{m^2 X_E}{\beta}$, the Sverdrup relation $\psi = \dfrac{1}{\beta}(x - X_E)$ is complete for $0 < x < X_E$ but the signal keeps propagating westward and an opposite gradient propagates away. The meridional velocity in the forcing region is balanced by an equal and opposite meridional velocity that continues to propagate westward.

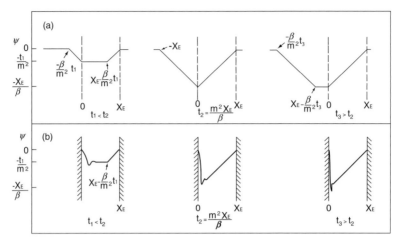

Figure 6.5. (a) Evolution of streamfunction with unit wind-stress forcing impulsively applied at $t = 0$ between $x = 0$ and $x = X_E$ on an unbounded plane. (b) Same as (a) but in a bounded basin. (From Cane and Sarachik, 1977.)

The fourth example (Figure 6.5b) is the same as the third, except between boundaries at $x = 0$ and $x = X_E$. Again the streamfunction grows within the basin until the signal arrives from the eastern boundary that there is no forcing to the east of the eastern boundary. The Sverdrup relation is set up within the entire basin at the time the signal crosses the basin, $t = \dfrac{m^2 X_E}{\beta}$, but now, in the absence of friction, the boundary layer on the western boundary of the basin continually thins according to Equation 6.15 and returns the meridional mass flux set up in the interior. In the presence of friction, the boundary stops thinning at the frictional boundary-layer scale r / β and the layer reaches a steady state.

6.3 Equatorial ocean dynamics: free waves

We return to the divergent shallow-water equations (Equation 3.66) with equivalent depth H_n. A single equation for v may be derived from this equation:

$$(v_{xx} + v_{yy})_t - \frac{f^2}{gH_n} v_t - \frac{1}{gH_n} v_{ttt} + \beta v_x = 0. \tag{6.22}$$

Exercise 6.2: Derive Equation 6.22 from Equation 3.66. (Hint: first replace Equations 3.66a and 3.66b with equations for $r = u + (gH_n)^{-1/2} p$ and $s = u - (gH_n)^{-1/2} p$.)

If we write $v(x, y, t) = V(y) \exp[i(kx - \omega t)]$, Equation 6.22 becomes

$$V_{yy} + \left(\frac{\omega^2}{gH_n} - k^2 - \frac{\beta k}{\omega} - \frac{f^2}{gH_n} \right) V = 0. \qquad (6.23)$$

6.3.1 f-Plane

On the f-plane, f is a constant; i.e. $\beta = 0$. Taking $V = \exp{(ily)}$, Equation 6.23 becomes

$$\omega^2 = f^2 + gH_n(k^2 + l^2). \qquad (6.24)$$

This is the dispersion relation for inertia-gravity waves – clearly from Equation 6.24, the frequency is larger than the local Coriolis frequency so that these represent relatively high-frequency motions. There is another solution to Equation 6.22 when $\beta = 0$:

$$\omega = 0.$$

This means the motion is completely independent of time so that setting the time-derivative terms to zero in the shallow-water equations (Equation 3.66) gives:

$$-fv + p_x = 0 \text{ and } fu + p_y = 0,$$

i.e. the balance is *geostrophic* and from Equation 3.66c, $u_x + v_y = 0$, i.e. the motion is *non-divergent*. These f-plane results should be compared with the beta-plane results obtained below.

6.3.2 Midlatitude beta-plane

At midlatitudes $f = f_o + \beta y$. It is customary to replace f in Equation 6.23 by $f_o =$ constant, making the equations considerably easier to solve. This is justified if the horizontal scale of the motions L is such that $\beta L \ll f_o$.

Then we may take $V(y) \sim \exp{(ily)}$ to obtain

$$\frac{\omega^2}{gH_n} - \frac{\beta k}{\omega} - \left[k^2 + l^2 + \frac{f_o^2}{gH_n} \right] = 0, \qquad (6.25)$$

which is the dispersion relation for the shallow-water equations on a *midlatitude beta-plane*.

It is not strictly consistent to replace f by f_o and yet retain all other terms in Equation 6.25. Assuming that $\beta L \ll f_o$ and $L \sim (gh)^{1/2}/f_o \equiv L_R$ (L_R is the Rossby radius of deformation), consider separately the two cases:

(a) $\omega \sim f$, i.e. frequencies are high so that the timescales are short compared to the inertial period f^{-1}. Then in Equation 6.25 the ratio of the second to the first term is of the order $\frac{\beta L}{f_0}$ so that the second term can be neglected and the dispersion relation becomes:

$$\omega^2 = f_0^2 + gH_n(k^2 + l^2), \tag{6.26}$$

which is the same as Equation 6.24, the dispersion relation for inertia-gravity waves on an f-plane. In other words, the effect of β on inertia-gravity waves is small.

(b) $\omega \ll f$ (slow, long-timescale motions). Then in Equation 6.25 the first term can be neglected and the dispersion relation becomes:

$$\omega \simeq -\beta k / \left[k^2 + l^2 + \frac{f_0^2}{gH_n} \right]. \tag{6.27}$$

The corresponding simplification of Equation 6.22 is

$$(v_{xx} + v_{yy})_t - \frac{f_0^2}{gH_n} v_t + \beta v_x = 0. \tag{6.28}$$

Equation 6.27 is the *Rossby wave* dispersion relation and Equation 6.28 is the linearized quasi-geostrophic potential vorticity equation. The quasi-geostrophic approximation filters gravity waves and, to leading order, the flow is non-divergent and geostrophic:

$$(u, v, p) \sim (l, -k, if_0)expi[kx + ly - \omega t]$$

with ω given by Equation 6.27. We will consider these motions at some length.

The dispersion relations for inertia-gravity waves, Equation 6.26, and Rossby waves, Equation 6.27, are plotted in the k,ω plane for meridional wavenumbers $0 = l_o < l_1$ in Figure 6.6.

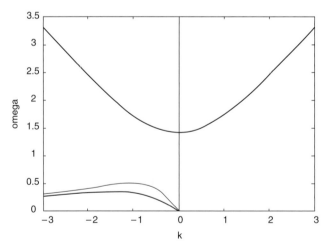

Figure 6.6. High-frequency (Equation 6.26) and low-frequency (Equation 6.27) dispersion relationships on a midlatitude f-plane. For the purposes of this diagram, omega (ω) in units of f_0, and length in units of $L_R = (gH_n)^{\frac{1}{2}}/f_0$. The thick line corresponds to $l=0.1$ and the thin line to $l=1$ (the $l=1$ gravity wave lies underneath the thick line and is therefore invisible).

Exercise 6.3: What is the maximum value of ω for all k and fixed values of l, f_o and H_n? What is the maximum value of ω for all k, l?

Note that the f-plane geostrophic modes with $\omega = 0$ have been replaced, on the midlatitude β-plane by the quasi-geostrophic Rossby waves with $\omega > 0$. There is still a gap in frequency between these low-frequency modes and the inertia-gravity waves with $\omega \geq f$.

6.3.3 Equatorial β-plane

At the equator $f_o = 0$, $\beta = 2\Omega/a$ and Equation 6.22 becomes:

$$(v_{xx} + v_{yy})_t - \frac{\beta^2 y^2}{gH_n} v_t - \frac{1}{gH_n} v_{ttt} + \beta v_x = 0 \qquad (6.29)$$

This equation was first analyzed by Matsuno (1966) in an atmospheric context and by Moore (1968) and Blandford (1966) in an oceanographic one.

There is a canonical scaling for the equatorial beta-plane: length is scaled by $L_{eq} = (gH_n)^{1/4}\beta^{-1/2}$ and time by $T_{eq} = (gH_n)^{-1/4}\beta^{-1/2}$. Some insight into this scaling can be obtained by considering how the midlatitude radius of deformation $L_R = (gH_n)^{1/2}/f$ would change if we use the equatorial value $f = \beta y$. As we approach the equator, L_R grows. At some value of $y = L$, the value of L_R becomes as large as L so that for further approach to the equator, an inconsistency would arise. This value obtains when $L = \frac{\sqrt{gH_n}}{\beta L}$ or $L = (gH_n)^{1/4}\beta^{-1/2}$. The value of the Coriolis parameter that corresponds to this value of $y = L$ is $\beta L = (gH_n)^{1/4}\beta^{1/2}$ so that the timescale is $T = 1/\beta L = (gH_n)^{-1/4}\beta^{-1/2}$. Velocities are then scaled by $L/T = \sqrt{gH_n}$.

With this scaling, the shallow-water equations, Equations 3.66, become:

$$u_t - yv + h_x = 0, \qquad (6.30a)$$

$$u_t + yv + h_y = 0, \qquad (6.30b)$$

$$h_t + u_x + v_y = 0, \qquad (6.30c)$$

and

$$(v_{xx} + v_{yy} - y^2 v)_t + v_x - v_{ttt} = 0 \qquad (6.31)$$

is the non-dimensional version of Equation 6.29. We have changed the notation a little, replacing the dimensional variable p by the non-dimensional depth h (dimensionally, $p = g'h$ for a reduced-gravity model). The boundary conditions on an infinite β-plane are u, v, h bounded as $y \to \pm\infty$. With these boundary conditions, Equations 6.30 are the basic equations for free waves for a given equivalent depth H_n. The solutions to Equations 6.30 for a given H_n do correspond to modes on a sphere for the same equivalent depth, but there is some geometric distortion.

There is a solution to Equations 6.30 that has $v \equiv 0$: the equatorial Kelvin wave:

$$u = h = \exp[-y^2/2]\exp[ik(x-t)] \tag{6.32}$$

i.e. $\omega = k$ with $c = 1$, or, dimensionally, $c = \sqrt{gH_n}$.

From Equation 6.30b with $v=0$, we see that the Kelvin wave's meridional-momentum balance is geostrophic while its zonal-momentum balance is that of a gravity wave (Equation 6.30a). The Kelvin wave travels eastward along the equator with the gravity wave speed c.

All other solutions are given by solutions of Equation 6.31. Let $v = V(y)\exp[i(kx - \omega t)]$ so that

$$V_{yy} - y^2 V + \left[-\frac{k}{\omega} + \omega^2 - k^2\right]V = 0. \tag{6.33}$$

The boundary conditions imply that the eigensolutions to Equation 6.33 are (see Appendix 2):

$$\psi_n(y) = \pi^{-1/4}(2^n n!)^{-1/2} H_n(y)\exp\left[-\frac{1}{2}y^2\right] \tag{6.34a}$$

with

$$-\frac{k}{\omega} + \omega^2 - k^2 = 2n+1 \qquad n = 0, 1, 2, \dots \tag{6.34b}$$

where the H_n are the Hermite polynomials of order n.

We can write Equation 6.33a as $V_{yy} + [2n + 1 - y^2]V = 0$ and define the turning latitude $y_T = (2n + 1)^{1/2}$. In this form it is easy to see that V is oscillatory equatorward of the turning latitudes $|y| < y_T = (2n + 1)^{1/2}$ and exponentially decaying poleward of the turning latitudes, $|y| > y_T$.

This β-plane approximation only represents modes on a sphere if the turning latitudes lie equatorward of the pole, i.e. if $y_T < y_{POLE}$ so that the boundary conditions can be satisfied. Dimensionally, $(2n + 1)^{1/2} L_{eq} \simeq 90°$ of latitude. For deep modes of the atmosphere $H_n \approx 10$ km, $L_{eq} \simeq 30°$ so only $n = 1, 2$ are good. For baroclinic modes of the ocean, $H_n \approx 0.6$ m, $L_{eq} \approx 300$ km $\approx 3°$ and the modes should be good for n of the order of 500.

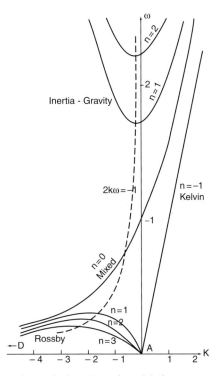

Figure 6.7. The dispersion relation Equation 6.34b on an equatorial beta-plane (to be compared to the f-plane version in Figure 6.6). (From Cane and Sarachik, 1976.)

The dispersion diagram corresponding to the dispersion formula of Equation 6.34b is shown in Figure 6.7.

Dimensionally we have

$$-\beta \frac{k}{\omega} - k^2 + \frac{\omega^2}{gH_n} = (2n+1)\, L_{eq}^2 = (2n+1)\frac{\beta}{(gH_n)^{1/2}} \qquad (6.35)$$

and the Kelvin wave

$$\omega = k(gH_n)^{1/2}.$$

For $\omega \ll T_{eq}^{-1} = \left((gH_n)^{1/4}\beta^{1/2}\right)$, Equation 6.35 becomes

$$\omega \approx \frac{-\beta k}{k^2 + (2n+1)\dfrac{\beta}{(gH_n)^{1/2}}}. \qquad (6.36)$$

These are Rossby waves. Comparing with the midlatitude expression, Equation 6.27, we see that the dispersion relation is similar but with $(2n + 1)\dfrac{\beta}{(gH_n)^{1/2}} = f^2(y_T)/gH_n$ replacing $l^2 + f_o^2/gH_n$ – the meridional wavenumber and radius of deformation term have been combined. (See Cane and Sarachik, 1976, pp. 631–632, for a discussion of the midlatitude extension of the equatorial beta-plane.)

For higher-frequency motions, $\omega \geq T_{eq}^{-1}$ Equation 6.35 is approximately

$$\omega^2 \approx gH_n[k^2 + (2n + 1)L_{eq}^{-2}] = f^2(y_T) + gH_n k^2, \qquad (6.37)$$

which is the approximate dispersion relation for inertia-gravity waves (compare to Equation 6.26). Note that in Equation 6.37, $n \geq 1$ so $\omega\sqrt{3}(gH_n)^{1/2}L_{eq}^{-1} = \sqrt{3}(\beta L_{eq}) = \sqrt{3}f(L_{eq})$. Again there is a frequency gap between the Rossby and inertia-gravity waves – though not as great as in the midlatitude case. Also, two modes cross the gap – the Kelvin wave and the $n = 0$ mode – the mixed Rossby-gravity wave. For small ω this behaves like a Rossby wave and for large ω like a gravity wave. Comparison with Moura (1976) (or Longuet-Higgins, 1968) shows that the equatorial beta-plane preserves many of the properties of the modes on a sphere. The group velocity $c_g = \dfrac{\partial \omega}{\partial k}$ is positive (eastward) to the right of $2k\omega = -1$ (where $c_g = 0$). For the Kelvin wave $c_g = 1$ (non-dispersive). For $\omega \ll 1$, $c_g \sim -(2n + 1)^{-1}$ for k small and $c_g \sim +k^{-2}$ for $|k| \gg 1$.

We can look at the structure of the free Kelvin and Rossby waves in terms of the previously defined solutions (6.34a):

$$\psi_n(y) \equiv \pi^{-1/4}(2^n n!)^{-1/2} e^{-y^2/2} H_n(y); \qquad n = 0, 1, \ldots$$

and in particular

$$\psi_o(y) = \pi^{-1/4} e^{-y^2/2}.$$

The free-wave solutions to Equation 6.30 with bounded solutions at infinity can be given for the Rossby, Kelvin and inertia-gravity waves directly. If we take

$$(u, v, h)^T = \exp i[kx - \omega_{n,j}(k)t]\mathbf{\Phi}_{n,j}(k, y) \quad n \geq 0, \qquad (6.38)$$

where the superscript T represents the transpose, then the free solutions are the eigenfunctions of Equation 6.30 rewritten in the form:

$$\boldsymbol{\Omega}(k, y)\boldsymbol{\Phi}_{n,j}(k, y) = \omega_{n,k}(k)\boldsymbol{\Phi}_{n,j}(k, y) \qquad (6.39)$$

where

$$\boldsymbol{\Omega} = \begin{pmatrix} 0 & iy & k \\ -iy & 0 & -i\dfrac{\partial}{\partial y} \\ k & -i\dfrac{\partial}{\partial y} & 0 \end{pmatrix}, \qquad (6.40)$$

with $j = 1, 2$ for the inertia-gravity waves and $j = 3$ for the Rossby waves. We will take $n = -1$ as a formal device for labeling the Kelvin wave – the utility of this device will become clearer below.

$$\boldsymbol{\Phi}_{n,j}(k, y) = \omega_{n,j}(k)\mathbf{W}_n(y) + k\mathbf{M}_n(y) - i[\omega_{n,j}^2(k) - k^2]\mathbf{V}_n(y) \qquad (6.41)$$

and the auxiliary vector functions $\mathbf{V}$, $\mathbf{M}$ and $\mathbf{W}$ are defined by:

$$\mathbf{V}_n = (0, \psi_n(y), 0)^T, \qquad (6.42a)$$

$$\mathbf{W}_n = (y\psi_n(y), 0, -d\psi_n(y)/dy)^T, \qquad (6.42b)$$

$$\mathbf{M}_n = (-d\psi_n/dy, 0, y\psi_n(y))^T. \qquad (6.42c)$$

The relations given in Equation A2.3 of Appendix 2 are useful in calculating the terms in Equations 6.42b and Equation 6.42c.

For the Kelvin wave,

$$n = -1; \; u = h = \frac{1}{\sqrt{2}}\exp[i(kx - \omega t)]\psi_o(y) \text{ and } v \equiv 0,$$

and we extend the definitions in Equation 6.42 to $n = -1$ as follows:

$$v_{-1} = 0 \text{ and } \mathbf{K} \equiv \boldsymbol{\Phi}_{-1} = \mathbf{M}_{-1}(y) = \mathbf{W}_{-1}(y) = \frac{1}{\sqrt{2}}(\psi_o(y), 0, \psi_o(y))^T, \quad (6.43)$$

Note the symmetries as a function of n: for n even, u and p are antisymmetric about the equator and v is symmetric. For n odd, u and p are symmetric about the equator with v antisymmetric.

Some valuable approximations can be obtained for low frequencies and long wavelengths (which is the regime where we will usually find ourselves): ω and k are both taken to be small and

$$\boldsymbol{\Phi}_{n,3} \sim k\left[\mathbf{M}_n - \frac{1}{2n + 1}\mathbf{W}_n\right] + 0(k^2). \qquad (6.44)$$

For ω small, but k large (short Rossby waves)

$$\Phi_{n,3} \sim k\mathbf{M}_n + ik^2\mathbf{V}_n + 0(k^3). \tag{6.45}$$

Exercise 6.4: Check the expressions in Equations 6.44 and Equation 6.45 for geostrophy.

6.4 Equatorial ocean dynamics: forced waves

We now begin to develop a method of calculating the response of an equatorial ocean to forcing. Let F and G be zonal and meridional sources of momentum such as wind stress, and Q be a mass (or heat) source. The *forced* shallow-water equations on an equatorial β-plane (i.e. the forced version of Equations 6.30) are

$$u_t - yv + h_x = F, \tag{6.46a}$$

$$v_t + yu + h_y = G, \tag{6.46b}$$

$$h_t + u_x + v_y = Q, \tag{6.46c}$$

where we have again scaled the equations with the equatorial length and timescales

$$L_{eq} = \left(\frac{c}{\beta}\right)^{1/2}; \quad T_{eq} = (c\beta)^{-1/2}.$$

Note that $c = L_{eq}/T_{eq}$ and $\beta = (L_{eq}T_{eq})^{-1}$.

Exercise 6.5: What are the dimensional versions of $[F, G, Q]$?

6.4.1 Scaling the equations

In Equations 6.44 and 6.45 we obtained low-frequency approximations for the *waves*: Now we consider low-frequency approximations $\dfrac{\partial}{\partial t} \ll 1$ (or equivalently $\omega \ll 1$) for the *equations*. Again we consider both large and small spatial scales, keeping the frequencies small.

For large scales: $\dfrac{\partial}{\partial x} \ll 1$ or equivalently $k \ll 1$ we can scale

$$\frac{\partial}{\partial t} \sim 0(\varepsilon), \frac{\partial}{\partial x} \sim 0(\varepsilon), \frac{\partial}{\partial y} \sim 0(1);$$

and

$$u, h \sim 0(1), \quad v \sim 0(\varepsilon).$$

This yields

$$u_t - yv + h_x = \varepsilon^{-1}F, \tag{6.46a'}$$

$$\varepsilon^2 v_t + yu + h_y = G, \tag{6.46b'}$$

$$h_t + u_x + v_y = \varepsilon^{-1}Q, \tag{6.46c'}$$

which indicates that the meridional wind stress will force less effectively than either zonal wind-stress forcing or mass (buoyancy) forcing. Note that for free solutions, Equation 6.46b implies that we should see geostrophy in the meridional direction but not in the zonal direction as we have already seen in the exercise for Equation 6.44. We note that eliminating the v_t term in Equations 6.46 is called the long-wave approximation.

Exercise 6.6: Put a factor a in front of the v_t term in Equation 6.46 and derive the analog of Equation 6.22. Then show that setting $a=0$ (the long-wave approximation) eliminates the inertial-gravity waves from the dispersion relation.

The dispersion relation for small k and ω is $\omega = -k/(2n+1)$ for the Rossby wave and $\omega = k$ for the Kelvin wave. In this low-frequency approximation, the Rossby waves and the Kelvin wave are non-dispersive and the group velocities are $c_g = -(2n+1)^{-1}$ for the Rossby waves and $c_g = 1$ for the Kelvin wave.

For low frequencies and small scales: $\dfrac{\partial}{\partial x} \gg 1$ or equivalently $k \gg 1$, we can scale

$$\frac{\partial}{\partial t} \sim \varepsilon \qquad \frac{\partial}{\partial x} \sim \varepsilon^{-1} \qquad \frac{\partial}{\partial y} \sim 1$$

so that

$$u, h \sim \varepsilon \qquad v \sim 1.$$

This yields

$$\varepsilon^2 u_t - yu + h_x = F, \tag{6.47a}$$

$$v_t + yu + h_y = \varepsilon^{-1}G, \tag{6.47b}$$

and

$$\varepsilon^2 h_t + u_x + v_y = Q. \tag{6.47c}$$

We see that in this $\omega \ll 1, k \gg 1$ case, if there is no mass forcing, the system is non-divergent and a streamfunction exists. The free solution would be geostrophic in the x direction (as we have seen in the exercise attached to Equation 6.45). The dispersion relation is $\omega = -k^{-1}$, which is highly dispersive, the phase velocity is $c = -k^{-2}$ and the group velocity is $c_g = +k^{-2}$.

6.4.2 A simple example of our method

The method we use to solve Equations 6.46 follows Matsuno (1966) in using an expansion in eigenfunctions where the eigenfunctions used are the free waves of the unforced problem. Though some sophisticated mathematics lies behind it, the method is not difficult to understand. The context of the equatorial β-plane adds complications which may obscure the fundamental idea. So we illustrate with a simple example, a non-rotating, forced shallow-water system. The equations are

$$u_t + h_x = F \qquad (6.48a)$$

and

$$h_t + u_x = Q. \qquad (6.48b)$$

The free solutions to this problem are gravity waves which propagate in the $+x$ and $-x$ directions (east and west, if one prefers), respectively. We can then define two modes: the $+$ mode propagates eastward and has

$$u^+ = h^+ = expi[k(x - t)],$$

while the $-$ mode propagates westward and has

$$u^- = -h^- = expi[k(x - t)].$$

Define two auxiliary quantities: $q = \frac{1}{2}(u + h)$ and $r = \frac{1}{2}(u - h)$.
Adding Equations 6.48a and 6.48b yields

$$q_t + q_x = \frac{1}{2}(F + Q) \equiv R^+, \qquad (6.49a)$$

while subtracting Equation 6.48b from Equation 6.48a yields

$$r_t + r_x = \frac{1}{2}(F - Q) \equiv R^-. \qquad (6.49b)$$

Consideration of the unforced case ($F = Q = 0$) shows that q is the $+$ mode and r the $-$ mode. Note that Equations 6.49 uncouple q from r. In contrast, u and h are coupled in the original set, Equations 6.48, which makes it more difficult to solve.

The simple wave equation, Equation 6.49, may be solved easily; the well-known solutions are

$$q(x, t) = q(x - t, 0) + \int_0^t R^+(x - s, t - s)ds,$$

and

$$r(x, t) = r(x - t, 0) + \int_0^t R^-(x - s, t - s)ds.$$

Then, in terms of these solutions, the original variables are simply given by $u = q + r$ and $p = q - r$.

We can restate this solution method in more general form. We wish to solve the system of Equations 6.48 which can alternately be written:

$$\mathbf{u}_t + \Omega \mathbf{u} = \mathbf{F} \tag{6.50}$$

where, with superscript T denoting transpose, $\mathbf{u} = (u, h)^T$; $\mathbf{F} = (F, Q)^T$ and the operator

$$\Omega = \begin{bmatrix} 0 & \dfrac{\partial}{\partial x} \\ \dfrac{\partial}{\partial x} & 0 \end{bmatrix}.$$

We know from our analysis of the free waves of the system that the eigenfunctions of Ω are $\mathbf{u}^+, \mathbf{u}^-$ given by

$$\mathbf{u}^+ = (u^+, h^+)^T = \frac{1}{2} e^{ikx} (1, 1)^T$$

and

$$\mathbf{u}^- = (u^-, h^-)^T = \frac{1}{2} e^{ikx} (1, -1)^T,$$

with eigenvalues $i\omega^+(k) = ik$ and $i\omega^-(k) = -ik$.

We then write

$$(u, h)^T = q\mathbf{u}^+ + r\mathbf{u}^-, \tag{6.51}$$

and noting that the inner product

$$(\mathbf{u}^+, \mathbf{u}^-) = 0,$$

we take the inner product of $\mathbf{u}^+$ and $\mathbf{u}^-$ with Equation 6.50 to derive a simple wave equation for q and r, respectively. After solving them to find q and r we sum as in Equation 6.51 to obtain u and h.

Exercise 6.7: Take $Q = 0$, and let a forcing

$$F = 1 \quad \text{for} \quad -1 < x < +1$$

and $F = 0$ otherwise be imposed beginning at $t = 0$. Suppose $u = h = 0$ at $t = 0$. Note that a steady-state response is $u = 0$ and $h_x = F$ or

$$h = h_0 - 1 \quad x \le -1;$$
$$h = h_0 + x \quad -1 < x < 1;$$

$$h = h_0 + 1 \quad 1 \le x.$$

Symmetry leads us to expect $p_0 = 0$. Calculate the time-dependent solution; does it go to this steady state? Examine its approach to a final state by sketching the solution at $t = 0.5$, $t = 1$ and $t = 5$. Note the role of the two wave modes.

Exercise 6.8: With the same forcing and initial conditions as above, imagine that the model is modified by a mean current U; (take $0 < U < 1$) so that :

$$u_t + Uu_x + h_x = F,$$

$$h_t + Uph_{xx} + u_x = 0.$$

Find the steady-state solution. Can you determine the unknown constants? Calculate the eigenvalues and eigenfunctions, i.e. the free waves and their wave speeds. How do they compare to the $U=0$ case? Calculate the evolution to a final state. Again sketch the solution at $t = 0.5$, $t = 1$ and $t = 5$. Note the role of the wave modes – now do they determine the constants in the steady solution? What happens if $U = 1$?

6.4.3 Calculating forced motions on an equatorial beta-plane

Assume that the x-dependence is of the form e^{ikx}; alternately, imagine that we have taken the Fourier transform from $x \rightarrow k$. Then Equation 6.46 may be written as

$$\frac{\partial}{\partial t}\mathbf{u} + i\Omega u = \mathbf{F}, \tag{6.52}$$

where

$$\mathbf{u} = \begin{pmatrix} u \\ v \\ h \end{pmatrix} \text{ and } \mathbf{F} = \begin{pmatrix} F \\ G \\ Q \end{pmatrix},$$

and Ω is given by expression of Equation 6.40.

Since the free-wave solutions

$$\mathbf{u}_{n,j} = \Phi_{n,j}^T \exp[i[kx - \omega_{n,j}(k)t]]$$

(again, superscript T denotes transpose) satisfy

$$\frac{\partial}{\partial t}\mathbf{u}_{n,j} + i\Omega u_{n,j} = 0,$$

it follows that

$$\Omega\Phi_{n,j}^T = \omega_{n,j}(k)\,\Phi_{n,j}^T. \tag{6.53}$$

We can expand the forced solutions of Equation 6.52 and the forcing both in terms of the free solutions of Equation 6.53:

$$\mathbf{u} = \sum_{n,j} a_{n,j}(k,t)\Phi_{n,j}(k,y), \tag{6.54}$$

$$\mathbf{F} = \sum_{n,j} b_{n,j}(k,t)\Phi_{n,j}(k,y), \tag{6.55}$$

where $b_{n,j}$ is determined by projecting the forcing $\mathbf{F}(k\ t,y)$ on the structures $\Phi_{n,j}(k,y)$, as calculated below.

Then

$$\frac{\partial}{\partial t}a_{n,j}(k,t) + i\omega_{n,j}(k)a_{n,j}(k,t) = b_{n,j}(k,t), \tag{6.56}$$

In the long-wave, low-frequency limit ($\omega, k \ll 1$)

$$\omega_{n,3} = \frac{-k}{(2n+1)} \tag{6.57}$$

for the Rossby waves, while for the Kelvin wave

$$\omega_{-1} = k. \tag{6.58}$$

Hence for these modes

$$\frac{\partial}{\partial t}a_n(k,t) - \frac{ik}{2n+1}a_n(k,t) = b_n(k,t) \tag{6.59}$$

(we have dropped the j because we are now concerned only with the low-frequency mode; so we know which j it is: $j = 3$).

Viewing this as an equation in Fourier-transform space, and recognizing $ika_n(k,t)$ as the transform of $\frac{\partial}{\partial x}a_n(x,t)$ allows us to write the last equations as:

$$\frac{\partial}{\partial t}a_n(x,t) - \frac{1}{2n+1}\frac{\partial}{\partial x}a_n(x,t) = b_n(x,t). \tag{6.60a}$$

Note that the left-hand side of Equation 6.60a is just a wave equation for a wave propagating at speed $-\dfrac{1}{2n+1}$; that is, westward for the Rossby waves ($n = 1, 2, 3, \ldots$) with speed $(2n+1)^{-1}$ and eastward for the Kelvin wave with speed $+1$:

$$\frac{\partial}{\partial t}a_K(x,t) + \frac{\partial}{\partial x}a_K(x,t) = b_K(x,t). \tag{6.60b}$$

The right-hand sides of Equations 6.60, $b_n(x,t)$ and $b_K(x,t)$, give the forcing.

In this low-frequency, long-wave limit, we have, for the Rossby wave:

$$\Phi_{n,3}^T \approx \mathbf{R}_n(y) \approx \frac{1}{2\sqrt{2}} \begin{pmatrix} (n+1)^{-1/2}\psi_{n+1} - n^{-1/2}\psi_{n-1} \\ 0 \\ (n+1)^{-1/2}\psi_{n+1} + n^{-1/2}\psi_{n-1} \end{pmatrix} \tag{6.61}$$

and, as ever, the Kelvin wave has $(u, v, h) = \frac{1}{\sqrt{2}}[\psi_o, 0, \psi_o]$.

Since b_n (and b_K) are the projections of (F, G, Q) on (u, v, p) of the free solutions given by

$$\frac{\int_{-\infty}^{+\infty} [Fu + Gv + Qh]dy}{\int_{-\infty}^{+\infty} (u^2 + v^2 + h^2)dy}, \tag{6.62}$$

it follows that the Kelvin projection is

$$b_K = \frac{1}{\sqrt{2}} \int_{-\infty}^{+\infty} (F + Q)\psi_o(y)dy, \tag{6.63a}$$

while for the Rossby waves

$$b_n = \frac{\sqrt{2}n(n+1)}{2n+1} \int_{-\infty}^{+\infty} \left[\frac{1}{\sqrt{n+1}}\psi_{n+1}(F + Q) + \frac{1}{\sqrt{n}}\psi_{n-1}(Q - F) \right] dy. \tag{6.63b}$$

Consider some examples of these projections:

(a) For mass forcing with simple Gaussian meridional shape: $F=0$ and $Q = \sqrt{2}\psi_o(y)S(x,t)$, then

$$b_K = S(x,t), \quad b_1 = \frac{4}{3}S(x,t) \text{ and } b_n = 0 \text{ for } n \geq 2;$$

the higher Rossby modes are not present because the forcing was chosen to have a shape that made their projections vanish.

(b) For zonal wind-stress forcing with simple Gaussian meridional shape: $Q=0$ and $F = \sqrt{2}\psi_o(y)S(x,t)$, then

$$b_K = S(x,t), \quad b_1 = -\tfrac{4}{3}S(x,t) \text{ and } b_n = 0 \text{ for } n \geq 2.$$

(c) For mass forcing chosen to be antisymmetric with respect to the equator: $F=0$ and $Q = S\psi_1 = \sqrt{2}y\psi_o$, then

$$b_K = 0, \quad b_1 = 0, \quad b_2 = S \text{ and } b_n = 0 \quad n > 2.$$

(There is also a b_o term which matters only in the forced region since the $n=0$ wave, the mixed Rossby–gravity wave, propagates so slowly.)

In case (a), for example, it remains to solve

$$\frac{\partial a_K}{\partial t} + \frac{\partial a_K}{\partial x} = b_K = S(x, t)$$

and

$$\frac{\partial a_1}{\partial t} - \frac{1}{3}\frac{\partial a_1}{\partial x} = b_1 = \frac{4}{3}S(x, t).$$

These equations can be considered in a more general context. For general forcings, having first found the bs, it remains to solve Equation 6.60 for the as; i.e. to solve the wave equation of the form:

$$\frac{\partial a}{\partial t} + c\frac{\partial a}{\partial x} = b. \tag{6.64}$$

The solution to Equation 6.64 may be found by the method of characteristics; e.g.

$$a(x, t) = a(x - c(t - t_0), t_0) + \int_0^{t-t_0} b(x - cs, t - s)ds. \tag{6.65a}$$

This form is appropriate for a forcing initiated at time t_0 and zero for $t < t_0$. If the forcing is zero beyond the point $x = x_0$, then the equivalent form below is more useful:

$$a(x, t) = a(x_0, t - c^{-1}(x - x_0)) + \int_0^{x-x_0} b(x - x', t - c^{-1}x')\frac{dx'}{c}. \tag{6.65b}$$

(Other useful forms may be obtained by a change of variable in the integrals: e.g. $s' = t - s$ in Equation 6.65a; $s' = x - x'$ in Equation 6.65b; etc. In Equation 6.67 below we use $s' = x - x'$.)

Suppose for example, the forcing is confined between longitudes $x = 0$ and $x = x_E > 0$. Then

$$a_K(x, t) = a_K(x = 0, t - x) + \int_0^x b_K(x - x', t - x')dx' \quad \text{for} \quad x > 0 \tag{6.66}$$

with

$$a_K(x, t) = 0 \quad \text{for } x < 0;$$

and

$$a_n(x, t) = a_n(x_E, t - (2n + 1)(x_E - x))$$

$$+ (2n + 1)\int_x^{x_E} b_n[s', t - (s' - x)(2n + 1)]ds' \tag{6.67}$$

for $x < x_E$
with

$$a_n(x, t) = 0 \quad \text{for} \quad x > x_E.$$

Once the as are determined, u and h are found by summing:

$$\begin{bmatrix} u \\ 0 \\ h \end{bmatrix} = \frac{a_K}{\sqrt{2}} \begin{bmatrix} \psi_0 \\ 0 \\ \psi_0 \end{bmatrix} + \sum_{n=1}^{\infty} a_n \mathbf{R}_n.$$

We can summarize the algorithm for calculating the forced response as follows: First: calculate the free waves:

1. Calculate $\psi_0(y) = \pi^{-1/4} exp(-y^2/2)$.
2. Calculate $\psi_{n+1}(y) \quad n = 0, 1, 2,$ from Equation A2.3:

$$\psi_{n+1} = \sqrt{\frac{2}{n+1}} y \psi_n - \sqrt{\frac{n}{n+1}} \psi_{n-1} \quad n = 0, 1, 2, \cdots$$

3a. Calculate the u and h meridional structure for the Kelvin wave,

$$\mathbf{K}(y) = \begin{bmatrix} u_K \\ h_K \end{bmatrix} \text{ with } u_k = h_k = \frac{1}{\sqrt{2}} \psi_0(y).$$

3b. Calculate u and h meridional structure for the nth Rossby wave from Equation 6.61 noting that $v_n = 0$ and

$$\mathbf{R}_n(y) = \begin{pmatrix} u_n \\ h_n \end{pmatrix} = \frac{1}{2\sqrt{2}} \begin{bmatrix} \frac{1}{\sqrt{n+1}} \psi_{n+1} & - & \frac{1}{\sqrt{n}} \psi_{n-1} \\ \frac{1}{\sqrt{n+1}} \psi_{n+1} & + & \frac{1}{\sqrt{n}} \psi_{n-1} \end{bmatrix} \quad n = 1, 2, \cdots$$

Second: for a specified forcing F and Q find b_n from Equation 6.63.

4a. Calculate $b_K = \frac{1}{\sqrt{2}} \int_{-\infty}^{+\infty} (F + Q) \psi_0(y) dy.$
4b. Calculate

$$b_n = \sqrt{2} \frac{n(n+1)}{2n+1} \int_{-\infty}^{+\infty} \left[\frac{1}{\sqrt{n+1}} \psi_{n+1}(F+Q) + \frac{1}{\sqrt{n}} \psi_{n-1}(Q-F) \right] dy \quad n = 1, 2, 3, \cdots$$

5. Solve Equation 6.60 for a_K and a_n using the forms in Equation 6.65 or other equivalent forms.
6. Then

$$\begin{bmatrix} u \\ h \end{bmatrix} = a_K(x, t) \mathbf{K}(y) + \sum_{n=1}^{\infty} a_n(x, t) \mathbf{R}_n(y).$$

Exercise 6.9:

(a) Use either Equation 6.59 or one of the later forms to find the as in the case of a spatially periodic forcing; i.e. $b \propto \exp(ikx)$.

(b) Using Equation 6.60 or Equation 6.75 or some other relations, derive the form for the as when the forcing and response are periodic in time; i.e. $b \propto \exp(i\omega t)$.

(c) Using unit-mass forcing, show $(0, 0, 1)^T = [\pi^{1/4}\mathbf{K}(y) + \sum_n (y)_n \mathbf{R}_n]$ where $(y)_n \equiv \int_{-\infty}^{+\infty} y\psi_n(y)dy$.

We now consider a forcing F, independent of x, switched on at $t = 0$ (initially over a resting ocean) and steady thereafter. Using Equation 6.65 with $t_0 = 0$ and $a(t_0) = 0$

$$a_n(x, t) = tb_n$$

with b_n a constant. In other words, u and h grow linearly in time:

$$u = Ut; h = Ht \tag{6.68}$$

with $U(y)$ and $H(y)$ independent of x and t. Substituting these forms into the zonal-momentum equation shows $v = V(y)$: while u and h grow linearly in time, the lower-order term v does not. We can interpret this linear growth of u and p in terms of the dispersion diagram in Figure 6.7: forcing at the origin ($k = 0, \omega = 0$) implies forcing on resonance and secular growth results. (We have already seen an example of this in the barotropic example in Equation 6.21a.)

Some examples of U, V and H in response to a forcing F are given in Figure 6.8. Note that, in Figure 6.8, even when the forcing is independent of latitude, the response is confined to the region of the equator: at higher latitudes $U, H \to 0$ and $V \to -F/y$. To calculate these pictures, use is made of the following formulas (Cane and Sarachik, 1981, p. 688). For

$$F(y) = \exp[-\tfrac{1}{2}\mu y^2] \tag{6.69}$$

$$b_{2n} = 0, \tag{6.70a}$$

$$b_{2n+1} = -\frac{2\pi^{1/4}a_{2n+1}}{(1+\mu)^{3/2}}\left[\frac{1-\mu}{1+\mu}\right]^n\left[\mu + \frac{1}{4n+3}\right], \tag{6.70b}$$

and

$$b_\kappa = \pi^{1/4}(1+\mu)^{-1/2} \tag{6.70c}$$

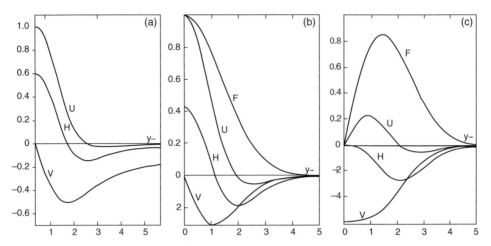

Figure 6.8. Unbounded baroclinic response to westerly wind-stress forcings. (a) $F = 1$ for all y. (b) $F(y) = \exp[-y^2/4]$. (c) $F(y) = y\exp[-y^2/4]$. (From Cane and Sarachik, 1976.)

with $\alpha_n = 0$ for n even and for n odd

$$\alpha_n = 1 \text{ and } \alpha_n = \left[\frac{n}{n-1}\right]^{1/2}\alpha_{n-2} = \left[\frac{n}{n-1}\cdot\frac{n-2}{n-3}\cdot\frac{n-4}{n-5}\cdots\cdots\frac{3}{2}\right]^{1/2} \quad (6.71)$$

or

$$\alpha_{2n+1} = [2^n n!]^{-1}[(2n+1)!]^{1/2}.$$

In the above, the α derives from

$$(y)_n \equiv \int_{-\infty}^{+\infty} y\psi_n(y)dy = 2\pi^{1/4}\alpha_{2n+1}.$$

(Note that the bs for $F=1$ are given by the above relations with $\mu = 0$.)

As a second example, assume $Q=0$, and for $t \leq 0$, $F = u = h = v = 0$. For $t>0$, suppose F is zero except in a narrow region near $x=0$; e.g.

$$F = \frac{1}{\Delta x} \quad \text{for} \quad 0<x<\Delta x \text{ with } \Delta x \ll 1 \quad (6.72)$$

and $F=0$ outside this region: $x<0$ and $x>\Delta x$. Note that in the limit $\Delta x \to 0$, $F \to \delta(x)$, a delta-function forcing.

In this limit, the response in Equations 6.66 and 6.67 is zero except that

$$a_K(x, t) = b_K \quad \text{for} \quad 0 < x \le t, \tag{6.73a}$$

$$a_n(x, t) = (2n + 1)b_n \quad \text{for} \quad \frac{-t}{2n + 1} \le x < 0. \tag{6.73b}$$

Exercise 6.10: Reproduce Figure 6.8 above for $F = 1$ and $F = \exp(-y^2/4)$. Find the solutions for F non-zero only in a region $0 < x < X_E$.

6.5 Equatorial ocean dynamics: adjustment

Whenever we assume a "balanced" state in any geophysical system, there is an implicit adjustment process that brought that fluid to that equilibrium. In particular, *hydrostatic balance* and *static stability* in the atmosphere and ocean are achieved by sound waves and buoyancy oscillations, respectively. The actual adjustment process is generally not explicitly computed; the balanced state is simply assumed. As an example, convective overturning, say by convective plumes or other eddies, will bring a statically unstable state to a statically stable one.

Lower-frequency quasi-geostrophic balance, including low-frequency behavior in equatorial regions, is effected by higher-frequency inertia-gravity waves. The adjustment to pure geostrophy on an f-plane is similar: this is the celebrated Rossby geostrophic-adjustment problem. The adjustment by gravity waves is often computed (inter alia by primitive equation numerical models) but we will not consider it in any great detail here.

The general method for calculating adjustment to impulsively started forcing may be stated as a sequence of steps:

(a) Calculate the unbounded response (Section 6.5.1).
(b) Calculate the inertia-gravity waves generated on the switch-on of the forcing (but not calculated in any detail here).
(c) Calculate the western-boundary response to (a) (Section 6.5.2) and (d).
(d) Calculate the eastern-boundary response to (a) and (c).

We have already seen an example of this method in the simpler context of the barotropic-vorticity equation in Section 6.2. In general, the long Rossby waves propagate energy westward. They do not impinge on the eastern side, and the role of the boundary there is just to cut off the forcing. They do carry energy into the western boundary, where they are reflected as short Rossby waves which make up a western-boundary current and a Kelvin mode if the Rossby signal has a symmetric part. We will see that the presence of the equatorial Kelvin waves makes the equatorial response quite different from midlatitudes, even though the Rossby wave behavior is similar.

6.5.1 Adjustment in the absence of boundaries

We consider the linear solution to a zonal wind-stress forcing constrained to a limited longitudinal extent but in the absence of boundaries. Therefore we must solve the set:

$$u_t - yv + h_x = H(t)T(0, x, X_E), \tag{6.74a}$$

$$v_t + yu + h_y = 0, \tag{6.74b}$$

$$h_t + u_x + v_y = 0, \tag{6.74c}$$

where the "top hat" function $T(a,x,b)$ is unity for $0 < x < b$ and zero otherwise. The solution will be the zonally unbounded solution, Equation 6.68, plus pieces at the edges of the forcing region $x = 0$ and $x = X_E$ needed to guarantee continuity. We can then write:

$$\mathbf{u} = \mathbf{u}_K^1 + \mathbf{u}_K^2 + \mathbf{u}_K^3 + \sum_n \mathbf{u}_{n,R}^1 + \mathbf{u}_{n,R}^2 + \mathbf{u}_{n,R}^3, \tag{6.75}$$

where subscripts K and R refer to Kelvin and Rossby modes and superscript 1 is the unbounded solution (as in Equation 6.68), superscript 2 is the response determined by continuity at $x = 0$, and superscript 3 is the response determined by continuity at $x = X_E$. There are two additional pieces to the solution in Equation 6.75 which will not be considered further: the inertia-gravity waves excited on switch-on, and higher-order Bessel functions which appear on the eastern side of the forcing discontinuities.

We can use the results following Equation 6.69 with $\mu = 0$ to write the solution within the forcing region directly:

$$\mathbf{u}_K^1 = \pi^{1/4} t \begin{pmatrix} 1 \\ 0 \\ 1 \end{pmatrix} \exp[-y^2/2] T(0, x, X_E) \tag{6.76a}$$

and

$$\mathbf{u}_{n,R}^1 = [ta_n\mathbf{R}_n - (2n+1)d_n\mathbf{V}_n]T(0, x, X_E), \qquad (6.76b)$$

where the a_n are given in Equation 6.73 and $d_n = 2\pi^{1/4}a_{2n+1}$.

The responses needed to guarantee continuity at the western and eastern edges of the forcing regions are, respectively:

$$\mathbf{u}_K^2 = -\pi^{1/4}[(t-x)]\mathbf{K}(y)T(0, x, t), \qquad (6.77a)$$

$$\mathbf{u}_{n,R}^2 = a_n[(t+(2n+1)x)\mathbf{R}_n + \mathbf{V}_n]T\left(-\frac{t}{2n+1}, x, 0\right) \qquad (6.77b)$$

and

$$\mathbf{u}_K^3 = \pi^{1/4}[t - (x - X_E)]\mathbf{K}(y)T(X_E, x, X_E + t), \qquad (6.78a)$$

$$\mathbf{u}_{n,R}^3 = -a_n[(t+(2n+1)(x-X_E))\mathbf{R}_n + \mathbf{V}_n]T\left(X_E - \frac{t}{2n+1}, x, X_E\right). \qquad (6.78b)$$

To see what this all means, recognize that within the forcing region, $0<x<X_E$, the solution will grow as t as long as signals from the edges of the forcing region have not arrived to inform a given point that there is no forcing outside the forcing region. The eastern boundary of the forcing region sends Rossby signals westward with speed $-(2n+1)^{-1}$ into the forcing region, while the western boundary of the forcing sends Kelvin signals eastward into the forcing region with speed $+1$. For definiteness, we plot the height field in Figure 6.10, with $X_E = 10$. In this case, it takes 10 units of time for an eastward-moving Kelvin signal to cross the forcing region and 30 units of time for the first westward-moving Rossby signal to cross the forcing region. (Please note that a positive height field corresponds to a deeper thermocline so the graphs are not ideographic – they must be reflected around the x-axis to become so.)

On the equator, the secular growth stops when the Kelvin signal arrives from the west or the sum of the Rossby signals arrives from the east. Figure 6.9 keeps only four Rossby modes in the computation.

If enough time is allowed for *all* the Rossby modes to pass all points in the forcing region (near the equator, $t=30$ will suffice) then

$$\mathbf{u} = \mathbf{u}_K^1 + \mathbf{u}_K^2 + \sum_n \mathbf{u}_{n,R}^1 + \mathbf{u}_{n,R}^3,$$

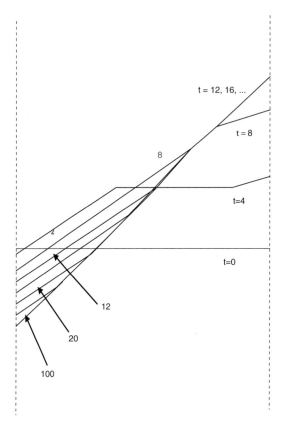

Figure 6.9. Sketch of the thermocline response with time to unit zonal forcing between $x = 0$ and $x = 10$ in an unbounded basin.

and using the solutions in Equations 6.76, 6.77a and 6.78b:

$$\mathbf{u} = x\pi^{1/4}\mathbf{K}(y) - \sum_n a_n(2n+1)(x - X_E)\mathbf{R}_n. \tag{6.79}$$

Note that expanding the forcing $(0,0,1)$ gives the identity:

$$(0,0,1)^T = [\pi^{1/4}\mathbf{K}(y) + \sum_n (y)_n\mathbf{R}_n], \tag{6.80}$$

so that the solution to Equation 6.79 in the forcing region becomes:

$$\mathbf{u} = x(0,0,1)^T - X_E \sum_n (y)_n\mathbf{R}_n$$

$$= x(0,0,1)^T - X_E[(0,0,1)^T - \pi^{1/4}\mathbf{K}(y)],$$

so that the final solution after all the signals have been heard from is, in the forcing region:

$$\mathbf{u} = (x - X_E)(0, 0, 1)^T + X_E \pi^{1/4} \mathbf{K}(y). \tag{6.81}$$

The first term has the correct tilt to be the final solution (to agree with $h_x = 1$) but the second term moves the tilting point: on the equator, from Equation 6.81,

$$h = x - X_E + .707 X_E,$$

which for $X_E = 10$ gives $h = x - 2.93$ which, as can be seen from Figure 6.10, is the final value for the height field within the forcing region. One might also notice that because the final solution, Equation 6.81, contains a Kelvin term, there will be a zonal current within the forcing region having the shape of a Kelvin mode.

Outside the forcing region, a sequence of height-field pictures looks like Figure 6.10.

If we sit on the equator at a point to the east of the forcing region, X_1 units from the eastern extent of the forcing, i.e. at $x = X_E + X_1$, the height-field will start increasing at time $t = X_1$ and will continue increasing for an additional X_E units of time until the signal from the western boundary of the forcing reaches the point. The height field will then stop increasing and the sloping height field, of width X_E, continues propagating eastward with Kelvin speed (in this case, unity).

A similar sequence of events happens to the west of the forcing region except that, in order to enforce the tilt $h_x = 1$ within the basin, the height-field perturbation is negative (the thermocline is closer to the surface) and the signals propagate with the

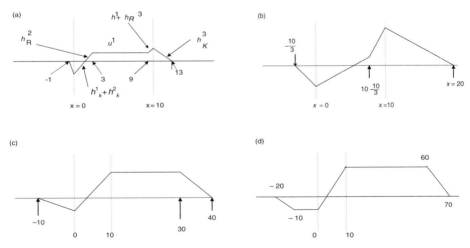

Figure 6.10. Sketch of height response to zonal forcing within $x = 0$ and $x = 10$. (a) $t = 3$, (b) $t = 10$, (c) $t = 30$ and (d) $t = 60$.

speed of the first Rossby wave, $-1/3$. Note that these pictures are almost correct on the equator, where the action of a single Kelvin signal and a single Rossby signal accounts for most of the response. As we move meridionally off the equator, the situation becomes more complex and, in particular, the Kelvin signal becomes less important and more Rossby modes must be included.

6.5.2 Calculating the effects of meridional boundaries

We now wish to describe how the ocean's response is modified by the presence of meridional walls at $x=0$ and $x = X_E$. We begin at the point where the forced motions in the absence of boundaries have been calculated. The task that remains is that of calculating the boundary response to these motions. That is, we seek the free solutions of the shallow-water equations, Equations 6.30, that are required to reduce the normal velocity to zero at the walls: in the case of meridional walls, $u=0$.

We may think of boundaries as modifying the unbounded forced response in two distinct ways. The first is as a barrier to incident motions: any part of the oceanic response bringing energy into a boundary must give rise to a reflection to carry the incident energy away from the boundary. The second is as a cutoff of the forcing: for example, a western boundary at $x=0$ has the effect of modifying the forcing by multiplying it by a step function $H(x)$, thus switching it off for $x<0$, as in the example shown in Figure 6.11. The unbounded solution for time t at a point $x>0$ will have to be modified insofar as it depends on motions that originated at points $x<0$, which are now outside the basin.

In the non-rotating example given earlier, group velocity and phase velocity are equal. Thus the $+$ mode carries energy into the eastern wall, where it is reflected into the $-$ mode. The amplitude of the reflection is determined by the condition $u=0$ at the wall (equivalently, by the condition that the reflected energy is equal to the incident energy). The roles of the two modes are reversed at the west, in the non-rotating example.

Exercise 6.11: Modify the non-rotating problem given earlier so that there are reflecting walls at $x=0$ and $x = X_E$ instead of those longitudes merely marking the extent of the forcing. What is the response for $t=0.5$, $t=2$ and $t \gg 2$? In the absence of friction is a steady state reached?

For the equatorial case – including waves of all frequencies – the energy reflected by incident waves can be reflected by other types of outgoing waves. Moore (see Moore and Philander, 1976) has given a method for calculating the reflection of an incoming wave. Some of the qualitative features of these reflections may be noted. Each incident wave excites a series of waves. A mode incident on a western boundary excites a response which is at least as equatorially confined as it, itself,

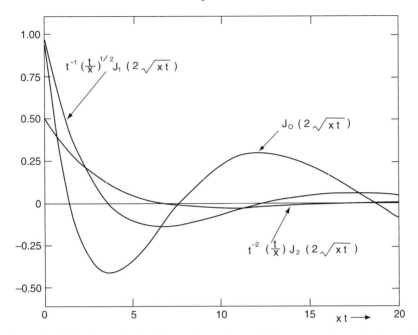

Figure 6.11. Functions giving the x, t dependence in the western-boundary layer. If the incident motion varies like t then $(u^B, h^B) \propto J_1$ and $v^B \propto J_2$. If it is independent of t after $t = 0$ then $u^B, h^B a \propto J_o$ and $v^B \propto J_1$. Note that the xt structure means that the shapes contract toward the western boundary as t increases. (From Cane and Sarachik, 1977.)

is. Unlike the midlatitude situation, a mixed mode or Kelvin wave will be part of the response. The latter propagates away from the boundary quickly; the former remains near the western side, though it shows some effects extending into the basin. A mode incident on an eastern boundary excites a response which is less equatorially confined than itself. The parts of this reflection that are closer to the equator propagate away from the eastern boundary the most rapidly. This is a fundamental asymmetry of equatorial dynamics: western boundaries tend to concentrate the response on reflection while eastern boundaries tend to spread the response on reflection.

Moore's method generalized without modification to allow the calculation of the boundary response to any zonal velocity as long as it is oscillating at a single frequency. It may be extended to a motion with arbitrary time structure by analyzing this structure into its frequency spectrum, calculating the boundary response as a function of frequency and then synthesizing over all frequencies to obtain the time-dependent boundary response. In essence, one begins by taking the Laplace transform of the initial motion and finally obtains the time-dependent response by inverting the resulting Laplace-transformed form of the response (see Lighthill, 1969; Anderson and Rowlands, 1976a, 1976b).

However, as we are only interested in low-frequency motions, we may take advantage of special properties of the low-frequency modes to reach the same results in a different way. (This method also applies to meridionally bounded basins [Cane and Sarachik, 1979] and hence may be used in numerical models: e.g. Cane and Patton, 1984.)

a. Western-boundary response

The low-frequency incident signals can only be long Rossby signals, though motions may also be directly forced *at* the boundary. Say the zonal velocity from all such sources is $u_I(y, t)$ at $x = 0$. The reflection must be made up of waves carrying energy eastward. At low frequencies these are:

(a) Equatorial Kelvin waves, which propagate rapidly. If their amplitude at $x = 0$ is $a_K(0, t)$ then their amplitude at $x = x'$ is $a_K(x', t - x')$; cf. Equation 6.65.
(b) Short Rossby waves, which propagate very slowly – so slowly that in the presence of friction they will be unable to move very far from the boundary. The totality of short Rossby waves at a given time make up the western-boundary current. Denoting the sum of these waves by (u^B, v^B, h^B), we saw earlier (Equation 6.47) that in the low-frequency limit they satisfy

$$- yv^B + h_x^B = 0, \tag{6.82a}$$

$$v_t^B + yu^B + h_y^B = 0, \tag{6.82b}$$

and

$$u_x^B + v_y^B = 0. \tag{6.82c}$$

The last of these implies there is a streamfunction

$$u^B = -\psi_y^B; \quad v^B = \psi_x^B, \tag{6.83}$$

so that, taking the curl of Equations 6.82a and 6.82b yields

$$v_{tx}^B + v_x^B = 0$$

and

$$\psi_{xt}^B + \psi_x^B = 0. \tag{6.84}$$

The boundary condition $u = 0$ at $x = 0$ means that, for any incident-zonal velocity u_I at $x = 0$

$$u_I(y, t) + u^B(x = 0, y, t) + a_K(0, t)\frac{\psi_o(y)}{\sqrt{2}} = 0 \tag{6.86}$$

for all y and t.

It follows from the continuity Equation 6.82c and the two boundary conditions: $v^B = 0$ at $y = \pm\infty$ and $u^B = 0$ at $x = +\infty$, that, for all x and t:

$$\int_{-\infty}^{+\infty} u^B(x, y, t,) dy = 0.$$

Hence, integrating Equation 6.86:

$$-\int_{-\infty}^{+\infty} u_I(y) dy = \int_{-\infty}^{+\infty} u^B(0, y, t) dy + 2^{1/2} a_K \int_{-\infty}^{+\infty} \psi_o(y) dy = \pi^{1/4} a_K. \quad (6.87)$$

Once a_K has been determined from Equation 6.87 $u^B(0, y, t)$ is given by Equation 6.86.

Equation 6.87 reveals an important property of the western-boundary response: *to leading order, all the incoming zonal mass flux is reflected solely in the Kelvin mode.* This fact is crucial to the adjustment process within a closed basin. The boundary-trapped short Rossby modes provide no net zonal mass flux: these modes transport the incoming mass meridionally, connecting the incoming mass flux to the equator to be transported away from the western boundary as equatorial Kelvin signals. This makes it possible for the Kelvin mode to return the net incoming mass eastward regardless of the meridional extent of the incident zonal flow. Of course, it is possible to have a large zonal flow at some latitudes without having any net mass flux; for example, a westward flow south of the equator and an equal eastward flow north of it. In such a case the boundary motions provide the meridional transport needed to close these fluid circuits. This transport may be found, from the interior solution, without explicitly calculating the boundary-layer structure. That is, once u^B $(0,y,t)$ is known, then so is the transport ψ^B $(x = 0)$. If desired, the structure of the western-boundary layer, which is made up of short, slow, eastward moving Rossby waves, may be calculated from Equations 6.82 and 6.84.

b. *Eastern-boundary response*

At low frequencies the signals reflected from an eastern boundary are long Rossby modes; denote the u and h components of their sum at $x = X_E$ by u^R (y,t) and h^R (y, t). The only low-frequency mode that can be incident on the eastern boundary is the equatorial Kelvin mode; let us suppose its amplitude at $x = X_E$ is $a_K(X_E, t)$. Then the boundary condition $u = 0$ at $x = x_E$ means that for all y and t

$$a_K(X_E, t) 2^{-1/2} \psi_o(y) + u^R(y, t) = 0. \quad (6.88)$$

As we saw earlier, both the Kelvin modes and the long Rossby modes are geo-strophic in y:

$$yu + h_y = 0;$$

hence the sum of the Kelvin and Rossby modes is as well. Since $u^K + u^R = 0$ at $x = X_E$, it follows that

$$\frac{\partial}{\partial y}(h^K + h^R) = 0 \quad \text{at} \quad x = X_E$$

or

$$a_\kappa(x_E, t)2^{-1/2}\psi_o(y) + h^R(y, t) = A(t). \tag{6.89}$$

i.e. the height field is constant at *all* latitudes on the eastern boundary.

Writing Equations 6.88 and 6.89 together:

$$\begin{bmatrix} 0 \\ A(t) \end{bmatrix} = a_K(X_E, t)\mathbf{K}(y) + \sum_{n=1}^{\infty} a_n\mathbf{R}_n(y), \tag{6.90}$$

where a_K is known, but A and the a_ns are not. A may be found by projecting $\mathbf{K}(y)$ on both sides of Equation 6.90 and using orthogonality, i.e. the fact that $(\mathbf{K}, \mathbf{R}_n) = 0$. The answer is

$$A = \pi^{-1/4}a_K. \tag{6.91}$$

Projecting $\mathbf{R}_n$ on Equation 6.90 then yields

$$a_n = 0 \text{ for } n \text{ even}$$

and

$$a_n = a_K \cdot 2\alpha_n \text{ for } n \text{ odd}, \tag{6.92}$$

where a_n is again given by Equation 6.71. Note that at its highest point $(y = 0)$ the Kelvin mode height field is less than A by a factor of X.

Exercise 6.12: How big is X and why is this the case?

Since $n \to \infty$, all Rossby modes enter the sum in Equation 6.90, the height A is set up *instantly* all along the eastern coast out to $y \to \pm\infty$. This is a consequence of the long-wave approximation, since with the full system of equations the fastest signal travels north and south from the equator only at the speed of a coastal Kelvin wave. However, calculations with the long-wave approximation *appear* to show a Kelvin wave propagating poleward along the eastern boundary. This may be explained as follows. At a latitude y_M the widest mode present will be the one for which $y_M^2 = 2M + 1$: lower n modes have small amplitude at this latitude, while larger n modes travel more slowly and so do not extend as far to the west. This mode has group velocity $-(2M + 1)^{-1}$ so at time t it extends a distance $x_M = (2M + 1)^{-1}t$. Now if we move up the boundary from the equator with Kelvin wave speed $c = 1$, we arrive at latitude y_M at time $t = y_M$ at which time $x_M = y_M^{-1}$ – the local radius of deformation. Thus if we move up the boundary at the Kelvin wave velocity, we always see the wave front at the local radius of deformation. The response thus has some of the characteristics of a Kelvin wave, though no true Kelvin wave is present. Longer time integrations show that the reflection does in fact continue to propagate farther westward into the basin. (In nature and in numerical general-circulation models, the long-wave approximation is not made and real Kelvin modes will be traveling up the eastern boundary.)

6.5.3 Steady-state solutions

We will be particularly interested in how the time-evolving circulation approaches a steady state. Even though the forcing is steady these inviscid equations need not reach a steady state when started from a resting initial state. Nevertheless, we anticipate that the long time circulation will bear some special relation to the steady circulation, perhaps, for example, oscillating about it. The steady inviscid equations generally do not admit solutions satisfying $u = 0$ at both $x = 0$ and $x = X_E$. It is well known that the addition of viscosity permits a viscous boundary layer at the western side only (the arguments are similar to those given in Section 6.2.2).

Furthermore, as illustrated in Figure 6.12d, the retention of the time dependences in the forced shallow-water equations, Equations 6.46, permits a steady-state flow to be corrected by a time-dependent boundary layer at the western side. We therefore envision a "steady-state" solution to Equations 6.46 as actually consisting of a steady interior solution plus a time-dependent boundary-layer correction at the western side. Hence we follow Sverdrup (1947) and impose the condition $u = 0$ at the eastern side $x = X_E$ on the interior solution. Applying this to the steady form of Equation 6.46 yields

$$u = - \int_{X_E}^{x} [G_x - F_y]_y \, dx + \int_{X_E}^{x} [yQ_y + 2Q] \, dx, \tag{6.93a}$$

$$v = [G_x - F_y] - yQ, \tag{6.93b}$$

$$h = \int_{X_E}^{x} \{y[G_x - F_y] + F\} \, dx + \int_{-\infty}^{y} G(x = X_E) \, dy - \int_{X_E}^{x} y^2 Q \, dx + h_0, \tag{6.93c}$$

where h_0 is independent of x and y (see below).

If $Q \equiv 0$ then the circulation is purely wind-driven, in which case Equation 6.93 reduces to the Sverdrup solution. If the curl of the wind stress is zero then there is no steady motion and the sea-surface setup balances the wind stress:

$$h_x = F, \quad h_y = G.$$

Such a solution satisfies all boundary conditions without the need of a western-boundary layer. For a thermally driven circulation ($F \equiv G \equiv 0, \quad Q \neq 0$) Equations 6.93 say that the steady solution is geostrophic with the thermal source locally balanced by mass divergence.

The presence of an eastern boundary brings the ocean toward the steady-state Sverdrup balance. In the present case, the curl of the wind stress may be non-zero, showing that the way in which the spin-up takes place is not governed by the presence or absence of wind-stress curl. In this response to a zonal wind, a significant part of the mass redistribution required to reach the final state is accomplished by meridional currents.

6.5.4 Adjustment to the steady state in a basin

We now have all the elements necessary to fit together to describe adjustment to a steady-state solution in a basin. These elements are:

- The unbounded solution which, until the effects of the boundaries can be felt, grow linearly with time if some zonal or mass forcing is present (as in Equation 6.68).
- The effects of the eastern boundary in cutting off the forcing and in returning a series of long Rossby modes.
- The effects of the western boundary which returns a Kelvin mode to all incoming signals that have meridionally integrated zonal mass flux, and a western boundary layer consisting of short Rossby modes which can redistribute mass along the western boundary.
- The approach to the final steady state which occurs first along the equator and then gradually more and more poleward.

Figure 6.12 provides a schematic view of the non-inertia-gravity wave components of the solution. Since the zonal current described by the unbounded response is generally non-zero, additional motions are stimulated by the presence of boundaries. At the eastern side the additional motions are syntheses of long-wave Rossby waves ($\omega, k \approx 0$). These modes propagate relatively rapidly: the group velocity of the nth mode is $-(2n + 1)^{-1}$. Since the more equatorially confined lower n modes propagate faster, this response extends further into the basin near the equator and becomes narrower with increasing latitude. Only the first N modes travel fast enough to have reached longitude x by time t where $N(x, t)$ is the largest integer

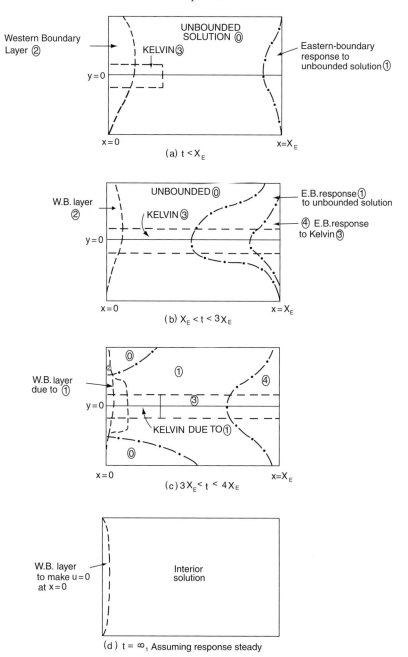

Figure 6.12. Schematic diagram of successive stages of adjustment. Description in text. (From Cane and Sarachik, 1977.)

such that $2N + 1 \leq t(X_E - x)^{-1}$. The resulting bulge in the eastern-boundary reflection is illustrated in Figure 6.12c with a dotted line indicating the wave front marking its farthest westward extent. Note that at time t the front can travel no further west than to $X_E - t/3$.

The western-boundary response consists of a Kelvin mode traveling away from the boundary with group velocity 1 and a boundary-trapped part that grows narrower and more intense with time, as indicated in Figure 6.11. The latter is a sum of modes that are a synthesis of short-wavelength Rossby waves with low group velocity so that these modes stay near the western boundary. Most of their energy is in the v component, which is in geostrophic balance. Since their zonal group velocity is so low, their energy density must be high in order for their energy flux to balance that of the incident motion. These features are qualitatively similar to the midlatitude case.

The asymmetry in the character of the eastward-and westward-propagating Rossby waves helps to explain why currents intensify on the western side of the ocean, as we saw in the barotropic context in Section 6.2.2. In addition, this reflection has features which are distinctly equatorial. Specifically, each incoming wave reflects as a whole series of waves, including the mixed mode or the Kelvin wave. Since the Kelvin waves carry energy away from the western boundary quickly, less of the incoming energy flux remains in the western-boundary current than is the case for midlatitudes. At $t = X_E$ the Kelvin mode from the western boundary arrives at the east and is reflected as a new series of long-wave Rossby modes; see Figure 6.12b. By $t = 3X_E$ the initial eastern-boundary reflection has crossed the basin and stimulates a new Kelvin wave as well as additional boundary-trapped motions. The significant difference from midlatitudes is the existence of signals that can traverse the basin rapidly.

The adjustment is most complete on and near the equator, since both Rossby and Kelvin signals have delivered their strongest messages here first. The bulging signal front in Figure 6.12c works its way westward across the basin and then, when it has passed completely across (the length of time depends on the meridional extent of the basin), adjustment is substantially complete, although wave fronts continue to bounce around until depleted by whatever inevitable dissipation mechanism is present. The adjusted case (Figure 6.12d) would then have the steady solutions, Equations 6.93, with a constantly thinning boundary layer on the western boundary. In the presence of friction, the thinning boundary layer would also reach a steady state.

The modes continue to bounce back and forth between the boundaries as in Figure 6.13, bringing the basin closer and closer to final adjustment, with the regions near the equator reaching adjustment before regions at higher latitudes.

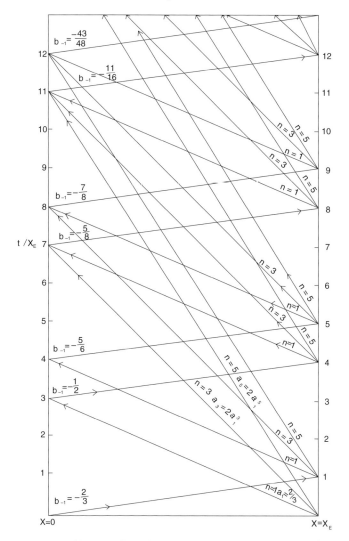

Figure 6.13. Mode diagram for adjustment (in this case to a meridional wind forcing of form $\psi_1(y)$). Kelvin modes reflect from Rossby modes impinging on the western boundary, Rossby modes reflect from Kelvin modes impinging on the eastern boundary. (From Cane and Sarachik, 1977.)

Our final comment on adjustment is the relevance of our assumed square basin to the real world. One might wonder what, in reality, plays the role of the western boundary. The western boundary serves mostly to reflect the Rossby modes into the Kelvin mode. Calculations by du Penhoat and Cane (1991) have indicated that the collections of islands forming the maritime continent is capable of reflecting most of the incident interannual Rossby motions (of the order of 80%) so that the maritime continent does form a reasonably effective western boundary.

6.6 Periodically forced motions

There is some measure of regular forcing at the annual cycle in both the Atlantic and Pacific Oceans. The Pacific also has an admixture of periodic forcing at the semi-annual period, while the Atlantic has very little semi-annual periodicity. A glance at the time series for NINO 3 (Figure 1.7) shows that there are epochs when warm phases of ENSO appear in an almost regular progression and there are epochs where lone warm or cold phases appear out of a featureless background. The question "Is El Niño sporadic or cyclic" is in fact the title of a discussion paper by Philander and Fedorov (2003). The basic conclusion of their discussion is that the actual time-dependence of the ENSO phenomenon sometimes shares characteristics of both.

We have explained the adjustment to forcings that are suddenly imposed, and have considered the time-dependent adjustment to this forcing. The response to periodic forcing is profoundly different in that we look only at the response at the forcing frequency (Cane and Sarachik, 1981). For periodic zonal wind-stress forcing where the wind stress is zonally uniform, we wish to solve the shallow-water equations:

$$i\omega u - yv = -h_x + F(y)\exp[i\omega t], \qquad (6.94a)$$

$$i\omega v + yu = -h_y \qquad (6.94b)$$

and

$$i\omega h + u_x + v_y = 0. \qquad (6.94c)$$

While we will not go into the solution in any detail (see Cane and Sarachik, 1981, with additions by Neelin *et al.*, 1998), we will indicate how essentially different are the solutions to a shallow-water ocean forced periodically at the surface and one adjusting to an impulsively applied surface forcing.

At each point of the ocean, the solution is a sum of all the equatorial waves oscillating at the forcing frequency ω that was forced by every other point of the ocean and by its reflections at both boundaries. For wind forcing uniform in the zonal direction and extending from the western coast at $x = 0$ and the eastern coast at $x = X_E$, a single parameter describes the entire linear response: $\phi = (\omega X_E/c)$, where c is the Kelvin wave speed for the baroclinic mode of interest. This parameter may be interpreted as the ratio of the zonal size of the basin to the distance a Kelvin wave travels in time ω^{-1}. Because the Atlantic is about one-third the length of the Pacific, the shallow-water response to annual forcing in the Atlantic is basically the same as the response to similar forcing with a 3-year period in the Pacific.

Figure 6.14 shows the response to annual forcing in the Atlantic (lower labels of the abscissa which give the distance directly in radii of deformation) or equivalently, the response to 3-year forcing in the Pacific (the upper abscissa labels give the fractional distance across the basin: $\xi = (x - X_E)/X_E$).

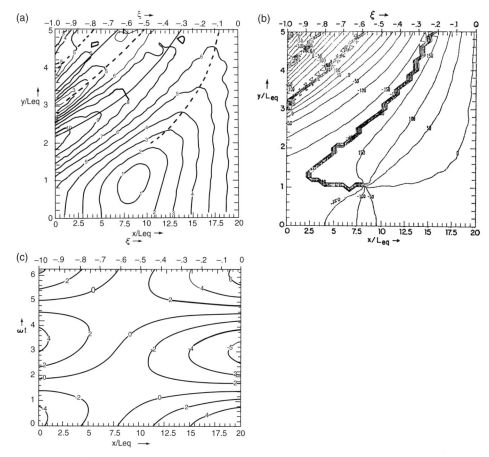

Figure 6.14. Response of the thermocline to annual forcing in the Atlantic or 3-year forcing in the Pacific ($\phi = 0.54$). (a) Amplitude of response. (b) Phase of response with respect to forcing (negative phases lead). (c) Depth of thermocline on the equator as a function of the non-dimensional time. (From Cane and Sarachik, 1981.)

The thermocline clearly does not pivot as a rigid see-saw. Within a radius of deformation of the equator, a minimum of amplitude (but not zero) exists about a third of the way from the western boundary. Were the thermocline to oscillate as a rigid see-saw, the phase would change from zero to 180° at the pivot point. Instead, the phase increases eastward near the equator. Since the apparent phase speed is

$$c_{phase} = \omega \left(\frac{\partial \chi}{\partial x} \right)^{-1} ,$$ there is apparent eastward phase propagation with effective

phase speed of about $\omega \pi / X_E$ or about 1/3 m/s. Since no actual wave propagates eastward with this speed (the Kelvin wave travels an order of magnitude faster),

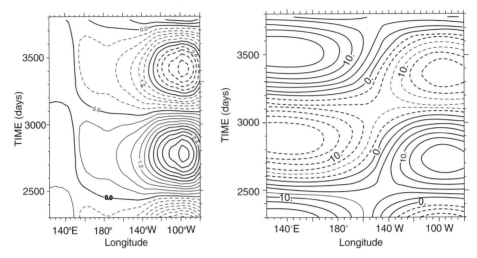

Figure 6.15. SST (Left panel) and thermocline variations (Right panel) from a linear coupled anomaly model exhibiting regular oscillations at period of about 3.5 years. (From Battisti and Hirst, 1989.)

the apparent phase speed is a result of the constructive and destructive addition of many different signals, all at the forcing frequency ω.

Figure 6.14 (Right) shows the motion of the thermocline on the equator. The thermocline motion *leads* the forcing at the eastern boundary with the lead increasing westward. That this simple periodic theory gives the general features of thermocline variation in coupled models that exhibit regular oscillations may be seen by comparing with Figure 6.15. The main difference between Figures 6.14 (Right) and 6.15(c) is that the wind field in the Battisti–Hirst simulation does not extend zonally over the entire basin but is confined roughly to the middle third, while the SST variability is confined to the eastern third.

6.7 The role of the ocean in ENSO

We have seen in Chapter 5 that the tropical atmosphere organizes convection into regions of persistent precipitation according to the warmth of the SST it responds to. In turn, the surface winds are driven by these same processes. It is the role of the tropical ocean to respond to surface fluxes of heat and momentum from the atmosphere and change its SST. The change of SST in the tropics is mostly in response to surface-wind changes in the atmosphere – the changes in surface heat fluxes are predominantly damping and respond to SST changes rather than causing them. SST then changes in response to changes in the winds through horizontal advection, which depends on surface gradients of SST, but mostly through wind-induced changes of upwelling. The temperature of upwelled water depends on the depth of the thermocline, which also depends on changes in winds through signals having characteristics of equatorial waves.

We have seen that the atmosphere organizes surface winds according to the SST provided by the ocean. The ocean provides SST changes in response to surface-wind changes provided by the atmosphere. It is now time to examine the large-scale low-frequency motions that arise from the interactions of the atmosphere and the ocean in the tropics – this is the topic of Chapter 7.

7

ENSO mechanisms

In the early days of modern ENSO studies (the early 1980s), it was common to find papers that used observed time-dependent fields of tropical surface fluxes of heat and momentum to drive an ocean model and ascertain that the resulting time-dependent fields of SST resembled the evolution of ENSO. It was also common to find papers that used observed time-dependent tropical fields of SST as boundary conditions for an atmospheric model to ascertain that the atmospheric response resembled both the local ENSO fields of sea-level pressure, precipitation and surface fluxes and the remote teleconnections of ENSO to the rest of the globe. Both types of studies are consistency checks on the models used for explaining ENSO but, of course, neither is, by itself, an explanation for ENSO. An explanation for ENSO must tell us where the atmospheric surface fluxes used to force the ocean and the oceanic SST used as a boundary condition for the atmosphere come from.

The only way to do this consistently is to couple an atmospheric model to an oceanic model and see if the coupled model exhibits *both* the atmospheric and oceanic aspects of ENSO. Even if a coupled model does exhibit ENSO behavior, it is not clear that the reasons for this behavior can be given: correct simulation does not guarantee correct understanding. It is for this reason that we begin with very simple, and therefore understandable, coupled models that exhibit interannual variability. It should be kept in mind that not all interannual variability can be identified with ENSO. Indeed it is still true at the time of writing (2009) that, while many coupled models give more or less reasonable representations of ENSO, and while aspects of the evolution of ENSO can be reliably predicted two or three seasons in advance, the basic mechanism for ENSO, in nature, remains uncertain. There are a number of possibilities and it is almost certain that the correct mechanism is to be found among these possibilities, either singly or in combination. It is the purpose of this chapter to introduce these mechanisms and indicate the arguments for their plausibility. It remains for future researchers to pin down the precise mechanism from among the possibilities here presented.

In the previous two chapters, we have examined the basic tropical mechanisms for the atmospheric forcing of the ocean and the oceanic forcing of the atmosphere. We saw that, in the tropical Pacific, surface water warmer than about 29 °C organizes low-level moisture convergence and therefore produces deep cumulonimbus convection. On monthly (and longer) timescales, the timescales relevant for our discussion, we can take the relation between the SST and low-level convergence as direct and reliable, while on shorter timescales (a week or less), different organizations of convection can exist for the same SST. We also saw that the surface winds on monthly and longer timescales are a combination of those forced directly by the regions of deep cumulonimbus convection and those forced by the boundary-layer horizontal pressure gradients. The result of wind forcing on the ocean is then to change the depth of the thermocline which, in concert with horizontal advection and upwelling, changes SST. The heat fluxes at the surface oppose these SST changes.

In this section, we will meld together these processes in the atmosphere and ocean and examine the types of motion that can arise as a result of the interaction. We will see that for large enough coupling (the sense of "large" to be defined below) there will always be unstable coupled atmosphere–ocean modes that grow exponentially with time, with the precise form of these coupled modes depending on the thermo-dynamics of the situation. The most realistic thermodynamics of the simpler models is given by the Zebiak–Cane (ZC in the following) model. We will see that one interpretation of the ZC model can be given in terms of the delayed-oscillator paradigm which explains regular oscillations at interannual periods that resemble ENSO (although the real ENSO is hardly regular). The full ZC model is irregular and nonlinear and the reason for the irregularity in the model is reasonably clear, but the reason why nature's ENSO is irregular are still controversial.

While classic instability is one form of ENSO growth, there is another. Stable systems can exhibit the property of "non-normality" (Appendix 3) in which small disturbances can first grow (sometimes to rather large amplitudes) before they decay. Constantly exciting a non-normal stable system with small random perturbations offers another class of theory for ENSO, one that is intrinsically irregular.

A third class of ENSO mechanisms is simply a combination of the two given above: unstable ENSO modes can be made irregular either through nonlinear inter-actions and/or by perturbations by inevitable ambient noise; in this case, motions of sub-monthly timescales.

This chapter will begin with simple models to exhibit some general properties of atmosphere–ocean interactions, even though they have little relevance to ENSO in the real world. The chapter will proceed to the so-called intermediate models, in particular to the ZC model, and look at the mechanisms that have been diagnosed for this class of model. The delayed-oscillator mechanism is explored in some detail then some other conceptual models, including the recharge oscillator, are introduced and the relationships among these models are considered. We then discuss the role of nonlinearities and dissipation, and whether their place in nature can be determined. The chapter will conclude with an examination of ENSO in more complex, coupled atmosphere–ocean general-circulation models.

It is hoped that the reader comes away with the (correct) impression that although much has been learned, there still remain fundamental problems in modeling ENSO and understanding the fundamental mechanisms for ENSO.

7.1 Pioneers of the study of ENSO: Bjerknes and others

There were historical precursors to ENSO theory in both the atmosphere and ocean – while none were complete, each was important.

Jacob Bjerknes (1969) was responsible for the point of view presented in Chapter 1, and for the idea that there are two coupled atmosphere–ocean states that comprise the warm and cold phases of ENSO. One state has anomalously cold water in the eastern Pacific, strong westward trades, heavy precipitation moved westward, and anomalously low sea-level pressure in the west and anomalously high SLP in the east (i.e. a strongly positive Southern Oscillation). The other has warm water extending eastward well into the tropical Pacific, weaker than normal westward trades (i.e. anomalously eastward zonal winds), eastward expansion of the region of persistent precipitation and anomalously negative Southern Oscillation. Recognizing that these two states involved a cooperative interaction between the atmosphere and the ocean was an extraordinary intellectual achievement and provided the starting point for all future progress in tropical atmosphere–ocean interactions.

Recognizing that the atmosphere and the ocean work cooperatively to produce the warm and cold phases of ENSO is not a complete theory for ENSO, since no mechanism has been provided for the transition between states. Indeed, as expressed by Bjerknes, the extreme warm and cold states seem stable.

An important advance in the oceanographic aspects of ENSO was the observation by Klaus Wyrtki (1975, 1985a) that the sea level in the western Pacific rises before and during warm phases of ENSO and declines as the warming approaches its peak. Wyrtki (1985a,1985b) developed a hypothesis for ENSO, a forerunner of what is now known as the "recharge oscillator": the aftermath of a warm event leaves the thermocline along the equator shallower than normal (i.e. equatorial heat content is low and SST is cold; this is the "La Niña" phase). Over the next few years the equatorial warm-water reservoir is gradually refilled. Once there is enough warm water in the equatorial band, the rapid (for the ocean) equatorial Kelvin and Rossby signals allowed by linear equatorial ocean dynamics can move enough of the warm water to the eastern end of the equator to initiate the next event. Wyrtki viewed this as working together with the feedbacks described by Bjerknes to create the ENSO cycle; it is ocean dynamics that provide the means for the never-ending transitions between warm and cold states. Wyrtki's ideas were derived from observational data for wind, SST and especially sea level, an indicator of the depth of the warm-water layer above the thermocline. At the same time, Cane and Zebiak (1985) proposed a strikingly similar picture, based primarily on the behavior of the intermediate Zebiak–Cane (1987) model, discussed below. Neither Wyrtki nor Cane and Zebiak expressed these ideas in the form of a simple set of equations, and neither were specific about just how the equatorial recharge was accomplished.

7.2 Simple coupled models

The first simple model of coupled atmosphere–ocean instability was that of Philander *et al.* (1984). They used the set of shallow-water equations for the atmosphere (essentially the Gill model) coupled to shallow-water equations for the ocean with temperature anomaly in the eastern Pacific set proportional to the thermocline-depth anomaly. They found *eastward*-propagating SST anomalies on interannual timescales. While this is not very realistic (we saw in Figure 2.17 that the anomalies tend to grow in place without much propagation) it was the first model to indicate that interannual variability could arise without special artifice, solely from the coupling of the atmosphere and ocean.

We will closely follow the work of Hirst (1986) because this includes the results of Philander, Yamagata and Pacanowski (and others, notably Anderson and McCreary, 1985) as a special case and because the presentation is very lucid and educational. The virtue of the model, its simplicity, is also one of the difficulties with the model: the interannual variability derived from the model bears scant resemblance to the actual ENSO. The resolution of this dilemma will be identified in this section, but addressed in the next section when we discuss the next in the chain of increasingly more realistic models, the Zebiak–Cane model. We emphasize that the presentation is ordered didactically rather than historically.

7.2.1 Formulation

The atmospheric model is akin to that of Gill (see Section 5.5.2):

$$U_t - \beta y V = -\varphi_x - AU, \tag{7.1a}$$

$$V_t + \beta y U = -\varphi_y - AV, \tag{7.1b}$$

and

$$\varphi_t + c_a^2 (U_x + V_y) = -B\varphi + Q. \tag{7.1c}$$

The Kelvin wave velocity c_a in the atmosphere is taken to be 30 m/s and the resulting equatorial radius of deformation $\left(\dfrac{c_a}{\beta}\right)^{\frac{1}{2}}$ is 1200 km. A and B are dissipation parameters for the horizontal velocity and geopotential, respectively, and Q is the heating.

The ocean model is based on a model of Anderson (1984) and Anderson and McCreary (1985). The ocean model assumes the layer above the thermocline, of depth $\bar{h} + h$, is well mixed, that there are no mean currents, and that the lower layer is at rest (as in the 1½-layer model of Chapter 3). The mean density of the upper layer is $\bar{\rho}(x)$, its average value is ρ_0, and the lower-layer density, $\rho_0 + \Delta\bar{\rho}$, is unchanging. Anomalies are linearized: $\rho = \rho_0 \alpha T$ and $\Delta\bar{\rho} = \rho_0 \alpha \Delta \bar{T}$ with α taken to be constant (which is a good approximation for the upper layer of the tropical ocean).

$$u_t - \beta y v + \alpha g \Delta \bar{T} h_x + \frac{1}{2}\alpha g \bar{h} T_x + \frac{1}{2}\alpha g \overline{T}_x h = -au + \frac{\tau^x}{\rho_0 \bar{h}}, \tag{7.2a}$$

$$v_t + \beta y u + \alpha g \Delta \bar{T} h_y + \frac{1}{2}\alpha g \bar{h} T_y + \frac{1}{2}\alpha g \overline{T}_y h = -av + \frac{\tau^y}{\rho_0 \bar{h}}, \tag{7.2b}$$

$$h_t + \bar{h}(u_x + v_y) = -bh. \tag{7.2c}$$

The temperature equation will be taken as various limits of

$$T_t + u\overline{T}_x = K_T h - dT. \tag{7.3a}$$

The Kelvin wave speed $c_o = (\dfrac{\Delta\rho}{\rho_0} g\bar{h})^{\frac{1}{2}} = (\alpha\Delta\bar{T}g\bar{h})^{\frac{1}{2}}$ is about 2.5 m/s and the equatorial radius of deformation 250 km.

The surface-wind stress and latent heating of the atmosphere are (linearly) parameterized respectively by:

$$\frac{\tau}{\rho_0 \bar{h}} = -K_S \mathbf{U} \tag{7.4}$$

and

$$Q = K_Q T. \tag{7.5}$$

The values of the coupling parameters K_S, K_Q and the dissipation parameters a, b, d and K_T are estimated from observations with the details given in Hirst (1986).

An equation for the atmospheric-perturbation energy, $E_a = \dfrac{1}{2}\left(U^2 + V^2 + \dfrac{\varphi^2}{c_a^2}\right)$, can be derived directly from Equation 7.1 where $<E_a>$ is the value integrated over the region of interest:

$$\frac{d\langle E_a\rangle}{dt} = \frac{\langle\varphi Q\rangle}{c_a^2} - A\langle U^2\rangle - A\langle V^2\rangle - B\frac{\langle\varphi^2\rangle}{c_a^2}. \tag{7.6}$$

We see from Equation 7.6 that the only generation term is the first one on the right-hand side (the other three terms are dissipation terms) so that energy can grow only when heating takes place in a region of high thickness or geopotential: i.e. only when latent heating heats where it is already warm. Since heating occurs over the warmest waters, this is automatically true.

The ocean energetics depends on which precise form of the temperature equation (Equation 7.3) is used. The simplest is Hirst's case I, where the temperature anomaly is simply proportional to the thermocline depth:

$$T = \kappa h. \tag{7.3b}$$

In this case, the ocean-perturbation energy is

$$E_o^I = \frac{1}{2}\left(u^2 + v^2 + \frac{\alpha g \Delta \overline{T}}{\overline{h}} h^2\right) \tag{7.7}$$

and the rate of change of perturbation energy is:

$$\frac{d\langle E_o^I\rangle}{dt} = \frac{\langle\mathbf{u}\cdot\boldsymbol{\tau}\rangle}{\rho_0 \overline{h}} - a\langle u^2\rangle - a\langle v^2\rangle - \frac{\alpha g \Delta \overline{T}}{\overline{h}} b\langle h^2\rangle \tag{7.8}$$

so that, again, the only generation term is the first one on the right-hand side of Equation 7.8 and it says that perturbation energy can grow only when the wind stress works on the perturbation currents, i.e. is in the same direction as the surface currents; for example when a westerly perturbation wind stress acts on a weaker westward mean current (eastward-current anomaly). It will be true for all the models derived from Equation 7.3 that perturbations can grow only under the two conditions already given: that, on average, perturbation latent heating occurs where the air is anomalously warm and perturbation wind stress works positively on the anomalous currents.

The other models derived from Equation 7.3 are:

$$\text{Model II:} \qquad T_t + u\overline{T}_x = -dT \tag{7.3c}$$

and

$$\text{Model IV:} \qquad T_t = K_T h - dT. \tag{7.3d}$$

The model using the full Equation 7.3a is denoted as Model III.

Because the forcing terms $\boldsymbol{\tau}$ and Q on the right-hand sides of Equations 7.1 and 7.2 are expressed in terms of the variables themselves by Equations 7.4 and 7.5, the full set of equations can be put in the form of an eigenvalue problem. First, Fourier transform the basic equations in x so that the variables are in the form $U(y) \exp[i(kx - \sigma t)]$ with a specified k real and σ possibly complex. The Equations 7.1, 7.2, and whatever form of Equation 7.3 is used are then discretized in the y direction and a single eigenvalue problem can be found of form $\mathbf{M}\xi = i\sigma\xi$ where ξ is a vector of the seven state variables $U(y), \varphi(y), \mathbf{u}(y), h(y), T(y)$ in discretized form, $\mathbf{M}$ is the matrix obtained from the discretized forms of Equations 7.1, 7.2 and 7.3 and the boundary conditions specify that all variables go to zero at large $|y|$. Any solution to the eigenvalue–eigenfunction problem with $\mathrm{Im}(\sigma) > 0$ gives unstable solutions that grow exponentially with time – the corresponding eigenfunctions give the form of the growing coupled modes. $\mathrm{Re}(\sigma)$ gives the corresponding frequencies of the growing modes.

7.2.2 Stable and unstable coupled solutions

It is worth going through the solutions in some detail since much of our intuition about the general nature of coupled atmosphere–ocean modes has been developed from solutions to these kinds of simplified models.

It can be generally noted that for each of the four models, when the coupling gets strong enough, unstable modes inevitably occur. The coupling is measured by the product of $K_Q K_S$, so it does not matter which of the coupling constants in Equations 7.4 and 7.5 is increased, it is only the product that counts. The coupled modes that become unstable are different for each of the four models of SST change, I, II, III or IV. Figures 7.1 and 7.4 show the growth rate and frequency for models I and III as a function of coupling (for a wavelength of 15 000 km, about the width of the equatorial Pacific) with the dashed vertical line as best estimate for representative coupling. The right panels in each of these figures show frequency and growth rate as a function of wavelength at the representative coupling. The dashed lines in the right panels are the values in the absence of coupling.

Figure 7.1 shows clearly that as the coupling increases (left panels) a mode becomes unstable. The right panel shows that this mode nearly has the dispersion formula for the free oceanic Kelvin wave and this is verified by looking at the structure of the coupled mode in Figure 7.2.

The structure of the coupled Kelvin mode is basically that of the Kelvin mode in the ocean with the atmospheric pressure pattern arranging itself so that the maximum wind anomalies lie over the maximum surface-current anomalies, which for Kelvin waves is coincident with the deepest depth anomalies which, in turn, for Model I, is coincident with the maximum SST anomaly. The coupled mode moves almost with the ocean Kelvin mode speed and the winds are aligned with the ocean currents to have $\mathbf{u} \cdot \boldsymbol{\tau} > 0$ so that the necessary condition for instability according to Equation 7.8 is satisfied. For Model I, the coupled Rossby mode (not shown) has the

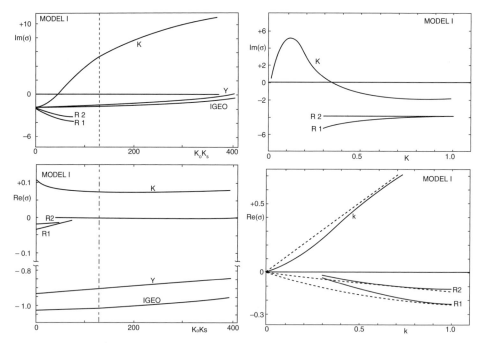

Figure 7.1. Solutions for Model I. Left panels give frequency $\mathrm{Re}(\sigma)$ and growth rate $\mathrm{Im}(\sigma)$ as a function of coupling for specified $k = 0.106$ corresponding to a wavelength of 15 000 km. Right panels give frequency and growth rate as a function of wavenumber for the coupling given by the dashed line in the left panels. (From Hirst, 1986.)

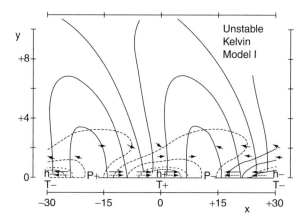

Figure 7.2. Structure of the unstable mode in Model I. In the ocean, the dashed contours give mixed-layer depth h anomalies, the dotted contours give SST (T) anomalies and the dashed arrows give ocean velocity anomalies. In the atmosphere, solid contours give pressure (P) anomalies and solid arrows give surface-wind anomalies. The meridional distance is measured in units of ocean equatorial radii of deformation – about 250 km. Contours of P and h are at 90%, 50% and 10% of extreme values (+ and −), while the contour of T is at 80% of extreme value. (From Hirst, 1986.)

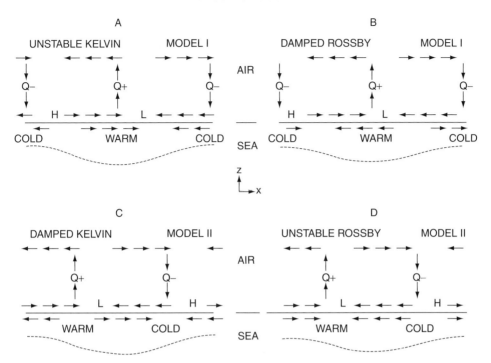

Figure 7.3. Summary schematic of damped and unstable coupled modes for Models I and II. (From Hirst, 1986.)

atmospheric surface winds in quadrature (opposite) to the ocean currents and the necessary energetic condition for instability is not satisfied – the coupled Rossby mode is damped for Model I – see Figure 7.3A and 7.3B. The atmospheric anomalies have a much larger meridional scale but travel with the oceanic Kelvin wave speed.

For Model II, at large enough coupling, the mode R1 goes unstable at long wavelengths while the Kelvin mode K is damped. Again, the internal oceanic structure is that of the first Rossby mode with the entire coupled mode moving approximately at the oceanic Rossby wave speed. The SST anomalies are confined to the equator and, because the atmospheric highs in Figure 7.3D are displaced in a manner characteristic of Rossby waves, the equatorial winds are in phase with the ocean currents and the condition for instability is satisfied. For Model II, the Kelvin-like coupled mode is damped (Figure 7.3C).

When we move to Model III, where all the terms in the thermodynamic Equation 7.3a are included, the coupled model again goes unstable for large enough coupling but now the coupled mode (denoted U) moves slowly eastward at such a slow rate that it cannot be identified with any free ocean mode (Figure 7.4).

The horizontal structure of the unstable U mode is shown in Figure 7.5, the ocean aspects of which bear no relation to a free ocean mode. It has long zonal

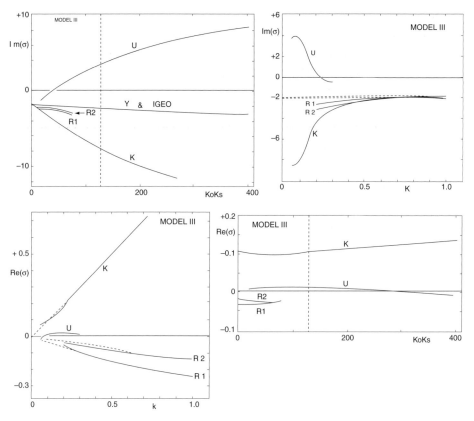

Figure 7.4. Solutions for Model III. Conventions as in Figure 7.1. U is the arbitrary label for the unstable mode. (From Hirst, 1986.)

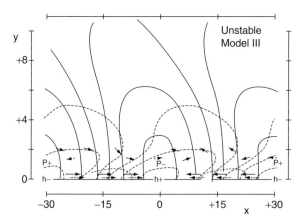

Figure 7.5. Horizontal structure of unstable mode U. Conventions as in Figure 7.2. (From Hirst, 1986.)

wavelengths and its internal ocean-depth anomaly and velocity anomaly, as well as the SST anomaly, are maximum on the equator. The depth anomaly leads the SST anomaly by about a half cycle. The atmospheric part of the mode has large meridional scales characteristic of the atmospheric radius of deformation, yet travels with the slow oceanic speed.

The U mode does not look much like the observed ENSO, in particular the anomaly propagates slowly eastward while the observed ENSO SST anomaly essentially grows in place (e.g. Figure 2.20a). Given that this simple coupled model is *not* a good representation for ENSO, what have we learned from this model? We can summarize the following points:

- Coupling of the atmosphere and the ocean near the equator can produce unstable coupled modes when the coupling is large enough. The instability is weakened by increased dissipation.
- Coupled modes will be unstable only if the mix of processes that change the SST can produce wind stresses that, in the net, work on the currents at the surface of the ocean, i.e. $\boldsymbol{\tau} \cdot \mathbf{u} > 0$.
- Coupling of the atmosphere and ocean can lead to modes of interannual period. Some of these modes do not involve the participation of the thermocline at all and, therefore, compared to observations (Figure 2.20b), cannot be a representation of ENSO.
- Depending on the mix of thermodynamic processes that change the SST, the modes can either resemble their ocean counterpart (Rossby or Kelvin modes), or not at all resemble them. In cases where the coupled mode looks like a free mode, the coupling causes the modes to travel slower than their oceanic counterparts.
- When the full complement of processes is included, the resulting coupled modes do not resemble free modes in the ocean.

Perhaps the lack of reality of the coupled mode U (the one with the most complete thermodynamics) is due to taking the beta-plane to be infinite, and therefore imposing boundaries at the east and the west will lead to a more realistic representation for ENSO. When Hirst (1988) did this, the result was surprising: the coupled modes did not seem to see the lateral boundaries of the ocean at all and were very similar to the modes on an infinite beta-plane.

Wakata and Sarachik (1991) identified the reason for the failure of these coupled modes to see the boundaries and, in the process, provided a link between the Hirst-type eigenvalue problems and the results of the Zebiak–Cane model discussed in the next section. Tracing the derivation of the SST Equation 7.3a shows that the parameters K_T and d both depend on the mean upwelling velocity $\overline{w(y)}$, which arises from the mean wind-driven divergence in the surface layer, and the mean thermocline depth $\overline{h(x)}$, which is tilted from east to west. The derivation shows that the correct interpretation of these parameters is:

$$d = \frac{\bar{w}\gamma}{H} + \alpha_s \text{ and } K_T = \bar{w} f(\bar{h}),$$

where the functional form $f(\bar{h})$ depends on the relation between the layer depth and the subsurface temperature and will be given below in Equations 7.14 and 7.15. The relevant point for this discussion is that the parameters d and K_T depend on the meridional extent of the mean upwelling velocity and the zonal tilt of the mean thermocline depth and therefore cannot be considered constant. By choosing these parameters as constant, Hirst had basically taken the friction parameter d as everywhere large and spatially constant, whereas, in fact, its large values are tightly meridionally confined to the equator and are relatively small, of value α_s, everywhere else. Having large frictional values everywhere damps the Rossby modes and thereby inhibits aspects of the problem involving westward propagation.

By solving the same Hirst linear eigenvalue problem for coupled modes in a basin, but now with these spatially dependent parameters, Wakata and Sarachik (1991) showed that a standing (i.e. not propagating) unstable coupled mode in a bounded model basin existed only when the East–West mean thermocline was 150 m deep in the west and 50 m in the east and when the meridional scale of the mean upwelling velocity was 150 km. This coupled mode is shown in Figure 7.6.

We see that the coupled unstable mode gives a reasonable ENSO cycle in a basin, albeit one that has too short a period. It should be noted that linear stability models can never give the amplitude of the oscillation, only the spatial and temporal properties. We learn, therefore, three more points on our journey to a reasonable ENSO mechanism:

- Propagating signals on the thermocline of the ocean are an important part of any coupled ENSO response. This is the same point that Wyrtki established observationally from his work with tide-gauge data.
- The magnitude and spatial extent of the oceanic dissipation (especially through the mean upwelling term) helps determine the propagation characteristics of the ocean-propagating signals and therefore of the coupled modes. In particular, to achieve an almost standing mode (without propagation), as observed, the meridional extent of the mean upwelling has to be very small (of order 150 km) and the slope of the mean thermocline must be near-observed.
- Coupled modes in a basin with periods on interannual timescales reasonably describing aspects of ENSO can be achieved under constant mean conditions without the necessity of an annual cycle. (This is not to say that the annual cycle is not important for the understanding of the real ENSO cycle. An obvious question is why does the ENSO cycle SST anomaly peak in boreal winter?)

We turn now to what has proved to be a very fertile and useful simplified model of ENSO. It predates the stability calculations we have just described and its design

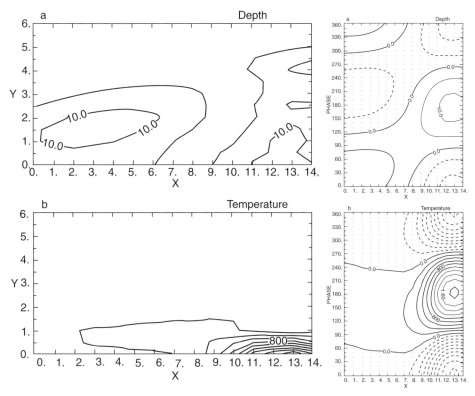

Figure 7.6. Left panel: a) Depth amplitude and b) SST amplitude, both as a function of latitude. Right panel: a) Thermocline on equator as a function of time moving upward and b) SST on equator. 360° in the right panel corresponds to the period of the mode, about 2.5 years. Because the modes are unstable, the exponentially growing part has been suppressed. (From Wakata and Sarachik, 1991.)

was drawn from a consideration of the data. We shall see that it incorporates all the lessons learned from the simple stability calculations described in this section.

7.3 The Zebiak–Cane model

The Zebiak–Cane model (Cane and Zebiak, 1985; Zebiak and Cane, 1987) is the simplest model that includes all the processes known to be important for ENSO and that incorporates the lessons we have learned from the simple linear coupled models in the previous section. Its major simplification is the formulation of the model as an anomaly model, where the anomalies are calculated relative to an annual cycle *specified* from observations. This simplification removes the necessity for simulating the mean climatic state and mean annual cycle: instead it requires that the mean and annual cycle in both the atmosphere and ocean be specified from observations. In retrospect, this has turned out to be crucial, since the annual cycle has proven particularly difficult to simulate correctly in more complex coupled models (see Section 7.8).

The model also simulates the other processes for the atmosphere and ocean that determine the SST anomaly at the surface. In particular, the ocean includes an explicit, if highly simplified, surface mixed-layer, which allows the mixed-layer processes of wind-driven convergence and divergence to be captured. The response of the thermocline to the winds is modeled by linear dynamics, and an approximate relation between the thermocline depth and the temperature of water entrained into the mixed layer is included. In the atmosphere, the effects of SST anomalies on the changes of the surface winds are included by the modified Gill-like model discussed in Section 5.7. The magnitude of the coupling of the wind to the stress is taken at a conventional value and the magnitude of the convergence for a given SST anomaly is adjusted to give reasonable magnitudes of the resulting surface winds.

To write the heat budget we need the surface advective velocity $\mathbf{u}_{sfc}$, the upwelling velocity w, and the temperature of the water entrained into the mixed layer, T_e, in addition to the physics of the ocean mixed layer. In most places the balance near the surface is just one-dimensional ocean mixed-layer physics (e.g. Gill and Niiler, 1973). In order to simplify the problem, the mixed layer depth h_m is taken to be constant at a mean tropical Pacific value of 50 m, and the temperature of the mixed layer has the same value as the sea-surface temperature.

Currents are more intense in the surface layer: to model them requires explicit consideration of the surface mixed layer. The simplest dynamical model is Ekman layer physics for the surface currents, $\mathbf{u}_{sfc}$. The currents also depend on the horizontal pressure gradient, so we need p_{sfc}. The temperature below the mixed layer T_{sub} depends on the vertical motion of the subsurface thermal structure: $\Delta T_{sub} \approx \dfrac{\partial T}{\partial z} \Delta h$ (Figure 7.7b) so we have to simulate the thermocline variations. The simplest model to determine these is the linear reduced-gravity model, with one active layer over an infinitely deep abyssal layer. We already saw this model in Section 6.2. Putting the reduced-gravity model together with a mixed layer gives the ocean component of the Zebiak and Cane (1987) model (Figure 7.7a). The earlier ocean models of Cane (1979) and Schopf and Cane (1983) were similar.

Summarizing this ocean model:

The dynamics are given by

$$\frac{\partial \mathbf{u}_1}{\partial t} + f\mathbf{k} \times \mathbf{u}_1 + \nabla p_1 = \frac{(\boldsymbol{\tau}_s - \boldsymbol{\tau}_I)}{h_m} \tag{7.9a}$$

and

$$\nabla \cdot (h_m \mathbf{u}_1) = w_e \tag{7.9b}$$

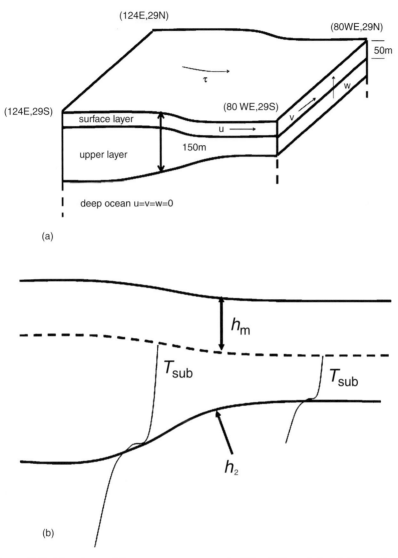

Figure 7.7. Structure of the ocean component of the ZC model. (a) Upper layer structure and (b) Relation of T_{sub} and thermocline depth.

for the upper (mixed) layer and

$$\frac{\partial \mathbf{u}_2}{\partial t} + f\mathbf{k} \times \mathbf{u}_2 + \nabla p_2 = \frac{\boldsymbol{\tau}_I - \boldsymbol{\tau}_B}{\overline{h}_2}, \qquad (7.10a)$$

$$\frac{\partial h_2}{\partial t} + \nabla \cdot (\overline{h}_2 \, \mathbf{u}_2) + w_e = 0, \qquad (7.10b)$$

for the lower layer extending from the bottom of the mixed layer to the (active) thermocline. $\boldsymbol{\tau}_s$ is the surface-wind stress, $\boldsymbol{\tau}_I$ is the interfacial stress at the bottom of the mixed layer and $\boldsymbol{\tau}_B$ is the interfacial stress at the thermocline. Since h_m and $\bar{h}_2$ are constant, the equations are linear. The horizontal currents in the surface mixed layer are driven by wind stress (Ekman processes) and by pressure gradients.

To recover the shallow-water equations: let $\mathbf{u}^T = \dfrac{[h_m\mathbf{u}_1 + \bar{h}_2\mathbf{u}_2]}{(h_m + \bar{h}_2)}$ and, since the stratification *within* the active layers is weak compared to $\nabla\rho/\rho$ (a measure of the density difference above and below the thermocline), we may assume $p_1 \approx p_2$.

Equations 7.9 and 7.10 can be combined to give equations for the total horizontal velocity (where r is the momentum dissipation):

$$\frac{\partial \mathbf{u}^T}{\partial t} + f\mathbf{k} \times \mathbf{u}^T + \nabla p = \frac{\boldsymbol{\tau}_s - \boldsymbol{\tau}_B}{\bar{h}} - r\mathbf{u}^T \tag{7.11a}$$

and

$$\frac{\partial h}{\partial t} + \nabla \cdot (\bar{h}\mathbf{u}^T) = 0. \tag{7.11b}$$

The equation for the Ekman velocity $\mathbf{u}_E$, which is just the difference $\mathbf{u}_1 - \mathbf{u}_2$ is:

$$\frac{\partial \mathbf{u}_E}{\partial t} + f\mathbf{k} \times \mathbf{u}_E \approx \frac{\boldsymbol{\tau}_s - \boldsymbol{\tau}_I}{h_*} \tag{7.12}$$

with $\dfrac{1}{h_*} = \dfrac{1}{h_m} + \dfrac{1}{\bar{h}_2}$. For the long timescales of interest $\dfrac{\partial}{\partial t} \ll f$ so the usual Ekman equations apply:

$$f\mathbf{k} \times \mathbf{u}_E = \boldsymbol{\tau}_s h_* - r_s\mathbf{u}_E,$$

where $\boldsymbol{\tau}_I = h_* r_s\mathbf{u}_E$, the interface stress at the bottom of the mixed layer, is *needed* at the equator. Zebiak and Cane (1987) argue that it is a stand-in for the nonlinear terms.

The thermodynamic equation for SST anomalies is given as:

$$\frac{\partial T}{\partial t} = -\mathbf{u}_1 \cdot \nabla(\overline{T} + T) - \mathbf{u}_1 \cdot \nabla T - [M(\bar{w}_s + w_s) - M(\bar{w})]\overline{T}_z,$$

$$- M(\bar{w}_s + w_s)\frac{T - T_e}{H_1} - \alpha_s T \tag{7.13}$$

where $M(x) = \max(x, 0)$ so that water from below is brought into the mixed layer only when the vertical velocity is upward. The first term on the right-hand side is the advection of the total SST by the anomalous currents, the second term is the advection of anomalous SST by the mean currents, the third term is the vertical advection of mean temperature by the anomalous vertical velocity (as long as it is positive), the fourth term is the total vertical advection (as long as it is positive) of temperature through the bottom of the mixed layer, and the final term is the surface flux that opposes the SST anomalies. While this is an equation for the temperature *anomaly* – the departure from the (specified) climatological temperature $\bar{T}$ – it is important to note that it is fully nonlinear and no terms are omitted. (It is most readily derived by subtracting the equation for the climatological temperature $\bar{T}$ from the equation for total temperature $T_{total} = \bar{T} + T$ and $\dfrac{\partial T}{\partial t} = \dfrac{\partial T_{total}}{\partial t} - \dfrac{\partial \bar{T}}{\partial t}$.)

The parameterization of the subsurface temperature in terms of the thermocline depth is crucial. The water entrained into the mixed layer comes from the entrainment zone and is a mixture of the mixed-layer water with temperature T and water in the ocean beneath it with temperature T_{sub}; with $0 < \gamma < 1$,

$$T_e = \gamma T_{sub} + (1 - \gamma)T. \tag{7.14}$$

The subsurface temperature T_{sub} is given empirically by:

$$T_{sub1} = T_1\{\tanh[b_1(\bar{h} + h)] - \tanh(b_1\bar{h})\} \text{ when } h > 0, \tag{7.15a}$$

and

$$T_{sub2} = T_1\{\tanh[b_2(\bar{h} + h)] - \tanh(b_2\bar{h})\} \text{ when } h < 0, \tag{7.15b}$$

where $\bar{h}(x)$ is specified.

The equations are solved in a rectangular "tropical Pacific" basin 15 000 km wide extending from 29°S to 29°N. After a small initial kick to ensure that the initial anomaly state is non-zero, the coupled model is allowed to run freely. The evolution of a warm event is shown in Figure 7.8.

This Figure shows that the evolution of SST anomalies looks realistic, the spatial representation is realistic, and the anomalies grow in place, essentially without propagation. The major discrepancy in evolution is that the warm phase of ENSO lasts too long – instead of dying away by northern spring, it extends into the northern summer. The winds to the west of the peak SST anomaly look realistic but the winds to the east of the peak are too strong and too zonal.

An important realistic feature is that the model ENSO events tend to peak at the end of the calendar year. Zebiak and Cane (1987; also see Blumenthal, 1991) show that the mean conditions in boreal summer and fall are particularly favorable for the positive feedback. The coupling weakens into the winter and anomaly growth ceases; one would expect the maximum SST anomaly to be attained when the rate of change is zero.

The evolution of the SST anomalies over time are shown in Figure 7.9 where the NINO 3 index from a thousand-year run of the model is shown. The evolution is clearly irregular, with decades of little activity and decades with relatively regular 4-year cycles of warm and cold phases. Since the anomaly-temperature equation is

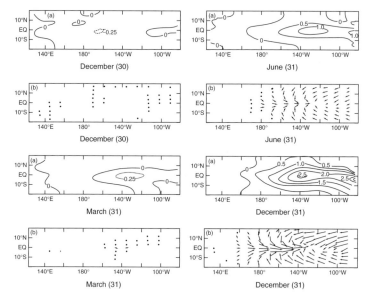

Figure 7.8. Evolution of warm phase of ENSO in 3-month intervals from December of model year 30 to December of model year 31. The upper part (a) of each month's display shows the SST anomaly for that month and the lower part (b) shows the wind field for that month. (From Zebiak and Cane, 1987.)

complete, including nonlinear terms, and the atmospheric heating is also nonlinear, the mean anomaly is not constrained to be zero. In fact, there is a small non-zero mean; the mean of the NINO 3 anomaly, for example, is about 0.3 °C. This is not an inconsistency in that nothing in the model formulation requires a zero mean, but it is an error insofar as the mean anomaly plus the specified basic state does not add up to the true climatology.

There is clearly a characteristic timescale of the order of 4 years. Since the longest instrumental record with which to compare it is of the order of 150 years (Figure 1.17) it is hard to know how realistic the long-term behavior of the model is, but parts of this long model run clearly look quite unlike the observations. For example, years 570 to 640 have an amplitude that is too large and the index is too regular. Years 780 to 820 look more like the observed NINO 3 record. The combination of Figures 7.8 and 7.9 shows that the model captures much of the correct near-surface behavior of ENSO.

The subsurface behavior is shown by the motion of the thermocline through the ENSO phases shown (along with the surface-wind stress) in Figure 7.10. We see that the thermocline starts deepening before the warming starts (in the late spring of year 31) and shows eastward-propagation characteristics similar to the observations in Figure 2.20. The heat content increases before the surface warming starts and starts

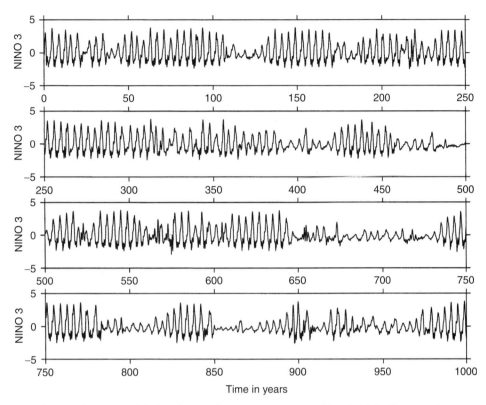

Figure 7.9. NINO 3 index from a thousand-year run of the Zebiak–Cane model. (From Cane, 1992.)

to decrease before the peak of the surface warming, again similar to observations. The wind-stress anomalies are roughly in phase with the SST anomalies.

What lessons can we take from this model; the first to attain broad agreement that it does, indeed, simulate the real ENSO cycle? The most important structural difference from earlier models is the inclusion of an explicit mixed layer, which enables the ZC model to represent all the processes that contribute to determining the SST, notably including "Ekman pumping"; the wind-driven surface-layer divergence that results in strong upwelling of colder thermocline waters along the equator. This is an essential link connecting thermocline movements to SST changes; a connection that lies at the heart of all our current ideas of how the ENSO cycle works. Zebiak and Cane (1987) show the results of numerical experiments where they artificially alter the effect on SST of changes in the zonal mean-depth anomaly of the thermocline. Doubling it changes the period from ~4 years to ~2 years and halving it changes the period to ~5–6 years. Most significantly, holding it fixed at zero eliminates interannual variability altogether.

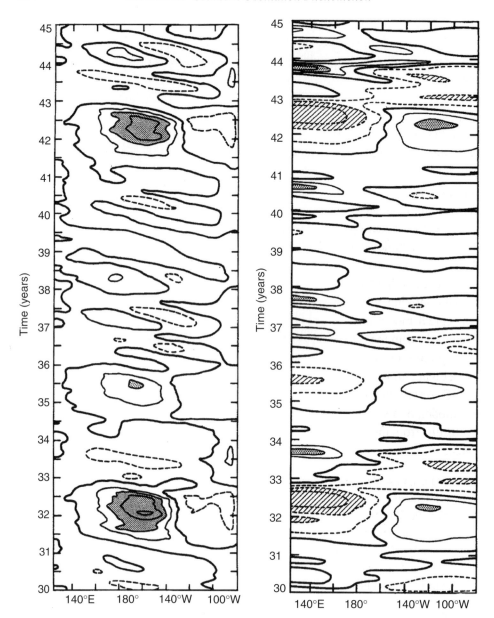

Figure 7.10. Wind-stress (Left panel) and thermocline-depth (Right panel) anomalies on the equator. Contour interval for the stress is 0.05 dyn/cm^2 with negative (westward) anomalies dashed and positive anomalies solid (eastward anomalies greater than 0.15 dyn/cm^2 stippled). Contour interval for the thermocline depth anomaly is 10 m, with positive (deeper) anomalies solid and negative anomalies dashed. (From Zebiak and Cane, 1987.)

More generally, the ZC model departs from the practice of coupling two shallow-water models together to form a conceptual model of tropical ocean–atmosphere interactions in favor of a model built from a consideration of what constitutes the essential physics of the tropical atmosphere and, most tellingly, the upper layers of the tropical ocean. The design took Bjerknes' feedback mechanism as the starting point, and took care to include the physics needed for that to function. It also took to heart the implications of the observational evidence presented by Wyrtki that variations in the volume of water in the tropics – the thermocline depth – are the central element of the ENSO cycle, an element that, fortunately, is adequately captured by the relatively simple dynamics of the linear shallow-water theory we elaborated in Chapter 6.

If the reader can envisage a time when no model, either a simple conceptual model or a coupled general-circulation model (GCM), was able to create an ENSO that an observationalist might acknowledge as akin to the real thing, it will be easy to appreciate another constraint that the ZC model met: a model would have to produce realistic-looking fields of SST, winds etc., with a realistic time dependence. A conceptual model in a few variables would not be accepted, even if it oscillated irregularly with the right mean period and was backed by a reasonable mechanistic description – even if the mechanism was the correct one. It was only after the ZC model that simple conceptual models were accepted as telling us something about ENSO. As we shall shortly see, all the contending ideas of how ENSO works use the ZC model, or a very similar model structure, as a touchstone.

Results from the model suggest the following hypotheses about the actual ENSO cycle:

- ENSO is an oscillation of the coupled atmosphere–(upper) ocean system.
- The interactions essential to creating and maintaining the cycle all take place in the tropical Pacific. No extratropical influences need to be invoked.
- That the surface layer of the ocean can respond strongly and swiftly to the atmosphere profoundly influences the character of ENSO.
- However, the basin-wide response of the upper ocean down to the thermocline is at the core of the interannual variability that defines the phenomenon.

To sum up, in the ZC model and, by implication, nature, the ENSO cycle is a combination of Bjerknes' hypothesis and linear equatorial-ocean dynamics. As Bjerknes envisioned it, a warm (El Niño) event results from a positive feedback. Warm SST anomalies in the eastern equatorial Pacific reduce the East–West temperature gradient and thus the atmospheric sea-level pressure gradient, decreasing the strength of the trades. The weakening of the winds reduces upwelling of cold water, reduces the eastward advection of cold water, and deepens the thermocline in the east, making the upwelled water warmer than before. All this increases the warm

SST anomaly and the positive-feedback loop is complete. A cold event (La Niña) has the same feedbacks but with opposite sign: colder SST results in strengthened trades, which further cool SSTs. The significant addition to Bjerknes' original hypothesis is the inclusion of the nonlocal modes of thermocline response that are part of the equatorial ocean's basin-wide response to the winds.

Given the model's structure, it is clear from the ZC model results that the interannual oscillation, the feature that Bjerknes could not account for, is a consequence of the equatorial ocean dynamics that control the displacement of the thermocline from its climatological state. In the next sections we will recount the still viable major ideas that have been put forward to explain the oscillation.

7.4. The delayed-oscillator equation

There is thus ample reason for a never-ending succession of alternating trends by air-sea interaction in the equatorial belt, but just how the turnabout between trends takes place is not yet quite clear.

(J. Bjerknes, 1969)

The current explanations for the perpetual turnabout from warm to cold states did not emerge until after the development of the numerical models, and so are properly regarded as one of the fruits of numerical modeling. While the explanation of ENSO can be made definite in some of the models, there are still major unresolved issues in the explanation of ENSO in nature.

The early ENSO explanations of Schopf and Suarez (1988), Battisti (1988) and Battisti and Hirst (1989) have linear equatorial ocean dynamics at their core. As in nature, let the main wind changes be in the central equatorial ocean while the SST changes are concentrated further to the east. Then the surface-wind amplitude, which depends on the East–West temperature gradient, varies with this eastern temperature. This eastern SST is largely controlled by thermocline-depth variations, not necessarily in phase. These variations are driven by the changes in the surface-wind stress according to the linear shallow-water equations on an equatorial beta-plane. If the eastern SST is warm (thermocline depth positive) then the wind anomaly will be westerly, forcing a Kelvin signal in the ocean to further depress the thermocline in the east, thus enhancing this state. Note that this ocean response extends eastward into regions remote from the wind changes.

As long as an initial warm perturbation of SST in the eastern Pacific leads to a westerly patch of wind anomaly further west in the Pacific, there are a number of other ways that the warming patch can grow. Currents forced by the wind stress can advect warm water eastward down the mean SST gradient to further warm the warm patch. Westerly wind anomalies can reduce the cooling due to upwelling to anomalously warm the warm patch. This surface-layer mechanism is very local, with the upwelling changes occurring only directly under the wind changes. Westerly winds can meridionally advect warm water toward the equator to further warm the warm patch – this is also largely local to the region of wind changes. Taken singly or in combination, the westerly wind anomalies warm the warm patch by this combination of processes and the anomalous warm patch grows and, with it, the westerly wind anomalies also grow.

However, this excess of warm water must be compensated somewhere by a region of colder water (shallower than the normal thermocline). The mechanism is summarized in Figure 7.11, which depicts the warming phase of ENSO in the models. Equatorial dynamics dictates that this be in the form of equatorial Rossby signals, which must propagate westward from the wind-forcing region. As we saw in Figure 6.10, a finite patch of westerly winds sends upwelling Rossby signals westward to the west of the wind patch and downwelling Kelvin signals eastward to the east of the wind patch. When the Rossby signals reach the western boundary they are reflected as "cold" equatorial Kelvin signals, which propagate eastward across the ocean to reduce the SST there. Thus, the original warm signal is invariably accompanied by a cold signal – but with a delay. This delayed-oscillator mechanism accounts for the turnabout from warm to cold states. The wraparound Hovmöller diagram from Schopf and Suarez (1988) in Figure 7.12 illustrates this in their model.

Positive Rossby signals in the sea level (i.e. deepening of the thermocline or downwelling) propagate westward (Figure 7.12a – note the reversal of east and west) and get reflected as positive Kelvin signals (again downwelling) at the western boundary. The wind-stress anomalies acts near the center of the basin (Figure 7.12c) and respond directly to the SST anomalies. When the wind-stress anomaly is positive (eastward) the Rossby signal deformations of the thermocline are negative (upwelling) and propagate westward to the boundary (Figure 7.12d – note the

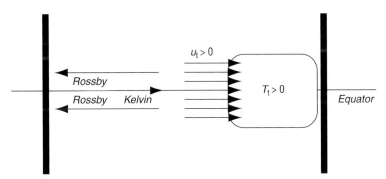

Figure 7.11. Schematic of growing warming phase of ENSO. (From Battisti *et al.* 1989.)

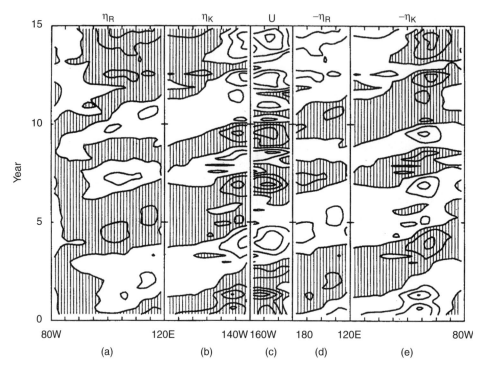

Figure 7.12. η_K is the dynamic height (an inverse measure of thermocline depth) on the equator and η_R the dynamic height off the equator between 5° and 7°N. Note that the East–West direction is reversed in (a) and (d) and that (d) and (e) plot negative sea level. Explanation of panels in text. Stippled is positive. (From Schopf and Suarez, 1988.)

reversal of east and west) and reflect as a negative Kelvin signal on the equator which then moves eastward (Figure 7.12e). Reflection at the eastern boundary spreads the signal, removing it from the equatorial zone, in contrast to the western boundary, which concentrates the reflection into the Kelvin signal.

To further appreciate the role of equatorial waves in sustaining the ENSO oscillation, consider the state of affairs when the eastern-thermocline and SST anomalies are near zero; for example, at the termination of a warm event. Then the wind anomaly must be near zero as well, so there is no direct driving to evolve the coupled system to its next phase. However, the previous warm event necessarily left a residue of cold Rossby signals in the western ocean, which eventually reflect at the west into a Kelvin signal which will reduce the SST in the east. The wind then becomes easterly and the cycle continues.

Suarez and Schopf (1988) summarize the results implied by Figure 7.12 by a simple "delayed-oscillator equation":

$$\frac{dT}{dt} = cT - bT(t - \tau) - dT^3, \tag{7.16}$$

where T represents the area average of the temperature of the warm patch in the eastern Pacific and τ is the travel time of the Rossby mode from the warm patch to the western boundary plus the travel time of the reflected Kelvin mode from the western boundary back to the warm patch: τ is thus the total time delay along the indirect route from the wind patch to the western boundary and back to the east. Since there are many Rossby waves with different speeds involved, τ is something of an average weighted by the contribution of all the Rossby modes. In Equation 7.16, c represents the local growth of the patch and b represents the effect of the returning Kelvin signal at the warm patch delayed by the travel time τ. Since the total mass of warm water is not changed by the winds, the amount going eastward in the Kelvin signals must be balanced by an equal amount of the opposite sign going westward in Rossby signals. This by itself would imply $b = c$, but since the higher-mode Rossby signals travel so slowly they would be substantially dissipated before reaching the west, implying $b < c$. Suarez and Schopf take $b < c$ on this basis.

The content of Equation 7.16 is mathematically straightforward: the cT term is simple unstable exponential growth with no periodicity. The $-bT(t - \tau)$ term is a lessening of the growth (since $b < c$) but growth is still occurring. The $-dT^3$ term is a nonlinear term needed to equilibrate the unstable growth at some finite amplitude. With $b < c$ and $c > 0$, oscillations occur only between the two fixed points

$$T_{\max} = \pm \left[\frac{c - b}{d} \right]^{1/2},$$

and we see that the nonlinear term is essential for a periodic solution.

Equation 7.16 was heuristic, devised by Suarez and Schopf to explain what they saw in Figure 7.12. On the other hand, Battisti and Hirst (1989) derive a simple linear delayed-oscillator model from a version of the Cane–Zebiak model. This version of the model, presented in Battisti (1988), is far more regular than the original ZC model, primarily for a reason identified in Mantua and Battisti (1995), namely that the mean climatology is weaker than the one used in the original ZC model. The weaker stress climatology in Battisti (1988) implies a weaker east–west mean temperature gradient on the equator and, not surprisingly, leads to an SST anomaly further east than in the ZC model (e.g. Figure 7.13) which naturally implies that the wind-stress anomalies are also further east than in the ZC model. Because the mean climatology is weaker, the Battisti (1988) model had to alter the relation between subsurface temperature and layer depth (Equation 7.15) to make it more

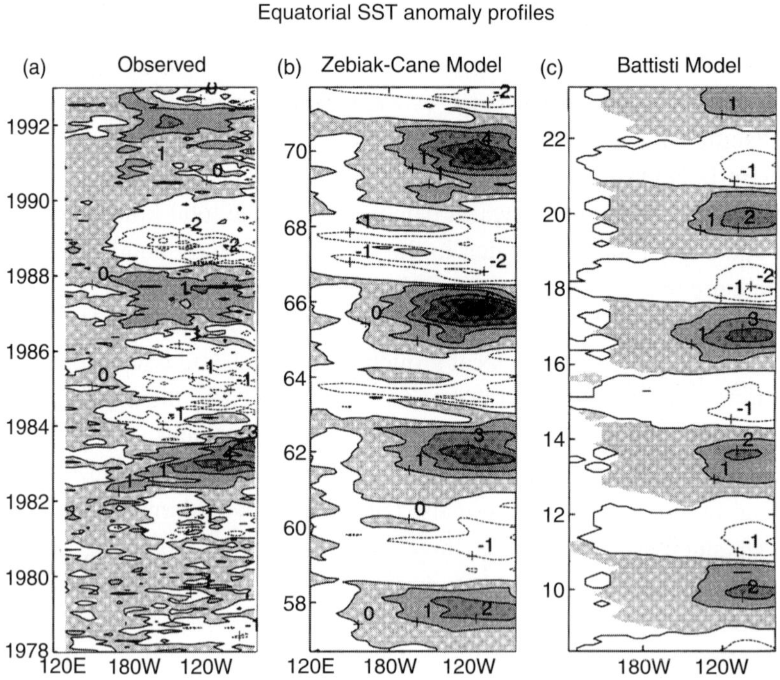

Equatorial SST anomaly profiles

Figure 7.13. SST averaged between 2°S and 2°N: (a) Observed from 1978 to 1993, (b) The ZC model and (c) The Battisti (1988) model. (From Mantua and Battisti, 1995.)

sensitive. The net result is that the instability in the Battisti model is greater than that in the ZC model, but the weaker mean climatology compensates to give an ENSO cycle of about the same magnitude. The major difference is that the Battisti model is regular, with the SST anomaly far to the east, while the ZC model is irregular, with the SST anomaly approximately as observed. A detailed comparison between the two models is given in Mantua and Battisti (1995).

The delayed-oscillator equation to be derived describes the regular oscillation in the Battisti (1988) version of the ZC model – Battisti and Hirst (1989) do extensive tests to demonstrate the ability of this equation to correctly emulate the results of the full coupled model.

The basic assumptions of the derivation are:

- τ^x and SST are in phase while thermocline depth h leads slightly. This is based on the ZC and Battisti model behavior. We shall see below that it is seen in the observational data. It is also a feature of the Cane and Sarachik (1981) analytic calculation of the response of a bounded linear equatorial beta-plane ocean to a periodic wind stress (see Section 6.6).

- The signal in both $\tau^{(x)}$ and SST are small west of 160°W. As discussed above, this assumption holds for the Battisti model because its weaker climatology puts the wind stress further east than in the CZ model – or in nature. Consequently, Rossby signals are able to propagate freely from 160°W to the western boundary.
- Equatorial ocean dynamics within ±5° of the equator are all that matter. The justification is that only the first few Rossby signals (and perhaps only the first) contribute substantially to the reflected Kelvin signal.

They further simplify as follows:

- The mean annual cycle is suppressed and all dynamics and thermodynamics are on the annually averaged mean state.
- The ocean thermodynamics and coupling are linearized (the ocean dynamics are already linear).

Now take the wind-stress average $<\tau>$ and the SST average $<T>$ over an eastern equatorial Pacific box, extending from 160°W to 80°W (the model South American coast) and from 5°N to 5°S. The linearized temperature-anomaly equation averaged over the eastern temperature box may be written (Battisti and Hirst, 1989, Eq. 2.3):

$$
\begin{aligned}
\frac{\partial \langle T \rangle}{\partial t} =& c_1 \langle \bar{u} \rangle \langle T \rangle - c_2 \left\langle \frac{\partial \bar{T}}{\partial x} \right\rangle \langle u \rangle - c_3 \langle \bar{v} \rangle \langle T \rangle - c_4 \langle M(\bar{w}) \rangle \langle T \rangle \\
&+ c_5 \langle M(\bar{w}) a(\bar{h}) \rangle h - c_6 \left\langle H(\bar{w}) \frac{\partial \bar{T}}{\partial z} \right\rangle \langle w \rangle a_s \langle T \rangle,
\end{aligned}
\tag{7.17}
$$

where the first term arises from mean zonal advection on the anomalous temperature gradient, the second term from the anomalous advection on the mean temperature gradient, the third term from mean meridional advection on the anomalous meridional-temperature gradient, the fourth and fifth terms from mean upwelling on the anomalous vertical temperature gradient, the sixth term from the anomalous upwelling on the mean vertical temperature gradient, and the final term is the local temperature dissipation by surface fluxes. As before, $M(x) = x$ and $H(x) = 1$ when $x > 0$, and both are zero otherwise.

The fourth and fifth terms come from the linearization of the anomalous vertical temperature gradient:

$$
\frac{\partial T}{\partial z} \propto (T - T_s) = (T - a(\bar{h})h).
$$

Each of the coefficients c_i may be evaluated from the model. Battisti and Hirst (1989) then take the dynamics as:

$$
h = h_{RK} + h_L = -a_w <\tau^x(t - \tau)> + a_L <\tau^x>.
$$

The first term is the thermocline variation due to the reflected Kelvin signal due to Rossby signals impinging on the western boundary. There is a delay τ from the time these waves were generated by the wind stress until the resulting Kelvin signal reaches the east. As with the Suarez and Schopf formulation, using a single delay time is conflating many Rossby waves, but here it is assumed that the $n = 1$ wave sets the delay time. The second term is the locally generated thermocline-variation Kelvin signal, directly forced by the local wind stress.

They further assume, and verify in the model, that the wind-stress anomaly is highly linearly correlated with the temperature anomaly, and that the upwelling anomaly and the zonal-current anomaly are highly linearly correlated with the anomalous zonal wind stress, which directly forces them:

$$\langle \tau^x \rangle \propto \langle T \rangle, \; \langle w \rangle \propto \langle \tau^x \rangle, \; \text{and} \; \langle u \rangle \propto \langle \tau^x \rangle.$$

As noted above, the last two relations are largely a consequence of mixed-layer physics, which requires that the wind forcing is local to the response region.

Since every term in Equation 7.17 is now proportional to the area-averaged anomalous temperature, this leads to the linear delayed-oscillator equation:

$$\frac{\partial T}{\partial t} = -bT(t - \tau) + cT. \tag{7.18}$$

Battisti and Hirst (BH) calculate the coefficients directly from the terms in Equation 7.17 based on the output of the Battisti (1988) model and thereby estimate:

$$c = 2.2 \, \text{yrs}^{-1}; \; b = 3.9 \, \text{yrs}^{-1}; \; \tau = 180 \, \text{d}$$

so in this model, $b > c$. The crucial difference between Equations 7.16 and 7.18 is that there is a local contribution to cT from terms c_1, c_2, c_3, c_4, c_6 and a_s in Equation 7.17, in addition to the thermocline-displacement term c_5, while the delayed contribution $-bT(t - \tau)$ comes only from c_5. The significant cancellations in the local term cT leads to a smaller contribution than from the delayed term, contrary to the argument of Suarez and Schopf cited above. It is the property that the delayed term is *larger* than the local term that allows the linear delayed-oscillator Equation 7.18 to have oscillatory solutions – nonlinearity is not essential for the oscillation.

We can now complete the description of the regular oscillation in the Battisti (1988) model referring to Figure 7.11: as the warming (say) patch grows in the eastern Pacific, the westerly wind patch to its west grows concurrently. The upwelling Rossby signal propagates to the west, reflects off the western boundary, and returns to the growing warming patch as an upwelling Kelvin signal. The warm patch is now warming due to unstable growth and simultaneously cooling due to the continuous action of the upwelling Kelvin signal which raises the thermocline and thereby allows the mean upwelling term to deliver cooler water to the surface. Because the delayed upwelling term is larger than the growth term, as indicated by the delayed-oscillator equation, the warm patch eventually becomes cold and the cycle continues but now with a growing cold patch in the eastern Pacific and an easterly wind-stress anomaly to its west. The growing easterly wind-stress anomaly sends downwelling Rossby signals to the west, which reflect as downwelling Kelvin signals which return to start warming the cold patch and ultimately turn it to warm, and the cycle continues.

The delayed-oscillator equation, Equation 7.18, by itself cannot be the correct paradigm for ENSO in any model, since it sustains growing solutions with no mechanism to limit the amplitude, which would therefore grow arbitrarily large: some nonlinearity is needed. But if the dynamics is essentially linear, with the nonlinearity acting mainly to limit the amplitude without changing the basic linear characteristics of the solution (as shown by Battisti and Hirst [1989] for the Battisti [1988] model), then the delayed-oscillator equation can be a useful analog and guide to the full model.

The linear Equation 7.18 has solutions $T = T_o exp(\sigma t)$ (with σ complex) when

$$\sigma = -be^{-\sigma\tau} + c,$$

or, taking the real and imaginary part,

$$\sigma_r = c - \frac{\sigma_i}{\tan(\sigma_i\tau)} \text{ and } \sigma_r = \frac{1}{\tau} \ln \frac{b \sin(\sigma_i\tau)}{\sigma_i}. \tag{7.19}$$

For a fixed τ, the real and imaginary parts of σ depend on the coefficients b and c. In particular, there will be oscillations when $\sigma_i > 0$, which obtains when

$$b > \frac{exp[c\tau - 1]}{\tau}.$$

For the parameters of the linearized Battisti model, $c\tau \approx 1$, so that the condition for oscillations becomes, approximately, $b > c$.

Exercise 7.1: Show from Equation 7.19 that the period of oscillation must exceed 2τ.

Exercise 7.2: Show that the condition for growth $\sigma_r > 0$ is, approximately, $b > \frac{\pi}{2\tau} - \left[\frac{\pi}{2} - 1\right] c$ which for $c\tau \approx 1$ is $b > c$.

Exercise 7.3: Show from Equation 7.19 that no growing oscillatory solution exists when τ becomes small. How small?

The full range of the solutions for fixed $\tau = 0.5$ years is given by Figure 7.14. Note that for $c = 2.2$/years, increasing b from zero first leads to pure exponential

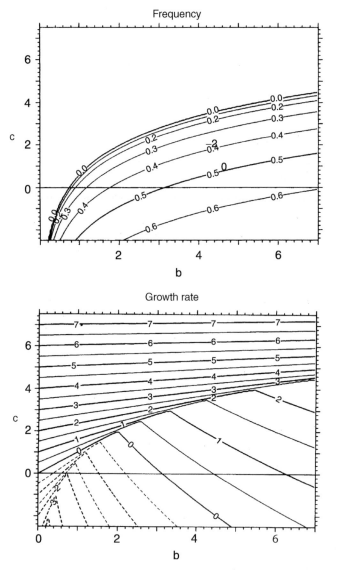

Figure 7.14. Stability properties of the delayed-oscillator equation. Bottom panel: The growth rate σ_r. Top panel: The frequency σ_i. Both as a function of b and c with units of inverse years. Here $\tau = 0.5$ year. (From Battisti and Hirst, 1989.)

growth with decreasing growth rate as b increases until $b = c$. Increasing b further increases the growth rate and the frequency.

Nonlinearities are of course necessary to obtain finite-amplitude solutions in a model that contains linearly unstable modes. Battisti and Hirst examine the nonlinearities in the Battisti (1988) intermediate model and conclude that the only nonlinearity that must be retained to capture the qualitative behavior of the model is the nonlinearity in vertical thermal structure $T_{sub}(h)$ as given by Equation 7.15. Their analysis yields a nonlinear version of Equation 7.18

$$\frac{\partial T}{\partial t} = -bT(t-\tau) + cT - e[T - rT(t-\tau)]^3.$$

The last term is derived by approximating $T_{sub}(h)$ as a cubic and noting that h depends solely on the wave-forced motions, direct and indirect, but not on the surface-layer physics terms.

Earlier we noted that the Suarez and Schopf (SS) delayed-oscillator equation *requires* the nonlinear dissipation term not only to limit the amplitude of the model ENSO, but to have an oscillation at all. We have seen that for the Battisti and Hirst (BH) delayed oscillator the oscillation depends on having $c < b$ while SS must have $c > b$ because in their model both c and b are based purely on linear ocean dynamics, which determine that the Kelvin coefficient c is larger than the retained Rossby components b. However, we can see that the two models may not be so different after all (i.e. they both allow regular oscillations) by linearizing the nonlinear term in Equation 7.16 about some non-zero temperature T_B:

$$\frac{dT}{dt} = cT - bT(t-\tau) - 3dT_B^2 T = c'T - bT(t-\tau).$$

Now

$$c' < b \Leftrightarrow c - 3dT_B^2 < b;$$

and since T varies between $\pm T_{\max} = \pm \left[\dfrac{c-b}{d}\right]^{\frac{1}{2}}$, write $T_B^2 = \alpha T_{\max}^2$ with $0 < \alpha < 1$, so that

$$c' < b \Leftrightarrow c - 3dT_B^2 < b \Leftrightarrow c - 3\alpha[c-b] < b \Leftrightarrow c(1-3\alpha) < b(1-3\alpha).$$

Since $c > b$ this will be true if $\alpha > 1/3$, a condition that is quite plausible. We may interpret the SS model as using the dissipation that occurs when $T_B^2 > \frac{1}{3}T_{\max}^2$ to meet the condition enabling oscillations. In the BH model, the magnitude of c is reduced by local wind influences, which, for example, reduce upwelling velocity and zonal advection of warm water when the trade winds slacken during an ENSO warm event. The positive feedback is thus reduced. These surface processes were left out of SS's Equation 7.16 – unless we interpret the dissipation term in this heuristic equation as a stand-in for them.

These processes are certainly part of the real ocean, and the real ocean's ENSO, but they may be overstated in the Battisti (1988) model, which has the wind changes too far to the east, increasing the direct effect of winds in the eastern box that Battisti and Hirst used to fit their conceptual model. The interaction of the surface-layer aspects of the ocean and the atmosphere are part of the "mixed" modes studied by Neelin (1991) (see also Jin and Neelin, 1993a,1993b; Neelin and Jin, 1993) as distinguished from what we will here call the thermocline mode, which involves wave dynamics and boundary reflections. Jin and Neelin examine various limits where the two modes are distinct, but indicate that the modes merge and both sets of characteristics are features of the ENSO cycle, in broad agreement with the analysis of Battisti and Hirst.

Our study of BH and SS has led us to the conclusion that local surface-layer processes are essential for the oscillation to occur. There is some irony in this since the essence of the "delayed-oscillator" theory is generally thought to be about equatorial wave dynamics, particularly the reflection at the west.

Contrary to the conclusion we reached above, it is possible to have growing, oscillating modes that depend solely on wave dynamics. Consider the situation where all the wind forcing is to the west of the region where the SST changes. This is a good description of real ENSO events until their later stages, when the wind changes reach far to the east. With the winds to the west, the changes in the SST region must all be transmitted by Kelvin signals from the forcing region and there could be no local wind influence on SST. Now a Kelvin signal directly forced by the wind, and a delayed Kelvin signal resulting from the reflection of Rossby signals at the west will, for the same amplitude, have exactly the same influence on zonal velocity, upwelling velocity and thermocline depth – all the factors that influence temperature. As with the SS model, this guarantees that $c > b$, but now we allow no dissipation or surface processes to provide a negative feedback and reduce the amplitude of c.

A model along these lines was constructed by Cane, Münnich and Zebiak (1990; CMZ hereafter). As with BH they take

$$\langle \tau^x \rangle \propto \langle T \rangle, \quad \langle w \rangle \propto \langle \tau^x \rangle, \quad \text{and} \langle u \rangle \propto \langle \tau^x \rangle,$$

which yields a temperature equation of the form

$$\frac{\partial T}{\partial t} = k_1 T - k_2 T_{sub}(h).$$

They then drop the time-derivative term on the grounds that k_1^{-1} is of order a month or two while the oscillations of interest have interannual timescales, so relative to these much longer timescales temperature adjusts rapidly: $T \approx (k_2/k_1)T_{sub}(h)$. This is the same simplification as in Hirst's Model I.

Following the method of Cane and Sarachik (1981) described in Section 6.6, CMZ solve – analytically – the shallow-water equations on an equatorial beta-plane bounded by meridians at $x = 0$ and $x = X_E$ forced by a zonal wind stress of the form

$$\tau^x = Af(x) \exp(-\mu y^2) e^{\sigma t},$$

where $\mu^{-1/2}$ is the meridional scale of the wind stress and

$$f(x) = \frac{1}{x_2 - x_1} \text{ for } x_1 \leq x \leq x_2 \text{ and } f(x) = 0 \text{ otherwise.}$$

Now $\langle \tau^x \rangle \propto \langle T \rangle$ so $A \propto \langle T \rangle = (k_2/k_1)T_{sub}(h)$ or $A = A(h) \propto T_{sub}(h)$. CMZ make one further simplification: the h in this expression is the average h in the eastern equatorial Pacific box and they replace it with h_e, the value of h at the eastern end of the equator, on the grounds that on interannual timescales this is representative of the box average. Hence $A = A(h_e)$. Here we will look primarily at the linear case $A = \kappa h_e$ with κ taken as constant, the "coupling strength" between ocean and atmosphere.

CMZ show that the results are not sensitive to the zonal width $x_2 - x_1$ of the wind patch, so we will only consider the simplest case, $x_1, x_2 \to x_c$ so that $f(x) = \delta(x - x_c)$, a delta function. For the present we also take the forcing to be at the center of the basin ($x_c = (1/2)x_E$). (Results are quite sensitive to the position of the forcing, as is true for the BH model. Moving this point changes the delay time for the reflected waves.) Then, with time scaled by the time it takes for a Kelvin wave to cross the basin (about 2 months for the first baroclinic mode in the Pacific) the dispersion relation is

$$\frac{\kappa^2}{2} = \cosh \sigma + \mu \sinh \sigma + \frac{\mu}{\sinh \sigma}. \tag{7.20}$$

This relation is plotted in Figure 7.15 for various values of the meridional width of the wind.

We see that that for all κ less than a certain value $\kappa_m(\mu)$ there are oscillating growing modes, with periods ranging from four times the Kelvin wave crossing time at $\kappa = 0$ (this is the free ocean mode of Cane and Moore, 1981) to infinite at $\kappa = \kappa_m$. For $\kappa \geq \kappa_m$ there is only non-oscillatory growth. The latter is the situation Bjerknes envisioned where the positive feedback is unchecked. In this model, that

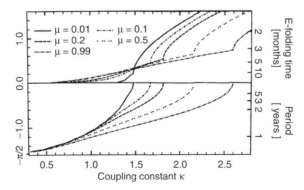

Figure 7.15. Dependence of the growth rate σ_r (Top) and frequency σ_i (Bottom) on the coupling strength κ for various values of μ where the meridional scale of the forcing is $L_y = \mu^{-1/2}$. (From Cane *et al.*, 1990.)

occurs when the coupling strength is so great that the reflected wave signal – the delayed signal – is never able to catch up with the directly forced Kelvin signal.

Let σ_m be the growth rate at $\kappa = \kappa_m$. Since κ_m is the minimum value of κ for real σ:

$$0 = \frac{\partial(\kappa^2)}{\partial \sigma}(\sigma_m) = \sinh \sigma_m + \mu \cosh \sigma_m - \frac{\mu \cosh \sigma_m}{\sinh^2 \sigma_m}.$$

Since $\mu < 1$ and we expect σ_m to be small we can expand the hyperbolic functions to obtain

$$\sigma_m \approx \mu^{1/3} - \frac{\mu}{3} + O(\mu^{5/3}); \quad \kappa_m = (2 + 3\mu^{2/3})^{1/2} + O(\mu^{4/3});$$

the smaller the meridional scale of the wind forcing the larger the value of κ_m and so the greater the range of coupling strengths that allow oscillating modes and the larger the growth rate for any fixed period. This behavior is evident in Figure 7.15.

Exercise 7.4: Note that for the form of wind stress we are using the value on the equator is independent of meridional scale, but $\int \tau^x dy \propto \mu^{-1/2}$. How would the results be changed if the wind stress were normalized so that the integral was independent of scale?

Expanding in a Taylor series about (σ_m, κ_m) yields

$$\kappa = \kappa_m + \frac{3}{2\kappa_m}(\sigma - \sigma_m)^2 - \frac{\mu^{-1/3}}{\kappa_m}(\sigma - \sigma_m)^3 + O(\sigma - \sigma_m)^4 \qquad (7.21)$$

and solving for σ by expanding in powers of $(\kappa - \kappa_m)^{1/2}$ yields

$$\sigma \approx \mu^{1/3}\left[1 + \frac{2}{9}\mu^{-2/3}\kappa_m(\kappa_m - \kappa)\right] \pm i\left[\frac{2}{3}\kappa_m(\kappa_m - \kappa)\right]^{1/2}. \qquad (7.22)$$

Comparing the approximate solution, Equation 7.22, to the exact solution, Equation 7.20, shows that these are good approximations to the full equation: quantitatively in the neighborhood of (σ_m, κ_m) and qualitatively for a great range of coupling strengths (Figure 7.16). It is obvious from Equation 7.22 that there are growing oscillating solutions for $\kappa < \kappa_m$ and pure growth for $\kappa > \kappa_m$. (However, unlike the exact solution, Equation 7.20, the approximate solution, Equation 7.22, has decaying solutions for κ sufficiently small.) As with the earlier models in this section, there are large variations in the period of the oscillation for small variations in the coupling strength (c and b in the models above). Since these numbers are not very precise, this theory does not tell us much about why the observed period is about 4 years.

We conclude that this version of the delayed oscillator does allow growing, oscillating solutions without any dissipation or consideration of the surface-layer

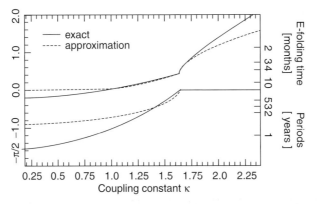

Figure 7.16. The approximate solution, Equation 7.22, compared to the exact solution, Equation 7.20. (From Cane *et al.*, 1990.)

processes involved in the fast SST mode. But, the reader might ask, where is the delay? CMZ show that if we regard σ as a (Laplace) transform variable and take $\kappa = A/h_e$ then the version of Equation 7.20 for any value of x_c transforms to

$$h_e(t) = (1+\mu)^{-1/2}\left\{ A[t-(1-x_c)] + \sum_{n=1}^{\infty}(a_n^- v^n - a_{n-1}^- v^{n-1})A[t-(4n-1)x_c-1]\right\}$$
$$+ \sum_{n=1}^{\infty}a_n^+ h_e[t-4n], \qquad (7.23)$$

where $a_n^{\pm}$ are the absolute values of the coefficients of x^n in the expansion of $(1-x)^{\pm 1/2}$ and $v = (1-\mu)/(1+\mu)$. The term in curly brackets is the response to the wind forcing; the first term is the directly forced Kelvin signal, and the terms in the sum are all of the reflected Rossby signals, terms simplified ad hoc into a single wave and single delay time in the earlier delay equations. The final term on the right is the contribution of signals reflected at the eastern boundary that propagate west and are reflected back to the east as Kelvin signals. This process is omitted in the SS and BH models. There is a nonlinear version of this model (Münnich *et al.*, 1991) where, as for BH's nonlinear model, the nonlinearity is due to the nonlinear relation of the subsurface temperature to the thermocline depth. Münnich *et al.* (1991) show that this model can exhibit aperiodic behavior, and that such behavior is favored by including the annual cycle or by including the asymmetry in the shape of the thermocline (as in Equation 7.15). Note that this equation is a pure-delay equation, in contrast to the differential-delay equations of the earlier models, although if we approximate

$$\frac{\partial T(t)}{\partial t} \approx \frac{T(t+\Delta t) - T(t)}{\Delta t},$$

then the BH Equation 7.18 becomes a pure-delay equation as well:

$$T(t + \Delta t) = -\Delta t b T(t - \tau) + (1 + \Delta t c) T(t).$$

(Note that Equation 7.23 will look like this if one drops the final sum [the eastern-boundary reflections] and replaces the sum of Rossby signals in the curly brackets with a single "representative" signal.)

While the delayed-oscillator mechanism for regular oscillations is clearly operative in the Battisti and Hirst model and does explain the parameter dependence of the linearized-coupled model that gives rise to these regular oscillations, it does not really explain the unique conditions under which the delayed oscillator is valid. CMZ note that the same setup in a model on a non-rotating plane instead of an equatorial beta-plane will have the same sort of delay – signals traveling west and being reflected east before they influence the temperatures that influence the winds – but does not allow growing oscillations. CMZ then show that the non-rotating model does allow growing oscillations if something breaks the symmetry: either having a mean eastward current so waves traveling east are faster than waves traveling west, or making the reflection at the east less effective than that at the west. Note that the delayed oscillator of Equation 7.18 does not seem to depend at all on the eastern boundary or signals emanating from it. This omission is made with the realization that the Rossby signals comprising the eastern-boundary reflections spread the signal poleward, dispersing the upper-layer warm (for El Niño) or cold (for La Niña) perturbation away from the equatorial zone. In contrast, western-boundary reflections (and the implied boundary layers consisting of short Rossby signals) concentrate the signal into Kelvin signals. Were this asymmetry not to exist the resulting coupled modes would not be oscillatory. In this view, the eastern boundary is essential for the delayed oscillator by what it does *not* do.

Though other mechanisms can give rise to unstable oscillations in coupled tropical models (e.g. Jin and Neelin, 1993a, 1993b), it is generally accepted that the delayed-oscillator paradigm accounts for the behavior of the numerical models discussed above, as well as that in the higher-resolution coupled GCMs which exhibit an ENSO-like oscillation. That is, the reflections at the west are essential for generating interannual oscillations. There is less agreement about precisely which other physical processes should be included as essential, and it is more difficult still to establish conclusively how it operates in nature.

The delayed oscillator is consistent with the refill idea described in the next section, which is supported by data (Wyrtki [1985a, 1985b] and the additional time series available in the Climate Diagnostics Bulletin of NOAA). Finally, the ZC coupled model, in which this mechanism is clearly operative, has demonstrated the ability to predict warm events a year or more in advance.

Experiments and analyses with ENSO models have demonstrated very strong sensitivities to rather small changes in parameter values. (Most of the references cited above provide examples, including Zebiak and Cane [1987], the first detailed description of the ZC model. A recent systematic study is Federov and Philander [2001].) In the anomaly models some of these changes are equivalent to changes in the mean background state. (Interesting examples of the effects of changes in equatorial heating due to changes in the Earth's orbital configuration may be found in Clement *et al.* [1999, 2000, 2001].) Since a greenhouse warming will alter this state, the implication of such sensitivity is that the characteristics of ENSO will be changed. There have been a few experiments to explore this possibility (e.g. Zebiak and Cane, 1991) but inferences must be highly tentative in deference to our limited confidence in the ENSO models and to the great uncertainties as to the nature of greenhouse-gas-induced changes. This area of research is likely to become quite active as climate modeling progresses.

7.5 The recharge oscillator and other conceptual models

Bjerknes developed the hypothesis that is still at the heart of all theories for ENSO, but it lacked an explanation for the oscillation. The earliest such explanation was not the delayed oscillator but the recharge oscillator first suggested independently by Cane and Zebiak (1985) and Wyrtki (1985a). This idea has been reinvigorated by Jin (1997a,1997b), who was the first to provide a simple equation at the level of Equations 7.16 and 7.18. Cane and Zebiak formulated their idea largely on the basis of the ZC model simulations, though they do note that it is in agreement with the scant data then available. They put the idea as follows:

This positive feedback is essentially the same mechanism proposed by Bjerknes, the most significant change being the inclusion of nonlocal modes of oceanic response. However, the feedback will not take hold unless a necessary condition for the instability of the coupled system is satisfied. Model results suggest that El Niño events will not develop if the zonally integrated heat content in the equatorial Pacific wave guide is lower than its average value. If conditions are favorable, an event may be triggered by a variety of perturbations, the most readily available being the bursts of westerly wind that occur with great frequency in the western equatorial Pacific. Mean conditions in the (northern) summer and fall are favorable to the positive feedback. Hence, once begun, ENSO anomalies will grow to large amplitude during those seasons. In the following spring the normal seasonal changes in mean conditions (reductions in trade winds, upwelling, and zonal temperature gradient) weaken the coupling between atmospheric and oceanic anomalies and the warm event can no longer be sustained. As the system relaxes, it overshoots the mean state in a manner characteristic of equatorial ocean dynamics, producing the cold SSTs and stronger than normal easterlies typical of the year following an El Niño event. At this time the heat content of the equatorial ocean is lower than normal. During the next few years the equatorial heat reservoir is refilled until the ocean is once again prepared to sustain a warm event.

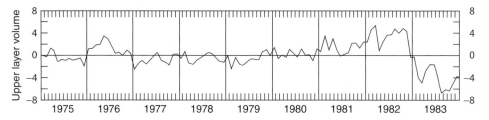

Figure 7.17. Anomalies of upper-layer volume of the tropical Pacific. (From Wyrtki, 1985a.)

Their account assigns a crucial role to equatorial ocean dynamics in generating the oscillation, but it is not specific about just what aspects of these dynamics are critical for allowing oscillations to occur. It also emphasizes the seasonal variation in the coupling strength, an idea subsequently verified by many others (e.g. Blumenthal, 1991).

The tide-gauge network that Wyrtki had deployed in the Pacific showed him (and the rest of us) that El Niño events involved a transfer of volume in the warm-water layer from west to east (cf. Wyrtki, 1979), but he was also able to construct a picture of the total amount of warm water in the equatorial Pacific (between 15°S and 15°N) showing that this volume decreased during a warm event, and then slowly refilled. This behavior is evident in Figure 7.17 and is confirmed by the more complete observational data now available. Wyrtki was also aware of the theoretical and observational work showing that the reflection of equatorial Kelvin waves impinging on the eastern boundary spreads the signal north and south, out of the equatorial zone. Here is his theory for the oscillation:

During periods when atmospheric circulation in the tropics is developed with normal strength, the trade winds push warm water toward the west and cause it to accumulate in the western Pacific both north and south of the equator. This process lasts several years until a significant amount of warm water is accumulated by a depression of the thermocline and by an increase of temperature in the mixed layer. There would be no such accumulation if there were no meridional boundaries in the ocean. Small, short fluctuations of the trade winds will have little effect on this long-term accumulation of warm water. Fluctuations of atmospheric circulation over the tropics will at some time lead to a relaxation of the trade wind field sufficiently widespread and long to allow the triggering of a Kelvin surge, namely a massive eastward displacement of the accumulated warm water along the equator … The warm water surging to the east is deflected by the coast of America to both the south and the north and is lost from the tropical ocean. This fact is evident from the sea level observations presented here and from direct observations of heat storage [White *et al.*, 1985]. Thus a complete El Niño cycle results in a net heat discharge from the tropical Pacific toward higher latitudes. At the end of the cycle the tropical Pacific is depleted of heat, which can only be restored by the slow accumulation of warm water in the western Pacific by the normal trade winds. Consequently, the time scale of the Southern Oscillation is given by the time required for the accumulation of warm water in the western Pacific. Its release is triggered by fluctuations of atmospheric circulation in the tropics. An El Niño–Southern Oscillation cycle represents a heat relaxation of the ocean-atmosphere system, in which heat stored in the tropical ocean is discharged toward higher latitudes.

Wyrtki is quite explicit about the roles of both eastern- and western-boundary reflections. In view of the results of CMZ demonstrating the dependence of the oscillation on the difference in these reflections, one could judge it to be a more satisfactory explanation than the delayed-oscillator models. On the other hand, it makes no mention of the surface-layer processes and it does not culminate in equations that could be used to calculate characteristics such as growth rate and period (though, as we have said, while the conceptual models we have considered do allow such a calculation, the dependence on model parameters is too great to say that they truly determine a period or growth rate).

Equations based on the recharge-oscillator idea were first developed heuristically by Jin (1997a). Figure 7.18 illustrates this paradigm. Jin (1997b) derived the same equations from a ZC-type model, which, as with BH, dictates parameter choices that allow the simple model to mimic the behavior of the intermediate model. Additional work, still ongoing, by Jin and collaborators has extended this recharge model to consider nonlinear and stochastic effects. Jin shows that the BH delayed oscillator is a particular case within the recharge-oscillator framework, one in which eastern-boundary reflections are eliminated by setting the reflection coefficients there to

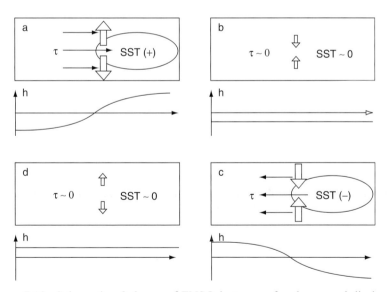

Figure 7.18. Schematic of phases of ENSO in terms of recharge and discharge. The large arrows give the mass and heat transport out of the equatorial strip. Note that positive thermocline depths are deeper: the net thermocline depth is shallower in b and deeper in d. (From Jin, 1997a.)

zero. He also shows that the same physics can be captured by either differential
equations or delay equations. We can make the same point by noting that the delay
Equation 7.23 was obtained as a transform of the exact Equation 7.20, while the
close approximation Equation 7.22 (cf. Figure 7.16) transforms into the differential
equation

$$\left\{\kappa_m + \frac{3}{2\kappa_m}\left(\frac{d}{dt} - \sigma_m\right)^2 - \frac{\mu^{-1/3}}{\kappa_m}\left(\frac{d}{dt} - \sigma_m\right)^3\right\}h_e(t) = A(t). \qquad (7.24)$$

(Dropping the cubic term gives a second-order equation more like Jin's. This
additional approximation is justified for σ close to σ_m in Equation 7.22. The reader
may easily verify that in the linear case $A(t) = \kappa h_e(t)$ this new equation admits
growing oscillating modes for $\kappa < \kappa_m$.) There are no explicit delays and no obvious
wave propagation: the periodic variations of the thermocline is given implicitly by
the net recharge and discharge.

The CMZ model is built on the theory developed by Cane and Sarachik (1981), so
it should be clear that the full treatment of periodic thermocline variations described
in Section 6.7 and utilized by CMZ contains both the delay physics and the
discharge and recharge of mass in equatorial strips, as in Jin (1997b). An application
of this theory to annual heat-content variations in the Atlantic, where the forcing is
predominantly periodic at a period of 1 year, is given in Cane and Sarachik (1983).
An unanticipated result is that the actual recharge and discharge is a small difference
of large terms: the direct interior discharge–recharge is almost cancelled by western-
boundary current transfers of mass into and out of the zonal strip. Any analog model
involving discharges and recharges would choose parameters giving the net transfers without a priori knowing the cancellations involved. Jin (1997a, 1997b) derives
the BH delayed oscillator from the recharge equations, and therefore regards it as a
special case of the recharge oscillator. On the other hand, our various exercises with
the CMZ equation ought to persuade the reader that the opposite is also true; that the
recharge oscillator is inherent in the delay model, and that it may be derived from it.
Given the rather amorphous notion of what constitutes either "the delayed oscillator" or "the recharge oscillator" – both have wave dynamics and reflection processes
at their core, but both consider surface-layer processes to be an ineluctable part of
the ENSO mechanism – it is not clear (to us at least) that there is any physical
difference between them, or any case where one paradigm applies and the other does
not. We can accept both as informative metaphors for a more complex reality.

The restriction of these models to the tropical Pacific region serves to bolster
Bjerknes' emphasis on this region. It is entirely forgivable that this simple paradigm
does not address the remote effects of ENSO, but it is troubling that it does not
capture all the tropical Pacific features associated with ENSO, notably the changes

in the western equatorial Pacific preceding the warming in the east. More generally, the Southern Oscillation (SO) is observed to exhibit some behavior distinct from El Niño, and this too is not reproduced. These tropical Pacific omissions leave open the possibility that connections essential to the ENSO cycle are not represented. The analog model of C. Wang, nicely summarized in Wang (2001) and references therein, includes specific processes in the western Pacific, in particular parameterizing changes in western Pacific wind stresses on the equator in terms of thermocline depth. He includes delayed effects and shows that the delayed oscillator and the discharge–recharge equations of Jin are obtained as special cases.

While these analogs are valuable and thought-provoking, it was the intermediate model of the ZC-type that was primary and the full characterization of time-dependent thermocline motion that was central to these analogs. None of the analogs are used to assimilate data or to make predictions.

The observed ENSO cycle is not regular, and some of the models share this feature. Nonetheless, the cause of the observed aperiodicity remains an unsettled issue. The results from model experiments using stable coupled models (e.g. Kleeman, 2008) suggest that it could be due solely to noise; that is, atmospheric or oceanic fluctuations distinct from the ENSO cycle. On the other hand, a correlation dimension test (Tziperman *et al.*, 1994, Figure 2) has clearly shown that the ZC model phase space is low-order, indicating that its aperiodicity is a result of chaotic dynamics. The simple conceptual ENSO model of Münnich *et al.* (1991), a version of the CMZ model, produces aperiodicity, doing so rather readily if a seasonal modulation is included.

7.6 Stochastically forced models

In linearly unstable model systems, a perturbation grows exponentially without limit. Something must, in reality, equilibrate the system at finite amplitude. Either some nonlinearity limits growth, or the system was not unstable to begin with. We have seen that the ZC model has Equation 7.15, reflecting the nonlinear profile of $T(z)$ in the ocean, as its basic nonlinearity: the coldness of the water upwelled into the surface layer to change SST as the thermocline shallows is limited by this relation. The other possibility is to explore a different parameter range, one in which the coupled modes are *not* linearly unstable. Initial efforts in this direction were made by Penland and Magorian (1993) and Penland and Sardeshmukh (1995) using Markov models constructed to mimic the statistics of observed SST. Since the observed time series of SST are already equilibrated, the modes are all decaying, and the only possible mechanism of growth must be transient and is connected to the system's inherent non-normality (see Appendix 3 on non-normality, which is a necessary preliminary to this section). Even earlier, Blumenthal (1991) similarly constructed a Markov model of the (necessarily equilibrated) output of the ZC model and similarly found the ENSO mode as the least decaying mode. In this section, we will examine non-normal growth in stable, coupled atmosphere–ocean models and, in particular, in linearized versions of the ZC model.

The issue can be stated succinctly as follows: for a coupled linear model of ENSO of the form

$$\frac{d\mathbf{u}}{dt} = \mathbf{Au} + \mathbf{f}, \tag{7.25}$$

where $\mathbf{u}$ is the state vector of quantities in the atmosphere and ocean, $\mathbf{A}$ is the linear evolution operator, and $\mathbf{f}$ some combination of nonlinearity and random noise; the linear stability properties of the system are determined by the nature of the eigenvalues of $\mathbf{A}$. If the eigenvalues of $\mathbf{A}$ have a positive real part, then there are exponentially growing modes and $\mathbf{f}$ must contain some nonlinear term in order to limit the amplitude of these modes. If the matrix $\mathbf{A}$ is non-normal (see Appendix 3), then even if the real parts of eigenvalues of $\mathbf{A}$ are negative, so that all the normal modes decay, there may be transient disturbances that first grow and then decay. If this is the case, then a purely random forcing $\mathbf{f}$ in Equation 7.25 may be sufficient to excite such a set of disturbances.

Thomson and Battisti (2000, 2001) made a model of this type that has the didactic advantage for this exposition of being a simple variant of the model described in Battisti (1988) based on the ZC model and already treated in the previous section. The Battisti model has a single unstable ENSO mode. Thompson and Battisti linearized the model and, by expanding in meridional parabolic-cylinder functions, as in Chapter 6, expressed the model in the matrix form, Equation 7.25. This allows the modes, adjoints and propagators to be easily obtained from the linearized evolution matrix $\mathbf{A}$.

The ENSO mode, i.e. the eigenvector of the matrix $\mathbf{A}$, is given in Figure 7.19. This mode was calculated in the presence of an annual cycle and has a growth rate of 1.8 per year and a period of 2.74 years, similar to the values calculated by Battisti and Hirst (1989). Because the matrix $\mathbf{A}$ is non-normal, the initial state that grows most rapidly into the mode is not the mode itself, but is rather the right singular vector of the propagator (or the optimal for short, see Appendix 3) and depends on the time to optimization. Because there is an annual cycle, the optimal also depends on start month, and the largest growth is attained for a 9-month optimal starting in May and peaking in January.

The optimal has an East–West tilt in SST and a feature in the south-east part of the basin. The thermocline component of the optimal has a trough all across the equator and a shallow feature in the south-east part of the basin. The optimization time is such that the optimal shown in Figure 7.20 is very close to the leading eigenvector of

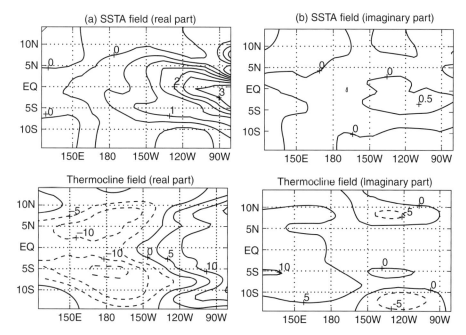

Figure 7.19. The ENSO mode in anomalies of SST (Upper) and thermocline anomalies (Lower). Because the ENSO mode is periodic with period 2.74 years, the real part represents the peak (warm) phase and the imaginary part represents the transition phase between peak warm and peak cold. (From Thompson and Battisti, 2000.)

the adjoint matrix $\mathbf{A}^+$. The optimal in Figure 7.20 grows into the mode in Figure 7.19. The optimal for other months and other optimization times looks very much like Figure 7.20 and the SST part of the optimal in Figure 7.20 agrees with optimals calculated from the SST only (Chen *et al.*, 1997).

Because the linearized $\mathbf{A}$ has unstable modes, Thompson and Battisti changed the parameters of the coupled model to stabilize the ENSO mode so that a statistically steady state may be attained when the system is forced by random noise. They created stable models by various combinations of reducing the coupling strength, increasing the dissipation and reducing the reflection coefficient at the western boundary. (Since the western boundary of the Pacific is punctuated by passages into the Indian Ocean, one might expect that some of the Rossby wave mass flux would not reflect into Kelvin waves. Du Penhoat and Cane [1991] calculate the reflection coefficient to be 0.8.) The NINO 3 index resulting from forcing a set of these stable models by random noise that is uncorrelated ("white") in both space and time is shown in Figure 7.21.

Figure 7.21 displays an interesting range of model behaviors. The slightly stabilized models, N.97 and T.97, may be described as regular oscillations with

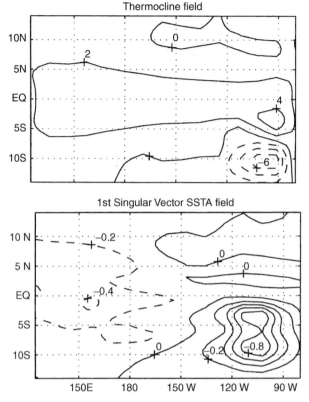

Figure 7.20. Optimal in SST anomaly (Top panel) and thermocline depth (Bottom panel) leading to largest non-normal growth. (From Thompson and Battisti, 2000.)

low-frequency amplitude modulation. As in the original Battisti model, the periods are shorter than the observed ENSO. Increasing the stability (the decay rate) increases the period; in the two examples here it is around 4 years. The heavily damped T.60 model is prone to produce large-amplitude La Niña events, while the more modestly damped T.80 seems to have the most realistic behavior of the four.

The amplitude of the non-normal disturbances excited by the stochastic forcing all tend to peak at times less than a year. The sustained responses in Figure 7.21 do not result from kicking off an optimal at $t = 0$ and then having it grow as if untouched by the random forcing. Rather, at each subsequent time, the optimal is increased by the part of the random forcing that furthers its growth, and soon stands out from the other modes, modes with less inherent ability to grow (i.e. with lower growth rates). The sum of the tiny bits grows into the ENSO mode (Figure 7.22).

The formal solution of Equation 7.25 with no initial perturbation ($\mathbf{u}(t=0)=0$) is

$$\mathbf{u}(t) = \int_0^t e^{\mathbf{A}(t-t')}\mathbf{f}(t')dt'.$$

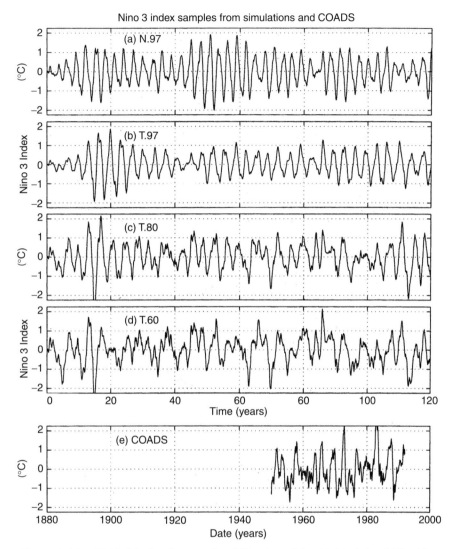

Figure 7.21. NINO 3 index for set of stabilized models. (a) Model stabilized by reducing the coupling constant but leaving the reflection and dissipation parameters alone. The 0.97 is the decay rate over the course of the year. (b), (c) and (d) Models stabilized by reducing reflection coeffecient and increasing dissipation rate to give the decay rates 0.97, 0.8 and 0.6. (e) Observed NINO 3 index for 40 years from COADS. (From Thompson and Battisti, 2001.)

It has been shown that there is a sequence $\mathbf{f}(t')$ that will generate the sequence $\mathbf{u}(t')$ with the largest amplitude at time t (or the $\mathbf{u}$ that is maximum in some other norm). This $\mathbf{u}$ is referred to as the stochastic optimal. Seminal work by Farrell and Ioannou (1993) in a fluid-dynamics context (reviewed in Ioannou and Farrell, 2006) and work applied specifically to the ENSO problem by Kleeman and Moore (1997;

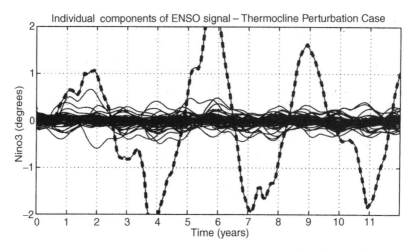

Figure 7.22. Individual monthly perturbations (thin solid lines) of the T.80 simulation adding up to the ENSO signal (dashed line). (From Thompson and Battisti, 2001.)

reviewed in Kleeman, 2008), and by Chang *et al.* (2004), all introduced the concept of stochastic optimals, the structures of the large-scale fields particularly sensitive to stochastic forcing. While all the coupled models used in these various works were linear and stable, the model differences meant that the stochastic optimals also differ.

Moore and Kleeman (1999), using meridionally symmetric physics, find that the large-scale intraseasonal forcing, in particular the Madden–Julian oscillation, is most efficient at producing interannual variance. It should be noted that the unstable ZC model, described above, is known *not* to have the Madden–Julian oscillation, yet the reality of its simulated ENSO and the skill of its prediction of the phases of ENSO rank with the best of any of the models. It is hard not to conclude that while the Madden–Julian oscillation may be important for describing the phases of ENSO accurately, it is perhaps not crucial for the fundamental existence of ENSO.

7.7 Noise or chaos? Stable or unstable? Linear or nonlinear? Does it matter and can we tell?

In the remainder of this section we discuss some issues that presently divide the ENSO community into those favoring noise and those favoring chaos. We conclude by asking if the distinction matters and whether we can tell the difference between the two.

7.7.1 The cause of irregularity

The literature offers noise and chaos as two distinct possibilities for the irregularity of ENSO. The argument for forcing by stochastic noise is straightforward: small perturbations grow either because the system is unstable, or because it is stable and non-normal, or because it is both unstable and non-normal. In each case, there is a sensitive dependence on initial conditions and random perturbations grow, implying irregularity. There is no question that there are small-scale, high-frequency (relative to ENSO) fluctuations in the atmosphere and in the ocean, but whether this stochastic noise is driving ENSO is open to question.

The ZC model (Section 7.3) does not have random noise and its source of irregularity must be different. Mantua and Battisti (1995) argue that the source of irregularity in the ZC model is the nonlinear interaction of the ENSO mode with the so-called mobile mode, a coupled westward-propagating Rossby mode. Zebiak and Cane (1987) show (their Figure 18) that with a background state fixed at July conditions interannual oscillations still appear, but become regular. They state that this holds for other months as well, albeit with different amplitudes and periods. They conclude that the seasonal cycle is responsible for the irregularity in their model, but do not rule out the possibility that there are plausible parameter sets that would allow irregularity even in the absence of a seasonal cycle. Jin *et al.* (1994) and Tziperman *et al.* (1994, 1995, 1997) also argue that the ENSO mode interacts with the seasonal cycle to produce irregularity and show that their rather different ENSO models all follow the universal quasi-periodic route to chaos as model parameters are varied. Tziperman *et al.* (1995) specifically study the ZC model and show, in agreement with the earlier analyses of Jin *et al.* and Tziperman *et al.* of simpler "toy" models, that the chaos is due to irregular jumping of the inter-annual oscillatory mode between different nonlinear resonances with the seasonal cycle. Loosely, if the nonlinearity is strong enough, the interannual mode tends to lock on to the seasonal cycle, quantizing its period to a multiple of the annual. However, the mode is indecisive about which period to choose and jumps irregu-larly between its choices (3 years or 4 years for the ZC ENSO), though there may be times when it sticks with one or the other for a number of cycles. This analysis also accounts for the tendency of ENSO to be phase-locked to the seasonal cycle and peak in boreal winter.

7.7.2 The cause of equilibration at finite amplitude

When the coupled interactions produce a stable mode, there is no need to explain the finite amplitude of the ENSO mode, since the only way to get growth is through non-normality: the ENSO mode grows out of small stochastic forcing and then decays. When the coupled interactions are strong enough for the coupled interactions to produce instability, the equilibration to a finite-amplitude ENSO mode occurs either through strong nonlinear terms (such as the T^3 in Equation 7.16) or through the nonlinearity implied by Equation 7.15, which limits the amount of warm and cold water available through upwelling.

7.7.3 Stable or unstable?

As we see, both the nature of irregularity and the cause of equilibration of the ENSO mode to finite amplitude depends on whether or not the atmosphere–ocean inter-actions are stable or unstable, which in turn depends on how strong the coupling is between atmosphere and ocean and how dissipative is the system, which includes how reflective the Rossby signals are at the western boundary.

The reflection of signals at the irregular western boundary is the best understood of these issues: it is generally agreed that the reflection coefficient at the western boundary is of order 0.8 (e.g. du Penhoat and Cane, 1991). The coupling depends on the relation between the stress and the wind (Equation 7.4), which is well known, and the relation between the SST and the heating (Equation 7.5), which is fairly well known, and the relation between the heating and the wind, which is fairly well known, and the relation between the wind and the SST, via thermocline displacements and otherwise, which again could be characterized as fairly well known. Unfortunately, all the "fairly well known" steps add up to considerable uncertainty.

This leaves dissipation. For the same coupling strength, if the dissipation is large enough, the coupled interaction will be stable, and if the dissipation is small enough, it will be unstable. The momentum-dissipation time in the ZC ocean model (Equation 7.11a) was taken to be $r^{-1} = 2.5$ years. Thompson and Battisti (2000) argue, on the basis of Picaut *et al.* (1993) that the dissipation is much greater; the dissipation time is between 6.5 and 8.5 months. Fedorov (2007), however, points out that the dissipation depends on timescale and the value given by Picaut *et al.* (1993) should be assigned to the annual cycle. For the interannual timescales appropriate to ENSO, he estimates that the correct dissipation time is 2.3 years, essentially the value used in ZC. This argues for the coupled ENSO mode to be unstable, even in the Battisti version.

7.7.4 Does it matter and can we tell?

We address the second queston first. Two structurally distinct models can exhibit many indistinguishable behaviors. We may illustrate this by a relevant construction. Blumenthal (1991) constructed a Markov model – a noise-driven stable model of the form in Equation 7.25 – from the output of the ZC model, which is clearly nonlinear and has been shown to be chaotic (Tziperman *et al.*, 1994, 1995). The Markov model was quite successful in simulating the behavior of ZC, and Blumenthal went on to analyze the optimal vectors of this noise-driven model, but obviously avoided any assertion that the ZC model must therefore be a stable noise-driven system. When the data being fitted comes from nature, there is no such check on the temptation to take a model's mimicry of a few aspects of the observations as proof that nature works just like the model.

Cane *et al.* (1995) asked what it would take to determine whether data came from the chaotic ZC model or from a noise-driven linear (Markov) model derived from time series of ZC fields. They concluded that it would be possible with 500–1000 years of observations of, say, NINO3 SST anomalies. This assumes the data is accurate – more accurate than one could expect of proxy data. This aside, the conclusion is unduly optimistic. The test they use to distinguish between the non-linear model and the linear knock-off depends on the more regime-like behavior of the chaotic nonlinear model, and it would be fooled if an external influence, such as solar-radiance variations or volcanic eruptions, was inducing persistent regimes, a possibility strongly suggested by Mann *et al.* (2005) and Emile-Geay *et al.* (2007, 2008). We are not aware of any other attempt to directly address the "can we tell" question. Perhaps there is a test with greater statistical power than that used by Cane *et al.* (1995), so we do not assert that it can never be answered affirmatively.

Does it matter? It depends. It surely matters as a point of intellectual curiosity: we would like to know just how the ENSO system functions. It probably matters if we wish to know if the system is likely to exhibit decades-long persistent regimes (no El Niños, many El Niños, persistent cold states, etc. ...) even in the absence of external forcing. Such regimes could have devastating consequences, such as persistent drought. A particular interest is the impact on predictability, which arises in Chapter 8. A short answer is that it matters in principle, but alas, in practice this difference is overwhelmed by the limits on our current predictions due to limited data and, more importantly, to errors in the models and to shortcomings of the schemes in using the data to initialize the models.

We all know that the real ENSO exists in a complex mix of nonlinearity and higher frequency "noise." Our inability to distinguish between the "noise" and "chaos" paradigms indicates that the real ENSO operates near the critical divide in "parameter space" where control passes from one to the other. There will be no profound behavioral differences between a state that is marginally stable and one that is slightly unstable. A more pressing issue at present is the failure of most of our complex coupled general circulation models to achieve a respectable simulation of ENSO.

7.8 Modeling ENSO by state-of-the-art coupled climate models

The basic idea of ENSO simulation is the same whether we deal with comprehensive coupled models or the simpler coupled models detailed in the previous sections.

The atmosphere is coupled to the ocean and, to the extent that the atmosphere determines the correct surface-wind stresses and heat fluxes and the ocean calculates the correct SST, the correct ENSO should arise spontaneously in the model tropical Pacific.

7.8.1 General concepts

The more comprehensive models have higher resolution; follow water vapor, liquid water, ice and snow explicitly; treat the radiative properties of aerosols and clouds explicitly; contain parameterizations for shallow and stratiform clouds; realistically define the ocean margins and bottom; have far more explicit and high-resolution treatment of the vertical ocean processes (especially mixing); and treat land processes explicitly. Further, the atmosphere and the ocean are treated on a global basis and, therefore, polar processes are also included. Some of the models have sophisticated physical, chemical and biological models for the uptake of carbon dioxide and other radiatively active gases (e.g. methane, nitrous oxide). There are now (2009) on the order of a dozen independent, complex coupled models in the world and most are used both for examining the current climate of the Earth and also for simulating the future response of the coupled climate system to the addition of radiatively active gases and aerosols to the atmosphere.

One might think that coupling more comprehensive, and therefore more complex, models of the atmosphere and ocean together will provide definite advantages over the simpler models when it comes to simulating ENSO. It turns out, however, that there are problems with these more complex models not only in simulating ENSO itself, but particularly in the correct simulation of the mean tropical conditions and tropical annual cycle. These are troublesome tropical biases that have persisted throughout the various upgrades of the comprehensive coupled models over the years. Since ENSO, and the tropics in general, play such a crucial part in all aspects of the global climate system, these biases are a serious limitation on the capacity of the climate models to give the correct response to the addition of these radiatively active gases and form one of the major obstacles to progress.

7.8.2 Simulation of the mean climate and annual cycle

There has been a tremendous amount of activity around the world in simulating the Earth's climate by comprehensive climate models in order to assess the climatic response to the addition of greenhouse gases to the atmosphere (especially IPCC, 2007). These comprehensive models use atmospheric resolutions of order 150 km (T.85 in spectral language), use state-of-the-art parameterizations of clouds and precipitation, couple atmosphere, ocean, land and cryosphere models together,

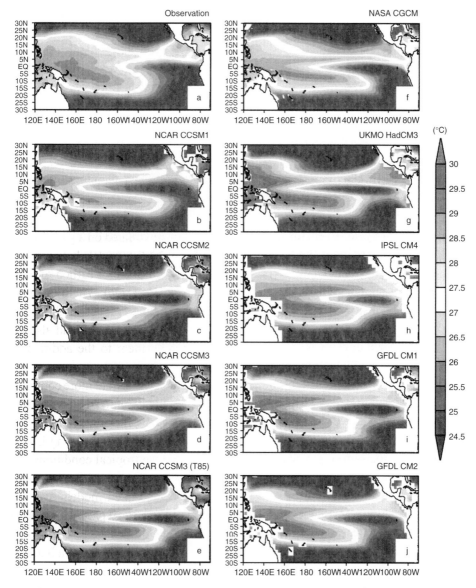

Figure 7.23. Annual mean SST from observations (Panel a) and from a number of comprehensive coupled-climate models used to simulate the response of the climate to the anthropogenic addition of radiatively active constituents to the atmosphere (Panels b to j). (From Sun *et al.*, 2006.)

and, all things considered, should provide a good simulation of the Earth's mean climate. Yet, for reasons still unexplained, significant biases remain in all the models.

Figure 7.23a shows the annually averaged SST in the tropical Pacific. The other panels show the simulation by a number of comprehensive climate models. *All* the models have the cold tongue extending too far to the west, and the South Pacific convergence zone (SPCZ) is too zonally aligned rather than pointing off to the south-east Pacific. The source of the westward extension of the cold tongue is that all the models have too strong easterlies extending too far westward. The net effect of this mean bias is that the region of persistent precipitation that lies over the warm pool in the western Pacific is too far west in the mean. Strong warm phases of ENSO, which tend to make the tropical Pacific a uniform warm temperature, therefore would have the SST anomalies extending too far to the west.

The annual cycle of SST is confined to the eastern Pacific, mostly to the east of 160°W. Here, the situation is mixed. Most models are not capable of giving a realistic simulation of the tropical annual cycle.

As we see from Figure 7.24, none of the coupled models analyzed simulates an accurate annual cycle. Even for those models where forcing only the ocean with climatological fluxes gives the correct annual cycle of SST, coupling to the atmosphere gives incorrect annual cycles of SST (Figure 7.25). This Community Climate System Model, shown in Figure 7.25, exhibits an annual cycle that is completely out of phase with observations – the cold season is in June and July rather than September–October. There is also some hint of biannual variability. Since the forced uncoupled ocean model gives the correct annual cycle, the bias in the climatology is a function of the coupling – its ubiquity indicates that its cause is recondite. We may note here that while the exact mechanism of the coupled annual cycle in the tropical Pacific is not well understood, the annual cycle is presumably not purely forced by the sun since, as we have seen in Figure 2.2 that the solar forcing is semi-biannual on the equator while the response is annual – the annual cycle is therefore a reasonable test of some of the same coupled mechanisms as ENSO itself. Each of these models is state-of-the-art, yet no one can yet say why one has a reasonable annual cycle and the other does not.

It is important to note that when the mean climatology has biases, the anomalies from the mean are necessarily suspect. Further, since the mean is incorrect, the heat sources that drive the predictable part of midlatitude variability are in the wrong places at the wrong times, and therefore give incorrect midlatitude variability. While ENSO prediction (see Chapter 8) is not immediately affected by mean biases, since the prediction is initialized by observations, the prediction evolves freely and therefore tends toward the wrong climate, eventually corrupting the forecast (this is an example of climate drift). A complete analysis of the annual cycle in all the models used in the IPCC (2007) is given in E. K. Jin *et al.* (2008) and in de Szoeke and Xie (2008) with results that are consistent: the mean and annual cycles are currently generally poorly done in coupled atmosphere–ocean models for reasons that are not presently known. The first of these papers attributes the inability of these coupled climate modes to correctly simulate and predict ENSO to this fundamental

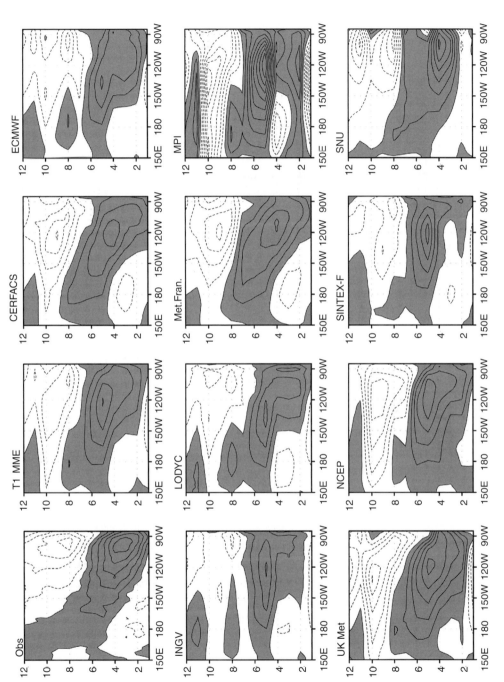

Figure 7.24. Annual cycle of Pacific SST on equator as anomalies from the annual mean. The multi-model ensemble is MME (After Figure 2 of E. Jin *et al.*, 2008 – courtesy Emilia Jin).

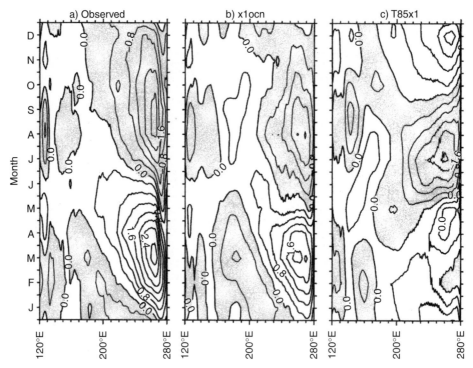

Figure 7.25. Simulations of the annual cycle (b) and (c) compared to observations (a); (b) shows the forcing of the ocean with observed climatological fluxes at the surface while (c) is the fully coupled CCSM. (From Large and Danabasoglu, 2006.)

problem in the models. Fixing these coupled model biases therefore becomes a very high priority for the next generation of comprehensive climate models.

7.8.3 *Simulations of ENSO*

As we pointed out in the previous section, all the current coupled comprehensive climate models have a bias that puts the annually averaged cold tongue too far to the west. It will therefore come as no surprise that the simulation of ENSO in these models has the ENSO SST anomalies also too far to the west. Figure 7.26 shows that all the comprehensive climate models exhibit this bias. In addition, all models have their periods too short compared with observations, and the North–South extent of the zonal-wind anomalies are too meridionally confined. These biases exist in the presence of thermocline simulations which can be either too shallow or too deep.

On the plus side, the magnitudes of the SST anomalies are approximately correct and the relationships between the thermocline anomalies, the zonal-wind anomalies and the SST anomalies are approximately correct, indicating the basic correctness of

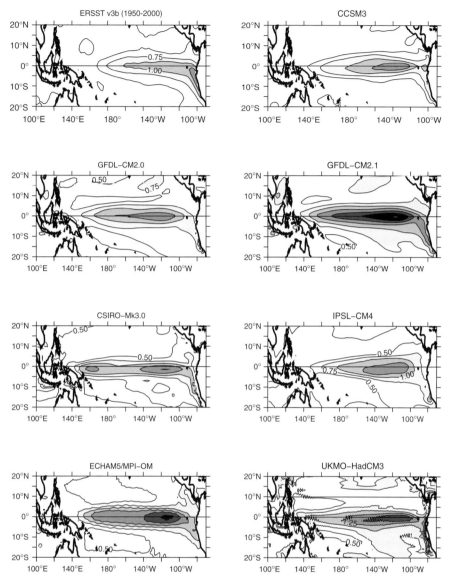

Figure 7.26. Upper left panel: Observations of standard deviation of interannual SST anomalies from the monthly climatology. Other panels show comparable standard deviation of comprehensive climate models from their own monthly climatologies. The contour interval is 0.25. (Courtesy of Daiwei Wang.)

the simulations according to the observations presented in Section 2.4. This basic correctness of the relationships illustrates continued progress over earlier assessments of the ability of comprehensive climate models to simulate ENSO.

It must not be thought, however, that the modeling situation using comprehensive climate models is acceptable. IPCC (2007) indicates that the response to the anthropogenic addition of radiatively active gases is only trustworthy on continental space scales, approximately 5000 km. Given that the time dependence of thermal forcing of midlatitudes by the tropics is poorly simulated, both spatially and temporally, it is not surprising that this is true. This is one more example of the indivisibility of climate: in order to simulate the correct spatial and temporal dependence of long-term climate, it is necessary to simulate all time and space scales that communicate with the time and space scales of interest. In practice, this means that no timescale, no space scale, no climate process and no small-scale process can be arbitrarily neglected.

8

ENSO prediction and short-term climate prediction

We begin by making some non-standard distinctions, solely for convenience in the following discussion. We will define "ENSO prediction" as the process of predicting the SST in the tropical Pacific a month to a year or so in advance. We will use "short-term climate prediction" to refer to the procedure of predicting the climatic conditions in the global atmosphere or over land away from the tropical Pacific a month to a year in advance. The utility of this distinction is that ENSO prediction can only be accomplished by coupled models, whereas short-term climate prediction, which depends in part on the thermal forcing due to the distribution of regions of persistent precipitation and is partly determined by the SST distribution in the tropical Pacific, can be accomplished by a global *atmospheric* model (with predicted tropical SST specified) but can also be accomplished by a fully coupled climate model. The distinction will become clearer in our discussion of one-tiered and two-tiered prediction below.

The possibility that coupled climate models, whether simple or complex, can predict aspects of the future evolution of ENSO is not at all obvious. The atmosphere is known to be of limited predictability, basically because it is chaotic: inevitable small errors in the initial conditions grow and, depending on the growth rate, eventually limit the skill of prediction after a given time. Since the error doubling time of the atmosphere is generally no more than a few days, the ultimate limit of prediction of the detailed state of the atmosphere is of the order of 2 weeks. No prediction of the weather beyond this limit can be made.

How then can we make predictions of the evolution of ENSO several months in advance? The answer lies in the nature of the coupling of the tropical atmosphere to the relatively sluggish tropical ocean. To the extent that the SST distribution in the ocean determines the statistical distribution (but not the instantaneous distribution) of cloud heating over the interval, say, of a month, the slow evolution of the ocean SST determines the evolution of the *statistics* of the atmosphere (the original argument was given in Shukla, 1981). Similarly, the statistics of the atmosphere applied as fluxes at the ocean surface determines the evolution of the ocean. It is the ponderous ocean component of the climate system involved with the evolution of the coupled system that permits long prediction times. What is forecast is the SST or, equivalently, the statistics of the atmosphere in equilibrium with the SST on time-scales of a month or so. Even if the atmosphere is chaotic, the SST can be predicted, and therefore the statistics of the atmosphere in contact with the ocean *can* be predicted. Note that this argument obtains only for the tropical regions, where the interaction of the ocean with the atmosphere is strong and where the atmospheric statistics on monthly or longer timescales are directly determined by the ocean. In midlatitudes, the state of the atmosphere is not determined by local SSTs and short-term climate prediction requires that remote tropical SSTs exert some control. It is likely to turn out that only the tropical SST is predictable a month to a year in advance and only those aspects of the global climate that depend on tropical SST can be foreshadowed with any skill at all.

Since the most useful type of future information is the probability distribution of future outcomes, "ensemble forecasting" has become the most useful prediction tool. Ensemble forecasting is based on the idea that the tropical ocean does not determine the instantaneous state of the tropical atmosphere, but does determine its monthly (and longer) averaged statistics. The exact state of the tropical atmosphere, i.e. the distribution of clouds, the height of the boundary layer, the exact instantaneous value of the wind stresses, is not, in general, known. Therefore, if a set of different forecasts can be accomplished, each with slightly different initial atmospheric conditions, all consistent with what is known about the atmosphere, and each compatible with the initial ocean SST, the distribution of the forecasts will serve as the probability distribution function (pdf) of future outcomes. In this way, the statistical aspects of the future atmosphere are limned out in the ensemble distribution of predictions. The peak of the pdf is the most likely future outcome and the width of the pdf gives an indication of how certain the forecast is – the more sharply peaked, the more certain.

Since there are many coupled models, each built independently by different modeling groups and, therefore, each presumably having different and independent biases and errors, it turns out, unintuitive though it may seem, that the combination of forecasts among different models gives a better forecast than the forecast produced by any individual forecast system. These combinations of forecasts into a multi-model ensemble also give a better idea of future probabilities and uncertainty than any individual model. A multi-model ensemble is limited only by the number of coupled forecast systems extant in the world. It may at some (distant) point in the future happen that one forecast system proves itself to be the absolute best and performs without systematic errors of any kind – in that case the multi-model paradigm may be abandoned. It will still be true that ensembles with slightly different initial conditions will be required to give an idea of future uncertainty, since it will never be true that the atmospheric initial conditions over the ocean will be observed with perfect fidelity at fine scales.

8.1 Weather prediction

While the focus of this chapter is on prediction a season to a year in advance, the comparison of the similarities and differences between weather prediction and short-term climate prediction proves illuminating. Aside from the obvious usefulness of the forecast information a few hours to a few days in advance, the twice a day (or in some cases four times a day) model-based *analysis* of the atmosphere gives a synoptic view of the atmosphere and forms the basis, through re-analysis (see Section 8.2), for the rational growth of the atmospheric record over long intervals of time.

Weather prediction proceeds by a number of standard steps (see e.g. Persson and Gravini [2005] for a very useful and complete review):

(a) Observations of the atmosphere, both direct and remotely sensed, are collected within a few hours of the initial time (i.e. within the initial time *window*). In general, the weather services of the world send their data to the global telecommunication system (GTS) which then makes the global collection of data available to all weather services.

(b) The observations are assimilated into a numerical model of the atmosphere by a data-assimilation procedure. This model-based *analysis* of the atmosphere is performed by combining the observations with the output of the forecast system for the initial time. Since the observations are imperfect and, by themselves, do not define the state of the entire atmosphere (especially in regions where no data exists or where the data is of such poor quality that the model gives a better estimate than the poor observations themselves), this combination of model with observations gives the best possible estimate of the state of the atmosphere. How the system knows the relative quality of imperfect observational data and model data is the essence of data assimilation. Sometimes model information at previous or future times is used to give the best estimate of state of the current atmosphere – the so called four-dimensional data-assimilation procedure. With a good data-assimilation procedure, the analysis should give the best possible estimate of the state of the atmosphere at the given initial time.

(c) The initial state for the forecasts is produced, essentially the model-based analysis at the initial time plus some subsidiary adjustments (removing gravity waves, adjusting the envelope of mountains, adjusting for shocks etc.).

(d) The model is run from the initial state out to n days, thereby providing forecasts for all times up to and including n days.

(e) As each real forecast time is reached, the forecast is compared with the analysis for that time in order to score the forecast.

(f) The forecast cycle is continually repeated and a series of forecasts is built up and verified by the series of analyses. The long series of forecasts is used to determine the overall skill, to analyze the dependence of skill on season and synoptic conditions and to examine the forecasts for persistent biases in specific regions.

The predictability of the atmosphere is limited to something of the order of 2 weeks, since the error doubling time is of the order of 2 days or so. This arises because the atmosphere is a chaotic system and inevitable errors in the initial conditions grow until the forecast accumulates so much error that it becomes valueless. The skill of forecasts has continuously increased over the years, partly due to the expansion of coverage enabled by satellite observations, partly by the increased ability of data-assimilation systems to deal with satellite data, partly by improved assimilation techniques for more standard data, and partly by general improvements in the atmospheric models used for weather forecasting. Much of this progress is attributable to being able to make more model experiments and forecasts at finer scales enabled by continuing increases in computing power. Improved forecasts can be expected as long as the observing system is, at the very least, maintained. Unfortunately, experience has shown that this cannot be taken for granted, despite the obvious benefits it enables.

8.2 Seasonal-to-interannual climate prediction

As we saw with weather prediction, the components of a prediction *system* are: observations, assimilation, analysis, initialization, forecast by model, and validation. The forecast model can range from simple statistical forecasting to the most complex coupled atmosphere–ocean forecast systems.

There were various statistical forecasts of the evolution of ENSO before 1986 (see e.g. Sarachik, 1990) with the first forecast using dynamical coupled atmosphere–ocean models made by Cane, Zebiak and Dolan (1986), forecasting the onset of the 1986–7 warm phase of ENSO from initial prediction time in the (northern) spring of 1986. The model used was the Zebiak–Cane model and, in the absence of ocean data in the tropical Pacific, the model was initialized by using the Florida State University (Legler and O'Brien, 1988) winds to force the ocean component of the model up to the initial time. The prediction proved to be correct and the era of ENSO prediction and short-range climate prediction was launched.

8.2.1 General concepts

There is no climate observing system, so the observations taken for weather prediction, oceanography, agriculture, hydrology, etc., must form the observational base of climate prediction. A climate observing system would satisfy the principles of climate observations and would be adequate to form a model-based analysis of the coupled atmosphere–ocean–cryosphere–land system. In the absence of such a climate observing system, and in the absence of ongoing analyses of the climate system, compromises must be made.

A measurement, once taken, is fixed in time and can never be retaken at precisely the same time. Some atmospheric measurements are taken and recorded, but for one reason or another, do not make it to the GTS in time. As these data taken at previous times are recovered, a new analysis of the atmosphere at these previous times can be performed by redoing the weather forecasting procedure at these previous times. Indeed, this can be done over the entire record using a single (best available) forecasting procedure and the record of stored and recovered data. If this is done with the best current models and data-assimilation techniques, then the best possible series of analyses of the atmosphere from the beginning of global observations to the present time can be obtained. This ongoing process is called re-analysis. It should be clear that re-analysis can never overcome inadequacies in the original measurements: it can, however, both correct for inadequacies in the original model (which may then possibly ameliorate problems with the original measurements) and can also use recovered observations which were not part of the original analyses.

One can conceive of a similar procedure for climate: data in the atmosphere, ocean, land and cryosphere is assimilated into a comprehensive coupled climate model (using the model predictions to the initial time as a first guess) to perform a comprehensive model-based analysis of the climate system. This analysis of the climate system would then form the basis for the initial conditions for the climate forecast. It would also be the optimal way to extend the observational climate record, since data in each part of the system would have some influence on the other and the climate analysis would be the best possible estimate of the state of the entire climate system. This procedure is presently not yet done.

An important question in all forecasting procedures is how to score the forecast: i.e. what constitutes "skill." Clearly, some comparison between the forecast at time t_n and the analysis of the observations at time t_n must form the basis of skill. The simplest possible measure of skill is the correlation of an index of a predicted quantity with its measured value and the root-mean-square (rms) difference of the amplitude of the predicted quantity with its measured value, both averaged over a long series of predictions. (Note that correlation alone measures the coincidence of phase without regard to amplitude, and by itself is not a good measure of skill.) "Persistence" is the correlation and rms error of the observed quantity as it evolves compared with its initial value, thus indicating how well the initial value of a quantity predicts the future evolution of the same quantity, again averaged over many realizations of the initial value. There is no general agreement about what

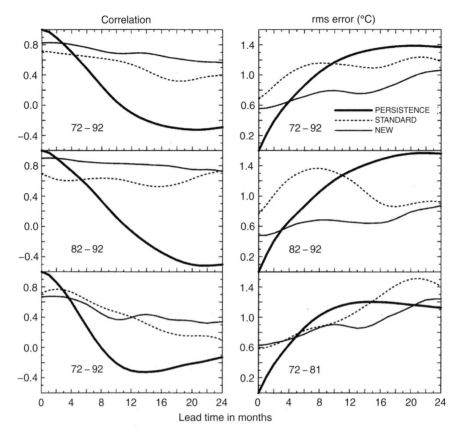

Figure 8.1. Skill scores for NINO 3 index for predictions using the Cane–Zebiak model. Left column: Correlation of predicted and observed NINO 3 index over the time interval noted. Right column: Root-mean-square difference between the predicted and observed NINO 3 index over the time interval noted. (From Chen *et al.*, 1995.)

level of skill is useful. It does seem clear, however, that at a minimum, skill that does not exceed the skill of persistence adds no value and, therefore, is useless.

Figure 8.1 shows an example of the skill so defined from the prediction system using the ZC model which, in the absence of long records of ocean data, is initialized by running the ocean model with the Florida State University winds to spin up to the initial state of the ocean. The Figure shows the correlation of predicted NINO 3 with observations, as a function of months of prediction (Left panel) and the growth of error of the NINO 3, again as a function of months of prediction (Right panel). Also shown is the persistence (heavy line), the forecasts by the original Cane–Zebiak scheme (as described in Cane *et al.*, 1986 – dotted line), and forecasts using a newer data-assimilation procedure that improves the initial state of the coupled model (Chen *et al.*, 1995). Note that in this prediction scheme, the predictions did not beat persistence at lead times less than 4 months. The source of this problem is the initial error, which is large for the original prediction system and is improved by the new adjustment procedure. Note also that prediction skill varies considerably from decade to decade (also see a longer set of forecasts in Chen *et al.*, 2004).

An important concept for all types of prediction is that of the probability distribution function of future outcomes. While it is certainly true that the state of the climate system as it evolves is unique, the prediction of the future state is necessarily imperfect. It helps to think about a *range* of future outcomes, expressed as the probability of each of the possible outcomes. Thus, even if the forecast is relatively sure about a certain outcome (say a winter in the upper ten percent of warmth) there may still be certain probability that the opposite of the outcome will occur (say a cold winter). The more certain the forecast, the more sharply peaked is the distribution about the forecast value. The less certain the forecast, the flatter is the probability distribution of outcomes. The basic problem of prediction therefore becomes the determination of the best method of determining the future probability distribution of outcomes. This is usually done by performing an *ensemble* of individual predictions that span the range of possible outcomes.

The reliability of the probability distribution can be tested in a hindcast mode – over many past years of data; ensemble retrospective forecasts are made and the predicted distribution of outcomes is compared with the actual distribution of outcomes over the entire record.

Predicting a probability distribution of outcomes complicates the scoring of the skill of prediction. At a very minimum, the actual future outcome should lie some-where within the predicted probability distribution. It should also be clear that while the skill can be determined objectively, there is no objective measure of the *usefulness* of a given level of skill – any information about the future is better than no information about the future and should be useful to someone who can take advantage of whatever skill is present.

As a final point, which we will return to in discussing the applications of predictions (Chapter 10), it will prove useful to a user of future climate information to know that a range of future outcomes is possible and that the user should act on climate information judiciously, keeping in mind that sometimes the opposite of what is most likely might just possibly occur. Since each possible future climate outcome implies an impact, a range of impacts, perhaps some beneficial, some malign, is implied. This judicious treatment of the range of future probabilities of impacts is the essence of climate risk management.

8.2.2 *One-tiered and two-tiered short-range climate prediction*

Ideally, the steps for short-term climate prediction are similar to those for weather prediction:

(a) Data is gathered in the atmosphere and ocean and at the land and ice surface, and assimilated into a coupled climate model.
(b) The data is combined with the forecast for the initial time (the so-called first guess) and an analysis of the whole climate system is made.
(c) This analysis, plus whatever practical adjustments need to be made, form the initial state of the forecasts. A number of possible perturbed initial conditions are produced for the construction of forecast ensembles.
(d) The coupled model is run into the future for each of these initial conditions.
(e) At each forecast time, the forecast is compared to the analysis at that time and statistics of skill are gathered.
(f) The cycle is continually repeated.

Because the climate evolves so slowly that it would be impractical to determine skill in real time, an additional step is needed:

(g) A series of retrospective forecasts is performed using the longest possible series of past analyses (or re-analyses) used both for initialization and for scoring (a retrospective forecast is one performed and scored on past data). Using this long series of retrospective forecasts, the overall skill of the forecast system is determined, the regional and seasonal stratification of skill can be assessed, and any systematic biases can be determined. Using the knowledge of biases obtained from the retrospective forecasts, forecasts can be corrected (so-called post-processing).

Because of the lack of data in the oceans and the huge amount of computer time it takes to run coupled models, compromises are often made. The original Cane–Zebiak model is itself a compromise: it predicts the SST anomalies in the tropical Pacific at modest cost in computer time, but it does not predict the effects around the globe since its active domain encompasses *only* the tropical Pacific. Subsurface data up till now has only been available in the tropical Pacific (see Chapter 2), so another common compromise is to allow the ocean to be an active participant only in the tropical Pacific and specify the ocean SSTs at their climatology or observed values elsewhere. Yet another compromise is to calculate the SST anomalies in the tropical Pacific from a simplified system (such as the Cane–Zebiak model or some other intermediate model) and use the resulting forecast SST distribution as boundary conditions for a relatively high-resolution atmospheric general-circulation model to determine the effect of the forecast SST anomalies on the global atmosphere. Calculating the SST anomalies with a coupled model and using the results as boundary conditions for a different higher-resolution global atmospheric model is called "two-tiered forecasting." Initializing the entire ocean and then performing the forecasts with a global coupled model is called "one-tiered forecasting."

The prime advantage of one-tiered forecasting is that the model climate evolves consistently throughout the model globe. The disadvantages are: first, that coupled models are expensive to run, and second, that there is no advantage to initializing the entire ocean if the data to do this are missing or otherwise inadequate in major parts of the ocean.

The prime advantage of two-tiered forecasting is that global atmospheric models are less expensive to run than fully coupled models. The disadvantage is that the coupled model used to generate the SST boundary conditions for the atmospheric model is usually regional and some other method must be used to generate SSTs elsewhere on the globe. This can be done by persistence or by some statistical method, but in any case, need not be fully consistent with the SSTs that would be generated by a global coupled model.

8.2.3 Ensemble prediction and probability distributions

In both one- and two-tiered forecasting, ensembles of forecasts are performed to give an idea of the probability distribution of future outcomes. In one-tiered forecasting, an ensemble of initialized ocean conditions is coupled to the atmosphere, where each initial condition defines one member of an ensemble of forecasts. In two-tiered forecasting, an ensemble of SST anomalies is generated from the coupled model and, for each of these SSTs, a number of atmospheric initial conditions consistent with each boundary condition are generated. These additional members of an ensemble are generated for the second-tier atmospheric model by using slightly different atmospheric initial conditions, each consistent with each SST boundary condition.

8.2.4 Multi-model ensembles

In a carefully controlled series of retrospective forecasts using common boundary conditions for two-tiered forecasting, the PROVOST project (PROVOST, 2000) found that different models gave different probability distribution functions of future outcomes. Since none of the models could be dismissed as clearly worst (i.e. each was best in some places at some times) only some combination of the different models would more correctly approach the true range of probability of outcomes. Further, the combination of models gives a better mean forecast. In a detailed follow-on program (Palmer *et al.*, 2004; DEMETER, 2005) using only coupled models, the detailed justifications, both empirical and theoretical, are given for multi-model ensembles (Hagedorn *et al.*, 2005; Doblas-Reyes *et al.*, 2005).

Our current understanding is that the best forecasts of mean values and the best possible probability distribution of future outcomes is given by combining ensembles from individual models into larger multi-model ensembles. The more members of the multi-model ensemble, the better the forecast and the more useful the probability distribution function of the forecast.

8.3 The current status of ENSO prediction and short-term climate prediction

There have been many excellent recent reviews of ENSO prediction and short-term climate predictability and prediction, both by single models and by multi-model ensembles (Latif *et al.*, 1998; Goddard *et al.*, 2001; the previously cited PROVOST and DEMETER volumes; Chen and Cane, 2008; E. K. Jin *et al.*, 2008).

A complete set of monthly ENSO forecasts of NINO 3.4 indices from the year 2002 to the present by a number of dynamical and statistical models is archived at the International Research Institute for Climate and Society website (www.iri.columbia.edu). As an example, Figure 8.2 shows summary forecasts of a large number of statistical and dynamical models over the years 2006 and 2007. This is a good period to look at because the ENSO state in the tropical Pacific was warm

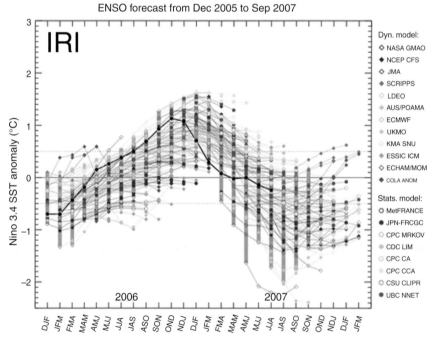

Figure 8.2. Forecasts of NINO 3.4 for a number of statistical and dynamical models for overlapping 3-month intervals from December 2005 to September 2007. The solid line is the observed value. (Downloaded from www.iri.columbia.edu/climate/ENSO/currentinfo/archive/200709/dynamical2.gif)

toward the end of 2006 and cold toward the end of 2007. Note that we will not be concerned with the performance of individual models (these can be found at the IRI website), but we note the following general features that have been cited elsewhere in the literature:

- At present, there are no clear winners among the classes of intermediate coupled models, full global dynamical models or statistical models.
- Every model busts sometimes.
- All the models have trouble with predicting the *amplitude* of warm and cold phases of ENSO.
- Forecasts initialized before the (northern) spring seem to go bad most often.
- Once spring has passed, most models tend to go in the right direction. It should be noted that the phases of ENSO are already developing in spring so that a cursory examination of observed values usually gives a reasonable forecast.
- While it is not clear solely from the time interval shown in Figure 8.2, examination of longer intervals indicates that stronger warm or cold phases of ENSO tend to be better predicted than weaker ones (Goddard and Dilley, 2005).
- The skill does not seem to have improved much since the first dynamical forecast in 1986 (Barnston *et al.*, 1999).
- Intraseasonal variability is not initialized and could account for the poor skill in forecasting the amplitude of ENSO since, as we have seen, intraseasonal wind variability at the surface can greatly enhance the effects of the pre-existing surface winds (Section 2.6). Since forcing from propagating intraseasonal thermal sources could force the midlatitudes some 2 weeks later, a possible source of long-range weather variability in midlatitudes is currently being neglected (e.g. Vecchi and Bond, 2004).
- As is clear from Figure 8.1, and from the experience of other forecasters, skill varies decadally for reasons that are at present poorly understood.

The skill of short-range climate prediction depends entirely on whether or not the region of interest is in the tropics or midlatitudes. In the tropical Pacific, accurate prediction of SST anomalies should enable accurate prediction of temperature and precipitation over specific regions: the west coast of equatorial South America, the Pacific Islands, the maritime continent, parts of Australia etc. For example, during warm phases of ENSO, the region of persistent precipitation expands into the central Pacific and away from the western Pacific. Upon a forecast of a warm phase of ENSO, it is therefore a good call that the maritime continent would be dry. Conversely, upon a forecast of a cold phase of ENSO, it is a good call that the maritime continent is rainy. Because there are considerable spatial variations over the maritime continent, more precise regional projections can only be made by more highly resolved models (e.g. Qian, 2008) used in a downscaled mode (i.e. the SST is predicted, a global atmospheric model is run in two-tiered mode, and using the output of the global atmospheric model as boundary conditions, a very high-resolution regional model is run). Within the tropical land areas, but outside the tropical Pacific, the area of dry conditions tends to be larger during warm phases of ENSO and the area of wet conditions tends to be larger during cold phases of ENSO (Lyon and Barnston, 2005).

8.4 Improvements to ENSO and short-term climate prediction

We can summarize the future road to progress in terms of improvements to each aspect of the forecast procedure: theoretical understanding, observations, assimilation procedures and models.

There are three basic theoretical issues that are presently unresolved and whose resolution would contribute greatly to our ability to improve the prediction of tropical SST and its effects. The first is the issue of the ultimate limit of predictability of tropical SST. If we are currently at the limit of predictability, then it would save a great deal of time and effort to know this, since the predictability of tropical SST could be improved only in accuracy, not in range. The limits depend on the mechanisms for ENSO: if ENSO is due to stable atmosphere–ocean interactions, then its predictability time could not be much more than the disturbance growth time, something of order of a year (Thompson and Battisti, 2001). If ENSO is due to unstable atmosphere–ocean interactions, then the fundamental limitation would be the length of time that motions on the thermocline retain their integrity (the so-called memory of the ocean) and the initialization of the thermocline could foreshadow the evolution of the system for much longer than a year. Chen *et al.* (2004) show that the large events are predictable more than 1 year ahead. Figure 8.1 indicates that for some periods, the predictability of ENSO is considerably greater than a year, while for some periods it is not.

This leads to the second theoretical issue: the decadal modulation of SST in the tropical Pacific and hence of all aspects of global climate influenced by SST. Since we know that these changes do impact climate, the prediction of ENSO alone is not a sufficient goal. Decadal variability is an area of active research as we write this, and at this time we cannot say whether the decadal variations we observe in the tropical Pacific are controlled by the same processes active for interannual ENSO variations, whether they are caused by something quite different or whether a mixture of the two is involved. In recent – and future – decades we would expect global warming to be involved, but just how is not yet clear (see Chapter 9). In either case, the timescales of global warming are intertwined with the timescales of decadal variability and we can no longer assume a stationary climatology, which calls the definition of anomalies into question. The point can be well illustrated by Figure 8.3, which compares anomalies from the mean of the record (Figure 8.3a) with anomalies from a decadal running mean whose origin, as we have pointed out, is still undeciphered. The continuing

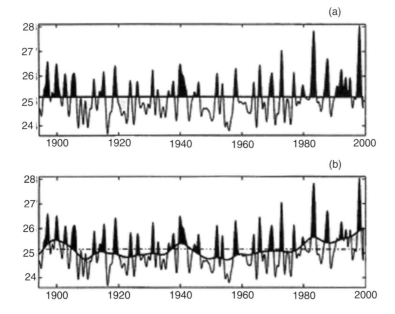

Figure 8.3. SST anomalies over 5°S to 5°N, 80°W to 120°W in the tropical Pacific. (a) Relative to the average over the entire record and (b) Relative to a decadal running mean of the record. (From Fedorov and Philander, 2001.)

warm phase of ENSO of the early 1990s, for example, so evident when taken with respect to a stationary mean, looks insignificant when taken with respect to the decadal running mean. It is possible that the decadal variability of ENSO predictability is due partly to this confusion of anomalies, but it is also probable that it is dependent on the specific nature of the mechanism for the decadal modulation of ENSO, which remains unknown.

The third theoretical issue is the nature of the propagation of signals from the tropics to the midlatitudes in the atmosphere. Even if we know the SST, and therefore the monthly distribution of heat sources in the tropics, this does not mean that we perfectly know the effects of these thermal sources on the higher latitudes. First, because these effects might be lost in the large natural variability of the midlatitude systems, and second, because the propagation of signals from tropics to midlatitudes through vertically and horizontally sheared winds is but imperfectly understood (e.g. Hoerling and Kumar, 2002; Liu and Alexander, 2007).

The observational situation is equally fraught. To the extent that the limitation on forecasting is due to our lack of understanding of decadal modulation, the ENSO observing system (Chapter 2) is inadequate. It has been in place only since 1995 and therefore cannot yet produce the long records needed to define the subsurface dynamics of decadal variability and, because it was designed to observe the waveguide (within 8° of the equator), it is not meridionally extended enough to define the observed meridional structure of the decadal signal in the tropics (Figure 2.33). On the other hand, to the extent that the basic inadequacy is the lack of a well-defined intraseasonal signal, the current weather-observing system in the tropics is weak on defining the structure, phase and amplitude of the synoptic intraseasonal oscillation.

The theory and practice of data assimilation in weather prediction is developing rapidly (Kalnay, 2003) but in a climatic context it is far behind. A coupled analysis of the climate system, consisting of data assimilation into a coupled climate model, is the only way to ensure that the data from the subsystems of the climate system (atmosphere, ocean, land, ice and snow etc.) are mutually consistent. A coupled analysis of the climate system is the consistent way to initialize coupled forecasts.

Finally, we have seen that the coupled models have specific biases (e.g. Figure 7.20 and Figure 7.21) and that these biases necessarily cause the coupled model to drift to the wrong climate state even if the forecasts are properly initialized. While the current causes of these biases are not known and are presumably subtle in origin (or else one of the coupled modeling groups would surely have already corrected them), the need to have the tropical forcings in the right place at the right time is a necessary condition for getting the global interannual variability correct. Thus, while coupled models are being run at higher resolution and with more realistic parameters, until the reason for these ever-present biases is understood, the limits on predictability are more tightly constrained than the ultimate limit on predictability allows.

The best way to improve all aspects of the prediction process is to commit to continuously and uninterruptedly provide the predictions, learn how to use them for specific applications in specific regions of the world, and complain mightily when the prediction skill is inadequate. The continuous confrontation between model predictions and reality and the ongoing attempt to learn to use the forecasts for the benefit of society is the surest way to scientifically learn the problems of the prediction procedures and to build the societal appetite to commit the resources to resolve these problems.

9

ENSO, past and future: ENSO by proxy and ENSO in the tea leaves

In this chapter we review what is known about ENSO as recorded in paleoclimatic proxies, and what is expected for ENSO as we enter a climate state altered by anthropogenic greenhouse gases. In neither case can we confidently construct a reliable picture from instrumental data; in one case we draw inferences from proxies to reconstruct what the climate was, and in the other we rely on imperfect models to foretell the future.

Our knowledge of ENSO in the paleoclimate record has expanded rapidly from the late 1990s. The ENSO cycle is present in all relevant records, going back 130 kyr (kilo-years) to the previous interglacial period (Hughen *et al.*, 1999). It was systematically weaker during the early and middle Holocene (the last 10 000 years), and, as we shall see, model studies indicate that this results from reduced amplification in the late summer and early fall, a consequence of the altered mean climate in response to boreal summer perihelion. Data from corals show substantial decadal and longer variations in the strength of the ENSO cycle within the past 1000 years; it is suggested that this may be due to solar and volcanic variations in solar insolation, amplified by the Bjerknes feedback. There is some evidence that this feedback has operated in the twentieth century.

All of us now anticipate a change in climate brought about by human activity. Among other things, we will have to adjust to a change in the year-to-year variations in climate. Will there be more El Niños, or more powerful El Niños? How will El Niño itself change in a greenhouse world? The short answer, to be expanded upon in Section 9.3, is that the best estimate at this time, which is based on the comprehensive general-circulation models used in the Fourth Assessment Report of the Intergovernmental Panel on Climate Change (IPCC, 2007), is that it will not change much at all, but we have very low confidence in this answer.

Another critical issue is whether the impacts of the ENSO cycle will change. For example, over the past century, the period for which we have instrumental data,

Much of the material in this chapter appeared in Cane (2005).

there is a statistically significant association between poor monsoons in India and El Niño events. This relationship seemingly broke down in the 1990s (Kumar *et al.*, 1999); monsoon rainfall was near normal during the powerful 1997 event. In contrast, during the very strong 1877 El Niño there was severe drought in India leading to widespread famine. Kumar *et al.* (1999) speculated that the change in the monsoon–ENSO relationship might be a consequence of global warming. However, the "normal" association has seemingly returned, as the moderate 2001–2 El Niño was accompanied by a weak monsoon.

9.1 ENSO past

9.1.1 ENSO in the Pliocene

The Pliocene is the geological period traditionally taken between 5.33 Mya (5.33 million years ago) and 1.8 Mya, though it would be better to put its termination with the onset of substantial northern-hemisphere glaciation at 2.73 Mya. What makes it so interesting is that it was the last time the Earth was as warm as it is about to become: ~3 °C warmer than the pre-industrial era. Atmospheric CO_2 levels are thought to have been ~400 ppm. Though there is great uncertainty in this estimate (it might well have been 350 ppm or 450 ppm), we have some confidence that it was something close to values in the year 2000 and less than twice pre-industrial CO_2 concentrations (see references in Haywood *et al.*, 2005; Pearson and Palmer, 2000).

The paleoproxy data for the period is most often interpreted as indicating a "permanent El Niño," although none has the temporal resolution to rule out the possibility of interannual variability with stronger or more frequent El Niño events than in the modern climate. A number of studies, but most importantly Wara *et al.* (2005), indicate that the thermocline in the eastern equatorial Pacific (EEP) was much deeper than in the modern ocean, and the EEP surface temperature was comparable to, and perhaps even higher than, that in the western equatorial Pacific. Molnar and Cane (2002) show that the global pattern of differences from modern climatology resembles the El Niño pattern, particularly the pattern of anomalies that accompanied the 1997–8 event (Molnar and Cane, 2008).

Philander and Federov (2003; see also Federov *et al.*, 2006) advanced the hypothesis that the permanent El Niño state was due to the thermocline being everywhere deeper than in modern times – too deep to allow the colder water within the thermocline to reach the surface, as it does in the modern eastern equatorial Pacific. The onset of major northern-hemisphere glaciations is attributed to the thermocline shallowing, though the reason for this change is not specified. They argue that the change in teleconnections to the high-latitude northern hemisphere with the demise of the permanent El Niño state triggers the growth of glaciers.

There are few coupled-model simulations of the Pliocene. Lunt *et al.* (2008) extended the work of Haywood *et al.* (2005, 2007) using the Hadley Centre coupled climate model (HadCM3) and an ice-sheet model. They conclude that the change from a permanent El Niño has very little effect on the growth of glaciers in Greenland, but that a decrease in atmospheric CO_2 consistent with (the very imprecise) reconstructions of the time history of atmospheric CO_2 is sufficient to account for the onset of glaciation. This model does not produce a permanent El Niño even with CO_2 elevated to 400 ppm. Taken at face value this would rule out the interpretation of the Pacific paleoproxy data as indicating a permanent El Niño, as well as the notion that the change in the tropical Pacific had a role in the onset of glaciation at the end of the Pliocene. However, the verisimilitude of the model is questionable and in some respects it seems at odds with proxy data from the Pliocene; for example, the data assembled by Molnar and Cane (2002) points to a cooler climate in the Gulf of Mexico region.

9.1.2 ENSO in the Holocene

There is good evidence that the ENSO cycle has been a feature of the Earth's climate for at least the past 130 000 years (Tudhope *et al.*, 2001; Hughen *et al.*, 1999). Figure 9.1 shows records from fossil corals collected on the Huon Peninsula in New Guinea, a location in ENSO's "heartland." In general, an oxygen-isotope signal reflects temperature, salinity and global ice volume. After correction for ice volume, in this location it primarily reflects variations in rainfall, which has a much greater range there than temperature. In any case, since greater precipitation and warmer temperatures occur together there, we can take $\delta^{18}O$ as an index of ENSO without troubling to disentangle the temperature and salinity signals. Every record shows oscillations in the 2–7 year band characteristic of ENSO. The records cover only a small fraction of the time since the last interglacial, so the possibility of some period without oscillations or with markedly different oscillations cannot be ruled out. However, there are enough records to be able to say that if there are such periods they cannot be common. We note that an ENSO model (Clement *et al.*, 2001) shows ENSO stopping only twice in the past 500 000 years: during the Younger Dryas, and about 400 kyr earlier when the orbital configuration was most similar to the Younger Dryas.

An earlier study of a laminated core from a lake in Ecuador (Rodbell *et al.*, 1999; also see Moy *et al.*, 2002) was interpreted at first as showing an absence of ENSO in the early and middle Holocene (Rodbell *et al.*, 1999; Federov and Philander, 2000, 2001). In this work the proxy for ENSO is the clastic sediment washed into the lake

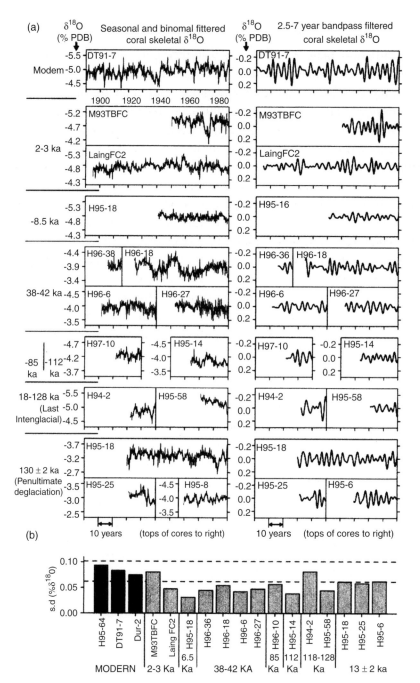

Figure 9.1. Paleo-ENSO variability from fossil corals. (a) Left-hand side: Seasonal resolution (thin lines) and 2.25-year, binomial-filtered (thick lines) skeletal ^{18}O records from fossil corals from the Huon Peninsula, with the record from modern coral DT91-7 shown for comparison. Right-hand side: 2.5–7 year (ENSO) bandpass-filtered coral ^{18}O time series. (b) Standard deviation of the 2.5–7 year (ENSO) bandpass-filtered time series of all modern and fossil corals shown in (A). An asterisk after the coral label indicates that the time series is < 30 years long. The horizontal dashed lines indicate maximum and minimum values of standard deviation for sliding 30-year increments in the modern coral records. (From Tudhope *et al.*, 2001.)

during the heavy rains that occur almost exclusively during El Niño events. This material is lighter in color than the usual lake sediments, so the number of El Niño events may be visually counted. It is more consistent with the Tudhope *et al.* (2001) record to suppose that although the ENSO cycle continues, there were few El Niño events during this period strong enough to wash material into the lake. In this view, ENSO does not start circa 5000 BP, but merely picks up strength. Because ENSO amplitudes can vary so much over a century (e.g. Figure 1.17), the fossil coral records are too short and too few to allow a confident statement that the early and middle Holocene were surely marked by a weakened ENSO cycle. The lake record, however, covers the whole period, and shows a systematic difference between the early–middle Holocene and the last 5000 years. The fossil coral records strongly suggest that the ENSO cycle was also weaker than at present during the glacial era, and of comparable amplitude to the modern during the last interglacial. More records are needed to establish that this description is indeed correct.

It is still not clear what changes in the mean state of the tropical Pacific (SST, SST gradients, upper-ocean temperature structure, location and abundance of rainfall, wind patterns) accompanied these marked changes in ENSO variance through the Holocene. In particular, a continuing middle-Holocene controversy is whether the mean state of the eastern equatorial Pacific was warmer or colder than today. On the basis of warm-water mollusk shells found on the coast of Peru at latitudes where they are not present today, Sandweiss *et al.* (1996) inferred that the mean temperatures were warmer – a persistent El Niño state. This is not consistent with other geological evidence or the proxy temperature record of Koutavas *et al.* (2002). Moreover, if an El Niño-like state prevailed, there should have been more rain at the lake sites in Ecuador. A possible resolution of this apparent discrepancy was offered by Clement *et al.* (2000), who suggested that the warm-water mollusks survive not because of a permanent warm (El Niño) state, but because the cold (La Niña) phase of ENSO was also weaker at this time of reduced variability.

Why was the behavior of ENSO so different in the early and middle Holocene? The likely cause is the difference in the Earth's orbital configuration at that time. Perihelion occurs in January at present, so 11 000 years ago, when it occurred in July, solar radiation was greater during boreal summer. Clement *et al.* (2000) imposed the perturbation heating due to orbital changes over the past 15 000 years on the ENSO model of Zebiak and Cane (1987). Figure 9.2 compares the number of strong warm events in the model simulation with the Ecuadorian lake record of Moy *et al.* (2002). As in the proxy record, the model simulation has a weaker ENSO cycle during the early and middle Holocene. The average period between events is not greatly different, but strong events are rare. An interesting feature of Figure 9.2 is the peak in the number of warm events about 1000 years ago present in both data and model. Clement *et al.* (2000) found that the general shape of this curve – few events

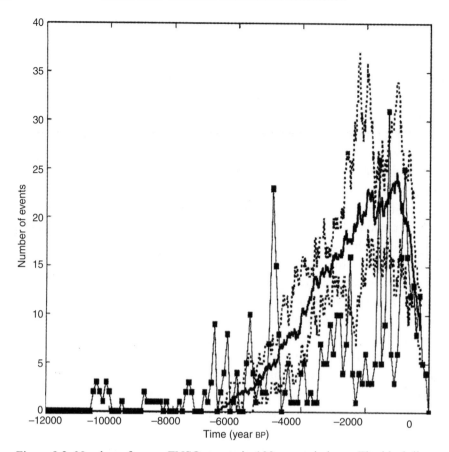

Figure 9.2. Number of warm ENSO events in 100-year windows. The black line with squares represents proxy data from a lake in Ecuador (Moy *et al.*, 2002). Warm ENSO events are defined as light-colored strata in the sediment record, which reflect pluvial episodes during large El Niño (warm) events. The solid line shows the ensemble mean of seven simulations with the Zebiak–Cane model forced by the orbital variations of the last 12 000 years. The dotted lines show the minimum and maximum values over the ensemble. Warm ENSO events are defined in the model as years in which the DJF SST anomaly in the NINO 3 region (5°N–5°S, 90°W–150°W) exceeds 3 °C. This event index corresponds to the middle of the rainy season in coastal South America during which large SST anomalies associated with ENSO events are capable of causing the ITCZ to move equatorward and bring large precipitation anomalies to the region. (From Cane *et al.*, 2006.)

in the early and middle Holocene, rising to a peak at ~1000 BP and then declining – was consistent across an ensemble of runs and concluded that these features were a consequence of the changing orbital forcing.

There is considerable sub-orbital timescale variability in the model runs, and even more in the lake record. While some of the latter is no doubt due to the nature of the "recording device," some is a feature of ENSO. The intermediate model has limited physics, and so has limited ways of generating internal variability. Moreover, external factors, particularly variations in solar radiance and in volcanic aerosol, might be expected to induce variations in the ENSO cycle. Asmerom *et al.* (2007) showed that a proxy record for ENSO derived from a speleothem from a cave in New Mexico was highly correlated with a proxy for solar irradiance. Emile-Geay *et al.* (2007) forced the Zebiak–Cane model with a proxy-derived solar irradiance time series for the Holocene and found a substantial and statistically significant enhancement of centennial-to-millennial scale ENSO variability. If the orbital variations were added to this forcing they found the same general pattern of weak ENSOs in the early and middle Holocene, as in earlier work.

In both the model and real versions of the modern climate, ENSO events amplify through a "growing season" that runs through the boreal summer and into the fall, after which growth ceases and anomalies begin to decay. (Thus El Niño and La Niña events peak around the end of the calendar year, when the rate of change is zero.) The growth is a consequence of the Bjerknes feedback; there is a positive feedback for only part of the year. In the model simulations of the early Holocene, the growth of anomalies ends around August, before the summer is over. This shorter growing season means that anomalies do not reach the peak values of today. The equatorial oceans received about the same annual solar radiation but its seasonal distribution was quite different. Northern hemisphere insolation was stronger in the late summer and fall, so the intertropical convergence zone, which tends to lie over the warmest water, was held in place in the higher tropical latitudes. A key link in the Bjerknes feedback is from SST to enhanced heating to changes in the winds, but the heating is associated with low-level convergence, and if the convergence cannot be moved on to the equator, the link is broken and the ENSO anomalies do not grow.

This analysis is based on a model of intermediate complexity, one that omits mechanisms that might alter the outcome, such as the advection of subsurface temperature anomalies. However, a number of studies with comprehensive coupled general-circulation models have also been shown to have a weak-amplitude ENSO cycle at 9 kyr BP and 6 kyr BP (Liu *et al.*, 2000; Otto-Bliesner *et al.*, 2003). Again, the Bjerknes feedback operates. They also point out that the stronger summer heating creates a stronger Asian monsoon, enhancing the trade winds, which leads to a weaker ENSO cycle.

9.1.3 *ENSO in the Pleistocene*

Thus, orbital changes alter the mean climate, and this in turn changes ENSO behavior markedly. The Tudhope *et al.* (2001) records also suggest that ENSO was weakened by glacial conditions at times when the model, which sees only orbital changes, maintains its strength (Clement *et al.*, 1999). The changes in orbital forcing and from modern to glacial are both strong perturbations, taking us far from modern conditions, so this period is a good test for models. A favorite glacial target for both observationalists and modelers is the last glacial maximum (LGM), the period circa 20 kyr BP when the last glacial epoch was at its coldest. Unfortunately, paleoproxy data do not yet give a clear picture of what the tropical Pacific looked like at this time. It has been suggested that the eastern equatorial Pacific was in an El Niño-like state (Koutavas *et al.*, 2002), but the picture that seems to fit the data best is that the ITCZ was closer to the equator, with the oceanic fronts also shifted to the south (Koutavas and Lynch-Steiglitz, 2005). Viewing the difference from the modern state as a North–South shift appears to be a better fit than the ENSO mold. We cannot say much from data about ENSO variability, and the coupled general-circulation models (CGCMs) give inconsistent results, with the National Center for Atmospheric Research (NCAR) model showing stronger ENSO variability (Otto-Bleisner *et al.*, 2003) and the Hadley Centre model (Hewitt *et al.*, 2001) showing little change.

9.1.4 ENSO in the last millennium: the response to solar and volcanic variations

Studies of ENSO over the last millennium provide examples of shifts in ENSO behavior without strong forcing. There are shifts in ENSO variance with timescales of decades to perhaps centuries, typically associated with changes in mean temperatures in the eastern equatorial Pacific. They could be a consequence of unforced natural variability, but we will make the case here that they are more likely a response to the variations in radiative forcing due to volcanic activity and changes in solar output. These forcings are not only much weaker than the orbital changes, but they have far less seasonal and latitudinal structure, so they provide more direct lessons for the greenhouse climate.

Decadal variations in ENSO are intertwined with Pacific-wide decadal variations in SST, sea-level pressure and winds. The Pacific Decadal Oscillation (PDO) has a pattern much like ENSO in the tropical Pacific (see Figure 2.33), but broader; it has its largest amplitude in the midlatitude north Pacific (Mantua *et al.*, 1997; Deser *et al.*, 2004). Recent work has shown that there are decadal variations in the south Pacific, strongly expressed in the movement of the South Pacific convergence zone, and that these are linked to the PDO (Garreaud and Battisti, 1999; Power *et al.*, 1999; Deser *et al.*, 2004). Power *et al.* (1999), noting that "PDO" is usually taken to be centered in the north Pacific, prefer "Interdecadal Pacific Oscillation" (IPO) to emphasize the basin-wide nature of Pacific variability. Having the signal appear in both hemispheres implicates the tropics as a likely source, and some of this work shows a direct connection in the data (see especially Deser *et al.*, 2004). How much of the basin-wide decadal variability is driven from coupled interactions in the tropical Pacific similar to ENSO, and how much is attributable to midlatitude sources is an area of active research. The IPO (or PDO) has been shown to affect the connections between ENSO and rainfall in Australia (Power *et al.*, 1999) and North America (Gershunov and Barnett, 1998). It appears that the total SST perturbation in (at least) the tropical Pacific must be considered to capture global impacts; ENSO alone is insufficient.

It is difficult to reach firm conclusions about decadal variations from the instrumental record, which is only long enough to provide half a dozen or so instances. The principal proxies able to resolve decadal variations are tree rings and isotopic analyses of corals. Both are at annual resolution and also resolve ENSO. The relevant tree rings are primarily proxies for precipitation in places where the influence of ENSO and the IPO is strong. They are thus indirect proxies for ENSO, subject to other large-scale climate influences, as well as the usual local and biological effects. This problem can perhaps be overcome by using multiple sites to extract the signal that corresponds to ENSO or the IPO; see Mann (2002) for a broad discussion. This approach has been used by a number of investigators to construct indices of the IPO going back several centuries, primarily using tree rings (Biondi *et al.*, 2001; D'Arrigo *et al.*, 2001; Gedalov and Smith, 2001; Villalba *et al.*, 2001), but also using both tree-ring and coral data (Evans *et al.*, 2000, 2001; Mann *et al.*, 2000) and corals alone (Evans *et al.*, 2002).

The corals from a given site used in these reconstructions have been from a single coral head, allowing records of a few hundred years or less. Cobb *et al.* (2003) overlapped shorter segments of fossil coral in a manner similar to the way in which tree-ring time series have been spliced together from individual trees. The result is displayed in Figure 9.3. Palmyra (6°N, 162°W) is in a prime location to provide an ENSO proxy, and Cobb *et al.*'s $\delta^{18}O$ record from modern corals correlates with the NINO 3.4 SST at $r = -0.84$ in the ENSO band, a value as high or higher than the correlation of any two commonly used ENSO indices (e.g. SOI, NINO 3, NINO 3.4) with each other. It is likely that the $\delta^{18}O$ signal primarily reflects rainfall and so correlates better with NINO 3.4 (and NINO 3) than with local SST (see Evans *et al.*, 2002).

Figure 9.3 also displays the results of a 100-member ensemble calculated by forcing the Zebiak–Cane model with a slightly updated version of the Crowley (2000) solar and volcanic forcing (Mann *et al.*, 2004). Given the ~ 1 °C variance in a single-model run, one would expect the variance of the mean of a 100-member ensemble to be ~ 0.1 °C if the variance in each member is independent, but if some of the variability is forced, then the variance of the mean will be higher. Regardless of whether it is noise-driven or a consequence of chaos, the variability of ENSO

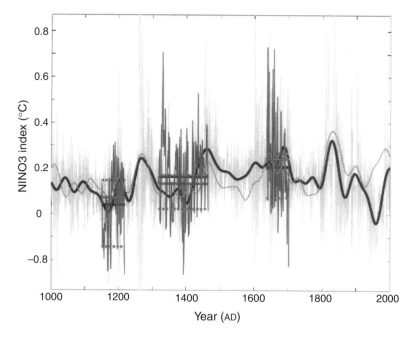

Figure 9.3. The annual mean NINO3 response of the Zebiak–Cane model to the combined volcanic and solar radiative forcing is compared with reconstructions of ENSO behavior from Palmyra coral-oxygen isotopes. The model is run over the interval AD 1000–1999; the coral reconstruction, shown as darker gray curves, is available only for the four intervals shown. The continuous, faint gray curve is the annual mean model NINO3 anomaly (in °C relative to the AD 1950–1980 reference period) averaged over a 100-member ensemble. Despite the averaging, considerable variability remains, largely due to the influence of volcanic eruptions. The heavy black line shows 40-year smoothed values of model NINO3. The coral data (darker gray curves) are scaled so that the mean agrees with the model (see Mann *et al.*, 2004, for details). Thick gray lines indicate averages of the scaled coral data for the three available time segments; the thick black lines are the ensemble-mean averages from the model for the corresponding time intervals. The associated inter-fourth quartile range for the model means (the interval within which the mean lies for 50% of the model realizations) is also shown (dashed gray lines). The ensemble mean is not at the center of this range, due to the skewed nature of the underlying distribution of the model NINO3 series. (After Mann *et al.*, 2004.)

makes it impossible that even a perfect model would agree in detail with the single realization present in the observational record. If the ENSO variability is forced, then it is possible for values averaged over a number of ENSO events to agree. Indeed, Figure 9.3 shows, for both model and data, cold SSTs in the mean in the late twelfth–early thirteenth centuries, moderate SSTs in the fourteenth–early fifteenth centuries, and warm SSTs in the late seventeenth century. In all three cases the means of the observations and the model ensemble are consistent within the ensemble sampling distribution (dashed lines in Figure 9.3). Moreover, the late-seventeenth-century warmth and the twelfth–thirteenth-century cold are well separated within the distribution of states from the model ensemble runs: one would expect the later period to be warmer than the earlier one in roughly seven out of every eight realizations. If these statistics carry over to reality, we would expect nature's single realization to be warmer in the later period, with close to a 90% probability. In both data and model there is also a systematic difference in the strength of the ENSO cycle in the two periods. There are numerous large El Niño events in the late seventeenth century and very few in the twelfth to early-thirteenth-century period. (This difference is statistically significant at the 0.1 level for both model and data.) Thus, both data and model show that for the last millennium more (less) ENSO variability goes with a warmer (colder) mean SST in the eastern equatorial Pacific.

The differences – in the model run, at least – are a consequence of Bjerknes' mechanism (Clement *et al.*, 1996). The result is, at first, counterintuitive: the warmer tropical Pacific temperatures occur at a time of increased volcanic activity and global cooling (Crowley, 2000; Jones *et al.*, 2001) and vice versa. If there is a cooling over the entire tropics then the Pacific will change more in the west than in the east because the strong upwelling in the east holds the temperature closer to the pre-existing value. Hence the East–West temperature gradient will weaken, so the winds will slacken, so the temperature gradient will decrease further – the Bjerknes feedback, leading to a more El Niño-like state. This chain of physical reasoning is correct as far as it goes, and the agreement between the data and the simulation with the simplified Zebiak–Cane model is evidence for the idea that the Bjerknes feedback holds sway in response to a change in radiation forcing. But the climate system is complex and processes not considered in this argument, such as cloud feedbacks, might be controlling.

9.2 ENSO in the twentieth century

Before turning to model projections of the future, we briefly consider what can be learned from the changes since the rise of CO_2 began in earnest in the late nineteenth century. Trenberth and Hoar (1997), noting that greenhouse-gas concentrations rose sharply in the past few decades, argued that the increase in the frequency and amplitude of ENSO events in the 1980s and 1990s was highly unusual, significantly different from the behavior in the preceding century, and thus attributable to anthropogenic causes. Rajagopalan *et al.* (1997) used a different statistical model to formulate their null hypothesis and concluded that the behavior was not significantly different from that in the earlier part of the instrumental record (also see Wunsch, 1999). The arguments are technical and inconclusive; the reader is invited to compare the last quarter of the twentieth century with the last quarter of the nineteenth century in Figure 1.17 and decide if the level of ENSO activity in the two eras is strikingly different. By some measures the 1877 El Niño was more powerful than any of the events in the twentieth century. Record drought in India, as well as severe droughts in Ethiopia, China, north-east Brazil and elsewhere, all contributed to what is fairly described as a global holocaust (Davis, 2001).

We noted that the data of Cobb *et al.* (2003) showed cooling in the eastern equatorial Pacific at times in the past when the global climate warmed due to increased solar radiation or reduced volcanism, a result reproduced in the modeling study of Mann *et al.* (2004) and explained by the Bjerknes feedback. However, this same relation does not seem to hold for the twentieth century, when radiative forcing and global temperatures increase. (Crowley [2000] found the greatest disagreement between global mean temperature and a model forced by solar, volcanic and greenhouse-gas variations in the early twentieth century.) Perhaps this change in behavior is due to the impact of atmospheric aerosol or perhaps there is something missed in our argument when the radiative increase is due to increased greenhouse gases. Another possibility is suggested by the results of Cane *et al.* (1997), whose plots of temperature trends from 1900 to 1991 are updated to 2000 in Figure 9.4.

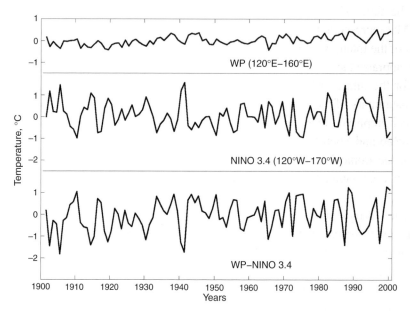

Figure 9.4. Time series of: (Top) The average SST anomaly in the WP region (120°E to 160°E; 5°N to 5°S); (Middle) Average SST anomaly in the NINO 3.4 region (120°W to 170°W; 5°N to 5°S); (Bottom) The difference WP-NINO 3.4, a measure of the zonal SST gradient. The least-squares estimate of the linear trends in the three time series (°C per century) are 0.41±0.06, –0.08±0.25 and 0.50±0.25, respectively. (Updated from Cane *et al.*, 1997: from Cane, 2005.)

The trend in the eastern equatorial Pacific is not significantly different from zero, but the East–West SST gradient does become significantly stronger over the century – as would be expected from the Bjerknes feedback. Complicating the picture, Vecchi *et al.* (2006) show that the sea-level pressure (SLP) gradient – the SOI – did not change, or perhaps weakened. Thus, the change in SST and the change in SLP are not consistent with Bjerknes' view of how the two should behave. At the time of writing this discrepancy is not resolved and it reduces our confidence in our ability to foretell ENSO's future behavior.

9.3 ENSO in the future

If we are to trust a model to predict ENSO in the greenhouse world, it is necessary, though perhaps not sufficient, that it reproduces the changes in prior centuries. In addition to simulating changes, it greatly increases our confidence if the model can simulate the defining features of the present ENSO cycle with some skill. Is the mean frequency close to 4 years? Is the largest warm anomaly where it is observed in the eastern equatorial Pacific? Does the model's cold tongue extend too far to the west, into the warm-pool region?

At the time of writing, the comprehensive coupled general-circulation models (CGCMs) representing the "state-of-the-art" are those used in the Fourth Assessment Report of the Intergovernmental Panel on Climate Change (IPCC, 2007). Many recent papers appraise the quality of these model ENSOs and attempt to assess their predictions for the future of ENSO; the paper by Guilyardi *et al.* (2009) is something of a meta-summary of this work, with many references. While the AR4 models' ENSOs are much improved over those of the previous generation of CGMs (see especially AchutaRao and Sperber, 2006), errors remain in their simulations of both the mean state of the tropical Pacific (e.g. Figure 7.23), the annual cycle (Figure 7.24), and ENSO-like interannual variability (Figure 7.25). The simulated rainfall patterns have not improved, and AchutaRao and Sperber (2006) find that "the quality of the El Niño composite precipitation rate anomalies is directly proportional to the quality of the boreal winter tropical precipitation rate." They also find that ENSO teleconnections to North America are better when the tropical rainfall pattern is improved.

In virtually all models, the equatorial cold tongue extends too far to the west, and the pattern of interannual variability follows suit. This means that heating anomalies driving teleconnections may be in the wrong place, and that aspects of ENSO physics, such as zonal advection of temperature, may be exaggerated (cf. Guilyardi, 2006; Capatondi *et al.*, 2006). The wind-stress response to eastern equatorial Pacific SST anomalies in the models is narrower and displaced further west than observed. Capatondi *et al.* (2006) note that the meridional scale of the wind stress can affect the amount of warm water involved in the recharge/discharge of the equatorial thermocline, while the longitudinal location of the wind stress anomalies can influence the advection of the mean zonal temperature gradient by the anomalous zonal currents, a process that may favor the growth and longer duration of ENSO events when the wind stress perturbations are displaced eastwards. Thus, both discrepancies of the wind stress anomaly patterns in the coupled models with respect to observations (narrow meridional extent, and westward displacement along the equator) may be responsible for the ENSO timescale being shorter in the models than in observations.

We note here that the same consequences follow from the linear periodic theory developed in Chapter 6, even without considering zonal advection (see Cane *et al.*, 1991).

The studies of the AR4 models typically involve about 20 different models and, using somewhat different measures, generally conclude that five or six of them produce reasonably good simulations of ENSO. These models have a peak in variance somewhere in the observed ENSO band of 2–7 years, are irregular, and have an amplitude not too far from the observed. They extend too far to the west, but do have the largest SST anomalies in the east, if not always as far east as they should be.

In most of the AR4 model runs for the twenty-first century the SST is projected to warm more on the equator than off it. Some models show a strengthened East–West gradient, but most show it weakening. If only the six or so "best" models are considered, then the change from present conditions is very small (van Oldenborgh *et al.*, 2005). Models do not agree on what will happen to the amplitude of ENSO; some decrease it, some increase it and some stay the same. On average it does not change much at all. The period between ENSO events decreases slightly in most models, a change that Merryfield (2006) attributes to the ($\sim 5\%$) increase in baroclinic mode wave speed that follows from the increased stratification in the ocean associated with a greater warming near the surface than at depth. He notes that this change is consistent with delayed-oscillator theory.

In summary, the models have not converged on a projection for the future of ENSO and the tropical Pacific. On average, they suggest that the change from present behavior will not be very great, but the spread among the models is too wide to be confident about what the future holds in store for the tropical Pacific and all that it influences.

9.4 Conclusions

Glendower: I can summon spirits from the vasty deep.
Hotspur: Why, so can I, or so can any man; but will they come when you do call for them?
(Henry IV, Part 1, Act 3 Scene 1)

ENSO variations impact climate world-wide because the changes in the heating of the tropical atmosphere they create alter the global atmospheric circulation. Changes in the mean state of the tropical Pacific would have similar impacts. Since societies and ecosystems are profoundly affected, we would like to know how ENSO and the mean state of the tropical Pacific will change in our greenhouse future. We must rely on models to make such predictions, since the past does not provide a true analog of the new climate we are creating. Our comprehensive coupled general-circulation models are impressive achievements, now able to simulate many features of the climate with striking verisimilitude. The ENSO cycle, however, is not their forte. Present attempts to summon the ENSO of the future bring forth a motley and uncertain set of responses. The paleoclimate record shows us that ENSO behavior is quite sensitive to climatological conditions, so it stands to reason that ENSO will behave differently in the future. But we cannot say how it will differ with any confidence. Indeed, the models' consensus estimate is that it will not change much at all.

There are reasons for optimism. The quality of ENSO simulations has improved dramatically in the past decade, and further progress is likely if computing power grows adequately. The paleoclimate record, almost devoid of information about ENSO only a decade ago, is expanding rapidly and even now provides enough information to test models under conditions substantially different from modern ones. Thus, there is hope that we can soon increase our confidence in forecasts of future variability. But at present the future of ENSO lies in depths of vast uncertainty, beyond our summons.

10

Using ENSO information

The problem of using ENSO forecasts is not at all straightforward. The basic difficulty arises from the fact that forecast information is probabilistic – our knowledge of the future is given imperfectly and we must learn to use this imperfect knowledge in an intelligent manner, especially when the skill is not high.

To illustrate the problem, we begin with an (admittedly fanciful) analogy. Suppose a stranger whispers in your ear that he is offering you a rare and unusual gift: a coin that looks and feels like every other coin of its type but will fall heads 55% of the time. The coin is yours to keep but it is up to you to find out how to make use of this gift.

The first problem is to find out if the stranger is telling the truth. So you flip the coin and it shows tails. This, of course, does not indicate that the stranger's words are fraudulent: one must flip the coin a very large number of times. So you flip the coin 100 times and 53 times it shows heads and 47 times it shows tails. This is promising, but it still does not prove that the coin is what the stranger said it is. So you flip the coin 1000 times and it falls heads 552 times and tails 448 times. Now it seems to be true that the stranger has told the truth – the coin is indeed a 55% heads coin. The more you flip the coin, the closer it comes to 55% heads.

How to use this coin? Clearly you would not go to a casino and bet a million dollars on heads: while the chance of winning this million dollars is slightly improved, the chance of losing a million dollars is 45% and this would be a disaster – to lose this amount of money you would be in debt for life. So you decide to bet a dollar at a time but do this for a large number of times. At the end of a day in which 2000 coin flips are made, i.e. 2000 one-dollar bets, you are likely to have won 1100 times and lost 900 times for a net gain of 200 dollars. While this seems highly inefficient, and progress seems abnormally slow, after a year of coin tossing, you have won $73 000 and this is enough to live on. The stranger has indeed given you a valuable gift: a living income for life. (The reader is urged to look at Lewis [1997] for a truly informative and entertaining guide to the unexpectedly relevant science of coin tossing.)

We see that a small advantage in probabilities can be used to advantage, not to get rich, but to get by. We know very little about the future in this example – only that the probability of a head is slightly higher than tails. The use of any prediction system that is probabilistic rather than deterministic has some of the same properties: it takes a large number of events to gain some idea of the true probability of occurrence and it takes a large number of uses of the probabilistic prediction to learn how to use the information to see if its use is worthwhile. Clearly, the higher the probability of a positive outcome, the more we know about the future, the faster one can determine the usefulness of the predictions, and the faster one can decide that its use is beneficial. The use itself must respect the probability of occurrence in order to be beneficial: this generally means that the gains are moderate and in proportion to the skill of the forecast. Any skill at all should lead to the possibility of beneficial use over a long enough time. Another way of saying this is that any information about the future, no matter how small, should be useful if properly approached. This approach is the problem addressed in this chapter.

This chapter will deal with using ENSO information. Physical science determines the nature of the information available, but social conditions determine whether and how the information is used. Attitudes, organization, participation and communication are crucial factors in determining the use of ENSO information. We start with some general considerations, move on to the use of past and present information, and give a general framework for thinking about the use of forecasts. Some extant and possible future applications are identified and discussed, and the general framework is exploited to reveal a general approach to overcoming the barriers to the use of ENSO information.

10.1 General considerations

There are three basic kinds of ENSO information: past ENSO information gathered by historical records and paleoclimatic proxies, current ENSO information ("nowcasts") diagnosed from the existing observational network, and future ENSO information, in the form of forecasts with lead times of 3 to 12 months. As we will indicate, each of these three types of information is useful to the extent it tells us something about the future. Note that we will not deal with the use of information about the future response of ENSO on decadal and global-warming timescales since information is lacking about both the scientific content of this issue (see Chapter 9) and the use of such information – this is not to say that this type of forecast would not be of great benefit if it were available.

We may characterize the general issue of using ENSO information in terms of six distinct questions which are commonly asked in the literature of knowledge utilization: 1. What is the information to be used? 2. Who will convey the message?

3. What is the medium by which the information is to be conveyed? 4. To whom? 5. For what action? 6. To what effect? We should consider these questions as interlinked, since, for example, the nature of the information required depends on the kind of action contemplated and the choice of those to convey the information may depend on the media available.

Only the first of these questions is the kind of physical science question that has been treated in this book. All the other questions are questions of social organization and policy and can be approached only by social science inquiry. It must be emphasized that the success in using ENSO information depends on successfully implementing these social issues (see National Research Council, 1999) but the detailed resolution of these social questions depends on the scale on which the information is to be used.

As the most direct small-scale example, we are all used to listening to the morning radio or television weather forecast in order to decide what to wear and whether or not to take an umbrella. No particular shaping of the forecasts is made (although the graphical displays on television vary from channel to channel) and the action taken is solely up to the initiative of the listener. The ongoing evaluation of the forecast is personal and informal and this subjective evaluation determines the action the listener decides to take.

When the scale becomes larger, the timescale longer, our experience with this new type of forecast more tenuous, and the stakes greater (perhaps involving large amounts of money, property, natural resources or lives) this passive approach no longer suffices. The user has to be identified, the method of communicating with the user (both to convey information and to determine the needs and situation of the user) has to be perfected, and the user has to be helped in acquiring, understanding and using the information. Ultimately, the sustainability of physical systems to observe, model and forecast the climate will depend on the user(s) being satisfied enough with the effort to insist that these prediction systems, expensive though they are, be maintained.

10.2 Using past ENSO information

If every year were climatically the same, and the seasons repeated exactly, we would, in the course of time, learn what actions would be optimal at each season. Farmers, for example, would learn to plant at an optimal time of year, fertilize at some other optimal time, and harvest at an optimal harvest time.

The fact that each year is different climatically from each other year, i.e. the climate varies interannually, means that there is a basic uncertainty in what is going to occur next year. Farmers then have to build uncertainty into their planning for the next year. They have to anticipate that some years will be very wet and they have to get the seed into the ground early and fertilize copiously to take advantage of this rainfall, while other years will be so dry that they will barely have a crop to harvest. In this latter case, they may want to carry insurance against total crop failure or belong to a cooperative to spread the risk. Knowing that each year is different from each other means either that an unvarying set of actions is designed for each time of the year to cover every contingency (i.e. to make the system resilient), or that adaptive actions are taken in accordance with the anticipated climatic conditions. Making the system resilient to a wide range of contingencies almost always means that many resources are devoted to resilience and the return on investment is necessarily less than if a more targeted approach to each year were possible. Until the advent of the type of short-range climate prediction described in Chapter 8, no information about the next year existed and only the resiliency approach was feasible. The basic function of adaptive response informed by climate information is to target the response only to the most likely contingencies, thereby conserving resources and increasing efficiency.

While it may seem that past climate information has little practical value, there are situations in which knowing how interannually varying climate behaved in the past will limit the basic set of actions to be taken in the future. Past climate information can tell us how different each year is from each other and how probable are longer runs of persistent conditions, e.g. droughts and pluvials.

Past ENSO information is of direct use for those Pacific locations directly under the influence of ENSO. For example, droughts and attendant forest fires obtain in and around Borneo when warm enough phases of ENSO occur for the region of persistent precipitation to move far into the central tropical Pacific and away from the maritime continent. Similarly, wet conditions obtain when cold enough phases of ENSO occur for the region of persistent precipitation to retreat to the region of the maritime continent. In order to be able to plan for these conditions, it becomes important to know the frequency of wet and dry conditions and, therefore, the frequency of very warm and cold phases of ENSO. If these conditions are rare, normal agriculture can take place with reasonable certainty that disruptions will be rare. If climate is slowly changing, so that periods of warm and cold phases of ENSO are becoming more common, the only way of really knowing this is having a past record of ENSO against which to compare.

As another example, streamflow in parts of the western USA depends on winter precipitation, which varies both with the phases of ENSO and with its decadal modulation. This streamflow is used for a variety of applications: transportation, recreation, drinking water, irrigation water, fish recruitment and hydroelectric-power generation. Clearly, if the precipitation was low for a significant amount of time, some of the basic assets that influence life and economy would be at risk. So if a region were drought-prone for a significant fraction of time, alternate mechanisms for acquiring energy and water would be required. Here, both the interannual and decadal modulation of precipitation are important to define the probability of dry conditions. A direct measure of past precipitation in a region can be obtained through combinations of proxy data for precipitation: for example, lake sediments and tree rings.

As a (famous) example of how this kind of past precipitation information can be used, we note that the Colorado Compact, signed in 1922, allocated the water of the Colorado river to various downstream states and to Mexico. The allocation was in absolute amounts: 7.5 million acre-feet of water was granted in perpetuity to each of the upper and lower Colorado basin and an additional 1 million acre-feet for the lower basin (which includes the water-hungry regions of Arizona and California). In addition, a treaty signed in 1944 guaranteed 1 million acre-feet to Mexico, and tribal nations within the Colorado basin have gradually been winning and exercising rights to more and more water from the river. Because the Compact was assumed to be based on robust flows in the Colorado, no provision was made for determining allocations in case of scarcity. The flow of the Colorado is known to be strong when warm phases of ENSO are current in the tropical Pacific, and are also known to be modulated on longer timescales by the Pacific Decadal Oscillation and on still longer-period timescales by longer-period variability in the Atlantic. Long-period droughts in the region are known to coincide with multi-year cold phases of ENSO (Cole *et al.*, 2002).

A glance at Figure 10.1 shows the basic problem with the Colorado Compact: the precipitation on which the Colorado Compact is based was abnormally high for the early part of the twentieth century, which led to high flows in the early part of

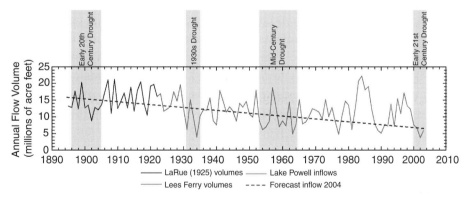

Figure 10.1. Flow at Lees Ferry, Arizona (which divides the upper and lower Colorado basin and therefore includes consumptive use in the upper basin and is essentially the amount of flow available to the lower basin). The dashed line is the overall trend and the shaded regions are periods of sustained droughts. (From USGS, 2004.)

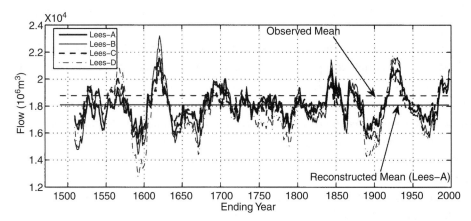

Figure 10.2. A long reconstruction of natural flow at Lees Ferry, Arizona. Different lines are slightly different reconstructions. (From Woodhouse *et al.*, 2006.)

the century – the flows have not been as high since. The Colorado Compact was based on abnormally high flows, which in the absence of long records, were assumed to be normal, and the river was therefore over-allocated – this has led to problems that persist to this day.

If, in 1922, a long-term reconstruction such as the one in Figure 10.2 had been available, it would have been realized that there are long-term variations in the flow and that the early-twentieth-century flows were not typical, and that the flow at the time of the Colorado River Compact was unusually large. The allocations then might have been given as percentages of current flow rather than as absolute amounts, and some scheme for allocation in the face of scarcity might have been written into the Compact.

Past ENSO information has another crucial use. Since warm and cold phases of ENSO recur relatively infrequently, it would take a very long time to develop a prediction system in real time. Obtaining past records of ENSO and its effects allows a series of predictions to be initialized, made and validated in a short amount of time compared to the evolution of the system. It is as if the stranger in Section 10.1 came with a long record of previous coin tosses so that the record could be perused, rather than generated, in order to verify that the coin was slightly biased.

10.3 Using ENSO nowcasts

We now have an observing system in place that allows us to know, in real time, the conditions in the tropical Pacific. No longer will warm and cold phases of ENSO progress without our knowing it. Knowing that there is a warm or cold phase of

ENSO in progress can be useful in a number of ways, basically the same ways that are implied by an ENSO forecast.

Strong warm and cold phases of ENSO become evident by the late summer to fall of the year in which the peak phases occur. In this sense, the evolution of the canonical ENSO (Figure 2.20) contains predictive information about the future evolution of the phases of ENSO, and knowing the current state of the evolution of ENSO allows a few-month prediction of future phases. For example, on knowing that a strong warm phase of ENSO was developing in the fall of 1997 on the basis of real-time observations and forecasts, an El Niño alert was declared by NOAA Climate Prediction Center and, knowing that heavy rains were one of the common consequences of warm phases of ENSO, water managers in California cleared ditches and canals, while Florida water managers lowered water levels in lakes and canals, both to avoid local flooding (Changnon, 2000). Individuals in California chose to fix their roofs sooner (rather than postponing this decision) to avoid water damage. Additional examples of the use of forecasts for the 1997–8 warm phase of ENSO are given in Pielke (2000).

10.4 End-to-end forecasting

The basic idea of applying ENSO forecast information is that the (probabilistic) forecast will eventually lead someone (or some organization) to make a decision that will lead to favorable outcomes, either in terms of money gained, time saved or natural resources preserved. That the requirements of such a process are not new led Hammer (2000) to argue for a systems approach to the decision support implied by ENSO forecasting. Sarachik (1999) proposed a similar approach and called such a system "end-to-end forecasting."

The steps in end-to-end forecasting are:

(a) Making the probabilistic ENSO forecast and making available the forecast and its uncertainty.
(b) Communicating the ENSO forecast.
(c) Elucidating the impacts due to ENSO (along with the impacts due to other low-frequency phenomena: Pacific Decadal Oscillation, SST in other basins, North Atlantic Oscillation etc.) in a region.
(d) Downscaling and shaping the forecast for regional use. Making resource forecasts.
(e) Assessing the uncertainties in the local climate or resource forecast.
(f) Examining the normal decision-making process of the potential user, including the decision calendar, the decision-making structures, the freedom to make decisions etc., in order to shape and target the ENSO forecast.
(g) Negotiating the forecast with potential users to better shape and target the forecasts for their benefit.

(h) Using the forecasts and other information to evaluate the possible outcomes of making a range of decisions. Taking action on this basis.

(i) Evaluating the benefits of the actions taken on the basis of the forecasts.

(j) Refining the entire set of procedures on the basis of the evaluation.

These steps are not to be considered a linear sequence to be performed in a defined order – the list is simply an identification of the steps needed, with some clearly depending on the others. A number of these steps need further discussion.

Probably the single most important factor in facilitating people and organizations to use the forecasts is making them relevant to the local region in which people live, work and plan for the future. This is done in steps (c) to (g) above. ENSO will have large-scale effects on temperature and precipitation in certain regions (as in Figures 1.4a,1.4b) but just identifying this level of impact is not adequate to define the true effects on a locality. The forecast has to be given specificity in three separate and interrelated ways: by downscaling the climate forecast in terms of local variables, by elaborating the effects of the regional climate forecast in terms of the effects on local resources, and by targeting the forecast to the specific needs of the user.

For example, the large-scale forecast might be downscaled to give the local probability of precipitation in a river basin, and a hydrology model of the local-river system might be coupled to this basin probabilistic precipitation forecast to give a probabilistic river-flow forecast. The negotiation in step (g) refers to potential users of river-flow forecasts, say the hydroelectric power industry, recognizing that precipitation forecasts might not be useful for their needs and requiring that flow forecasts, say 6 months in advance, are what they really need, entering into negotiation with the forecasters to agree on a product that is both useful to the user and possible for the provider. Clearly, the necessity for the user to make known specific needs and the forecast provider's wish to meet these needs implies a long period of sustained engagement while these contrasting issues get worked out.

Specific examples of resource forecasts abound in the literature. Cane *et al.* (1994) noted that maize yields in Zimbabwe correlated highly with the SST characteristic of the changing phases of ENSO (alternately with the Southern Oscillation index) so that if this high correlation is maintained, a skillful forecast of ENSO implies a skillful forecast of maize yield. Shaman *et al.* (2003) were able to use seasonal climate forecasts (mostly based on ENSO teleconnections) to forecast soil wetness in Florida and, because soil wetness correlates highly with human St. Louis encephalitis (SLE) (presumably through the intermediary of a mosquito vector), they were able to construct a forecast system for SLE (Shaman *et al.*, 2006). Stephens *et al.* (2000) used crop models coupled to seasonal climate forecasts (mediated by weather models appropriate to the forecasted seasonal climate) to forecast the wheat crop in Australia. A quasi-operational streamflow forecast was maintained for the western part of the state of Washington (Wiley, 2006) and for the entire western part of the USA (Hamlet and Lettenmaier, 2006) on the basis of downscaled and bias-corrected official seasonal outlooks issued by the US Climate Prediction Center.

The final stage of forecast use is making (or altering) a decision on the basis of the forecast. Just about every human enterprise has some weather or climate aspect and the climate forecast has to be presented in a way appropriate to the decision to be made.

It should be noted that beyond the purely physical-science steps of taking the data and making the forecasts, the success of end-to-end forecasting depends primarily on social factors. How to engage the user, how to communicate the forecasts, how to analyze the decision processes of a user, how to engage and sustain the engagement with the user – these are all matters for social science research.

10.5 Using forecasts – some potential examples

Except in the direct region of the tropical Pacific, useful forecasts of local precipitation or temperature usually have other requirements than simply predicting the phases of ENSO. In large regions of the western USA, monthly precipitation depends not only on the state of ENSO, but also on the state of the Pacific Decadal Oscillation and possibly the state of the Atlantic, so that a better forecast can be obtained if ENSO is forecast and the state of the PDO and the Atlantic SST known. In north-east Brazil, a combination of ENSO and the state of the tropical Atlantic determines the rainfall (Uvo *et al.*, 1998). In east Africa, the rainfall is determined by both the state of ENSO and by sea-surface temperature in the Indian Ocean (Goddard and Graham, 1999).

Because the forecasts are made in an ensemble sense, the forecasts (possibly downscaled) of rainfall or temperature in a region will have a range of probability centered around some central value. This probability distribution for the quantity of interest is the output of the prediction system.

The first dynamical forecast by coupled atmosphere–ocean models was made in 1986 using only winds as initial data and the ENSO observing system (Figure 1.16) was not put into place until 1995, so there has not been time for a complete end-to-end system to establish itself. We will content ourselves here with indicating what applications of ENSO forecasts have been suggested, recognizing that it will take many years to fully evaluate the use of the forecasts.

In agriculture, the decision can be what to plant, how much fertilizer to add or how much crop insurance to buy. In water-resource management, the decision can be how much water to spill from a dam, how much water to allocate to power generation, how much to fish survival, and how much to downstream municipalities for drinking water. In retail sales, the decision could be how much to order and when to stock winter apparel, and whether or not to buy natural gas futures to heat the stores and warehouses. Fishers may decide whether or not to buy additional boats or equipment to fish their traditional fishing grounds, whether or not to move to different grounds, and whether or not to sell fish stocks on future delivery contracts. International aid organizations may decide to stock emergency food aid early and in specific locations in response to forecasts of low agricultural output, and health organizations may decide to inoculate in anticipation of heavy mosquito infestations.

The private sector can use ENSO forecasts for staging, future hedging and regional allocation. Because much of the private-sector activity is conducted in competitive situations, information about uses and strategies are rarely made public. Changnon (2000), however, reports that power utilities were the most consistent users of the 1997 ENSO forecast and the uses included revamped maintenance schedules, revised buying on the natural gas spot markets, revised plans for stocking coal for the coming winter and altered strategies for futures contracts.

10.6 Improving the use of climate information

It has been noted in a number of different surveys of water management in the USA (Pulwarty and Redmond, 1997; Callahan *et al*., 1999; Rayner *et al*., 2005) that seasonal to interannual forecasts are rarely used in making decisions about next year's water. In the words of Callahan *et al*. (1999) "The barriers to managers' use of climate forecasts include low forecast skill, lack of interpretation and demonstrated applications, low geographical resolution, inadequate links to climate variability related impacts, and institutional aversion to incorporating new tools into decision making." Lest this seem like everything, Nicholls (2000a, 2000b) has pointed out that (Western) human thought seems to be subject to a number of well-studied cognitive illusions and biases which prevents even well-educated people from fully understanding probability and uncertainty, and this leads to what must be considered irrational decision making.

There is not a unique answer to overcoming these barriers to the use of climate information – every step in the process has to be addressed. If we use the series of steps given in Section 10.4, we can suggest the following improvements:

(a) Consistently and continuously improve the prediction system by a coordinated sequence of improvements to the observing system, the coupled models, the data-assimilation techniques and the suite of ensemble predictions.
(b) Shape the large-scale forecasts so that the best information can be consistently given from, preferably, a single authoritative source with the distribution of ensemble results

encapsulated in a forecast probability distribution. Choose the medium and communicators for maximum salience and credibility.

(c) For each region, have local organizations catalog, research and make generally available the impacts of ENSO variability and the percentage of variability explained. Add information about impacts due to longer-term variability including PDO, NAO, the Atlantic multi-decadal variability etc.

(d) Have regional organizations downscale the large-scale forecasts and make resource models to examine the range of resource variability in response to climate variability and in response to specific distributions of forecasts.

(e) Have these organizations catalog and make available the uncertainties in the local resource forecasts by means of the models in (d).

(f) Establish and improve research into the normal decision processes sector by sector for public institutions (private institutions will presumably do this themselves or hire other private institutions to do this).

(g) Design a participation process that allows the users to interact with the forecast providers to formulate a forecast that has value to the users and gives support to the providers.

(h) Design a set of simulation tools that allows the users to get comfortable with the use of climate information in making decisions in a simulated learning context where nothing is at risk. This allows the nature of various decisions to be considered and compresses the timescale so that many simulated decisions can be made in a relatively short time.

(i) Evaluate the consequences of each decision in (h) to build confidence in the system.

(j) Establish an organization (a Climate Service) responsible for the success of the entire system and its constant improvement.

Finally, it should be pointed out that every new idea or technology follows a known path as it works its way into society (Rogers, 2003). From early adopters, to general adoption, to adoption by laggards, the rate of adoption depends in an essential way on the properties of the new idea itself. There are five attributes of an innovation that helps determine its rate of adoption: relative advantage, simplicity, compatibility or fit, observability and trialability. From this point of view, a hand-held video game which is far better than Pac-Man, simple to learn and use, compatible with the leisure pursuits of teenagers and those of similar tastes, violating no cultural or religious norms, obvious to anyone looking over one's shoulder, and readily tried for oneself, is a perfect product that has all the attributes to rapidly diffuse into society. Climate prediction, however, while telling more about the future than no information, is difficult for the average person to understand, gives probabilistic information that conflicts with our need for certainty, is conducted by specialists out of the view of most people, and takes a very long time and special knowledge to test for oneself. It therefore has almost none of the attributes that makes it easy to diffuse into society, and one could therefore expect that it would take a very long time to be accepted as a common technology. The observing system and the coupled prediction and analysis system is expensive to maintain and improve, so it is a race to find and satisfy users of ENSO information lest the will to spend the money to maintain the observing and prediction system lags. Adoption of ENSO prediction is by no means a sure thing and has to be constantly worked at to succeed.

11

Postview

11.1 Looking back

We have examined, in some detail, the observations relevant to both the tropical Pacific and to ENSO, and the processes in the atmosphere and the ocean needed to explain ENSO. In the atmosphere, these processes are: the processes that anchor the regions of persistent precipitation to warm SST anomalies; the processes that determine the convergence of moisture over warm SST anomalies to maintain the regions of persistent precipitation; and the processes that determine the anomalous surface winds in terms of anomalous SST and its associated anomalous precipitation. In the ocean, the processes of interest are: the processes that change SST to produce SST anomalies; the processes that determine the depth of the mixed layer; and the processes that determine the time-dependent anomalous position of the thermocline in response to forcing by anomalous wind stresses.

Because the present complex coupled numerical models of climate do not yet simulate the tropical climatology or the phenomenon of ENSO with a sufficient degree of realism, we have concentrated analysis on the simpler ("intermediate") coupled models. These models are simple in that they have relatively few degrees of freedom, can be run for large numbers of cases and parameter changes, and can be analyzed relatively exhaustively. Rather than repeat what we have already discussed in Chapters 7 and 8, let us summarize what is known and not known about ENSO and its predictability.

We learned that the delayed-oscillator equation is a robust analog for regular ENSO oscillations. It provides a conceptual model for regular oscillations in an intermediate coupled atmosphere–ocean model, and in addition, is capable of correctly describing the changes of period when some of the basic model parameters are changed, in particular the size of the basin, the magnitude of the various couplings and the magnitude of the dissipation. (We note that the same might be said of the recharge-oscillator equation, which we argued embodies the same essential physics.) Moreover, we saw that the simplest fully nonlinear model, the Zebiak–Cane model, is not only capable of simulating the basic features of ENSO, it is also as good a prediction model as currently exists for predicting the occurrence and amplitude of warm and cold phases of ENSO a season to several seasons in advance. We saw that the ZC model had a basic atmosphere–ocean instability as its fundamental dynamics.

We also saw, again in a simplified context, an alternate formulation of ENSO which, while retaining the same underlying dynamics as the ZC model, appeared in a completely different guise. *Stable* coupled linear models driven by higher-frequency 'noise' simulate similar ENSO properties and have similar predictability properties up to a year in advance. In this stable case, a year seems to be at the limit of predictability. We argued that there is no current way of deciding the issue of stability or instability in the real ENSO system (as opposed to models of ENSO where the stability of the coupled interactions determining ENSO is known, or rather can be known) so that a more practical test is the range of predictability of ENSO. The current models are able to predict the SST anomalies characteristic of ENSO a few seasons in advance with retrospective forecasts, indicating that there are some epochs where the range of predictability is up to 2 years in advance. If we understood the reason for this decadal modulation of predictability, and if the 2-year range in some epochs was in concert with non-normal disturbances having their usual 1-year period of growth, then we could conclude the system was unstable. At the present time, the ifs have it, and we cannot come to any conclusions about the stability or instability of ENSO based on the range of predictions.

While we have a paradigm for regular oscillations, ENSO is not regular and the issue of the actual mechanism for irregularity has not been definitively settled. The two candidates are forcing by noise and scrambling by nonlinear interactions. Since the linear version of any atmosphere–ocean model is non-normal, small amounts of noise will force rather large amounts of variance when the linear system is either stable or unstable. Nonlinearities require interactions between different timescales and the obvious candidate to interact with the interannual timescale is the annual cycle. A number of simple models have been shown to produce irregularity through this mechanism but, again, there is no way to know what the actual mechanism is in nature. Perhaps it is some combination of the two possible mechanisms.

We conclude that the overall situation is less than satisfactory. While we have hints of mechanisms and of the ultimate range of predictability from a large number of simpler model studies, and while we can make useful predictions even without knowing the precise mechanism, we cannot presently know if there is more prediction skill to be mined at greater range or if the skill can be much increased at the current range of prediction. Perhaps the situation is somewhat analogous to the case of weather prediction some 40 years ago where forecasts were (futilely) attempted at far beyond the range of deterministic predictability. It took Lorenz' seminal insight that the system is chaotic, and that the ultimate range of predictability could be determined by understanding the growth rates of initial errors, before clarity could be attained.

11.2 Looking ahead

If we are to achieve the needed clarity in the future elucidation of ENSO, a number of prerequisites seem essential to us.

- We need longer records of the past behavior of ENSO in order to document the relationship between the changing mean climate and the behavior of ENSO. While the record of ENSO in the glaciers of the tropics is rapidly disappearing due to melting, there are still corals and other proxies to be discovered and there are real possibilities that a more comprehensive past record of ENSO can be obtained.
- We need to assure the future record of ENSO and the observations needed to interpret the decadal behavior of ENSO. This requires a commitment to the existing TAO/Triton array (Figure 1.16), to a meridional expansion of this array, and to the maintenance of satellite altimetry and scatterometer wind-stress measurements.
- We need to understand specific processes with a view to inserting them in comprehensive climate models with a good degree of accuracy. In particular, we need to understand how a single tropical heat source (i.e. region of persistent precipitation) forces the midlatitudes after emitting planetary waves that travel though the full three-dimensional wind field between the tropics and midlatitudes. We also need to understand the kind of low-level stratus clouds that lie over upwelling regions off the west coast of South America. Current models do not adequately simulate these clouds, leaving the upwelling regions of Ecuador and Peru – which should be cold due to upwelled water and shielding by stratus – too warm.
- We need to understand the annual cycle in the tropics and use this understanding for simulation. As we saw, the mean climate and the annual cycle are poorly done in all present comprehensive climate models. Because the annual cycle is poorly done, anomalies with respect to the model climatology are poorly defined and the location of the heat sources in the tropics are in the wrong place at the wrong times, thereby misplacing forced variability and limiting the accuracy of midlatitude variability. The single most important obstacle to the present simulation of variability, including ENSO, in climate models is the poor simulation of the climatology.
- With the climatology fixed in comprehensive climate models, we need to experiment with these models to better elucidate the nature of ENSO. In particular, it will be of great interest to see if the set of comprehensive coupled climate models, each built out of the best possible components and the most tested parameterizations for clouds, mixing, and surface fluxes of heat and momentum, give robust and consistent results for the stability or instability of ENSO. Whichever it is, it will also be possible to experiment with these models to test whether the amount of internal noise is correct and to experiment with

different amounts of external noise added to whatever internal noise is present, in order to elucidate the role of noise in mechanisms for ENSO. It will also be possible to examine the life cycle of ENSO in these models and to check whether or not the Madden–Julian oscillation is important by examining the nature of ENSO with the MJO and then with the MJO artificially suppressed. Finally, if these comprehensive models have the correct decadal patterns, it should be possible to elucidate the nature and cause of decadal modulation of ENSO.

- We need to advance the state of data assimilation in coupled models so that the initialization of ENSO prediction is advanced and so that a complete model-based analysis of the climate system becomes possible. Currently, data taken in either the atmosphere or ocean does not consistently constrain the system, since data assimilation is done in each system separately. A coupled data assimilation would guarantee that the data is dynamically consistent and allows an optimal estimate of the state of the entire coupled system, necessary for both initialization for prediction and analysis for archiving. A model-based analysis of the climate system performed regularly and systematically is the only way to consistently grow the climate record.

- We need to continue to explore and demonstrate the beneficial use of ENSO predictions so that users become more sophisticated in understanding the basis of the forecasts and hopefully, on the basis of their positive experiences, demand more and better climate information. Adaptation will undoubtedly become one of the major themes of the twenty-first century and crucial tools of adaptation are predictions a year in advance and information about how climate variability, in particular ENSO, will change as the climate warms.

During the last 20 years of the twentieth century, ENSO studies concentrated on the simpler intermediate coupled models of ENSO. We expect twenty-first century ENSO studies to concentrate on simulation with comprehensive climate models. When the bias problems involving the climatology have been solved, we expect that these comprehensive climate models can be used to solve some of the pressing ENSO problems involving the precise mechanism for ENSO, the role of stochastic noise, the ultimate limit of ENSO predictability, the response of ENSO to a changed mean climate and the nature of decadal modulation of ENSO. We see enormous opportunities in the future in solving a set of problems that are currently stymied by the lack of comprehensive enough tools to address them; in particular comprehensive climate models and the sustained observations needed to elucidate decadal variability in the Pacific. ENSO is *not* presently a solved problem – perhaps some day one of the readers of these words will rectify this situation.

Appendix 1: Some useful numbers

Radius of Earth:	$6370 \, \text{km}$
Area of Earth:	$0.51 \times 10^{15} \, \text{m}^2$
Solar constant:	$1367 \, \text{W/m}^2$
Area covered by oceans:	$0.36 \times 10^{15} \, \text{m}^2$
Heat capacity of water:	$C_{pw} = 1 \, \text{cal/gmK} = 4.19 \times 10^3 \, \text{J/kgK}$
Heat capacity of dry air:	$C_{pa} = 0.24 \, \text{cal/gmK} = 1.0 \times 10^3 \, \text{J/kgK}$
Density of air (at surface):	$1.23 \, \text{kg/m}^3$
Density of water:	$1 \, \text{gm/cm}^3 = 10^3 \, \text{kg/m}^3 = 1 \, \text{tonne/m}^3$
Planetary-vorticity gradient:	$\beta = \dfrac{df}{dy} = 2.28 \times 10^{-11} \cos \phi_0 \, m^{-1} s^{-1}$ where ϕ_0 is latitude.
	$\beta = 1.62 \times 10^{-11} \, m^{-1} s^{-1}$ at $45°$
1 year	$3.15 \times 10^7 \, \text{s}$ (which can be remembered as $\pi \times 10^7 \, \text{s}$)
Latent heat of water:	$L = 2.5 \times 10^6 \, \text{J/kgK}$
Universal gas constant:	$R = 8.31 \times 10^3 \, \text{J/Kmol}$
For dry air:	$R = c_p - c_v, \; c_p = \dfrac{7}{2} R, \; c_v = \dfrac{5}{2} R$

Derived quantities

The mass of a water column of $1 \, \text{m}^2$ area and $10 \, \text{m}$ deep is 10 tonnes.

The mass of the total air column exerting $1020 \, \text{hPa}$ at the surface is about 10 tonnes.

$50 \, \text{W/m}^2$ into a column of area $1 \, \text{m}^2$ of water $50 \, \text{m}$ deep heats the column 1 K in 50 d ("the 50–50–50 rule").

$100 \, \text{W/m}^2$ into a unit column of water $100 \, \text{m}$ deep heats that column 0.6 K/mo.

$100\,\mathrm{W/m^2}$ into a unit column of air to the top of the atmosphere heats the air column $0.8\,\mathrm{K/d}$.

It takes $29\,\mathrm{W/m^2}$ to evaporate $1\,\mathrm{mm/d}$ of water from the surface.

For the mean temperature of surface of the Earth, $T = 15\,^{\circ}\mathrm{C} = 288\,\mathrm{K}$, $\sigma T^4 = 390\,\mathrm{W/m^2}$.

For $T = 300\,\mathrm{K}$, $\sigma T^4 = 459\,\mathrm{W/m^2}$.

Appendix 2: The parabolic-cylinder functions

Consider the equation:

$$\frac{d^2\psi}{dy^2} + (a - y^2)\psi = 0 \tag{A2.1}$$

on an infinite plane: $-\infty < y < \infty$. The solutions to Equation A2.1 for which $\psi \to 0$ as $y \to \pm\infty$ exist only when $a = 2n + 1$. The normalized solutions are (Gradshteyn and Rizhik, 1965, Sections 7.37 and 7.38)

$$\psi_n(y) = \frac{1}{\pi^{1/4}} \frac{1}{(2^n n!)^{1/2}} \exp(-\frac{y^2}{2}) H_n(y), \tag{A2.2}$$

where $H_n(y)$ are the Hermite polynomials:

$$H_n(y) = (-1)^n e^{y^2} \frac{d^n}{dy^n} e^{-y^2}.$$

The normalized solutions to Equation A2.1 are orthonormal:

$$\int_{-\infty}^{\infty} \psi_n(y)\, \psi_m(y)\, dy = \delta_{nm},$$

where $\delta_{nm} = 0$ if $n \neq m$ and $\delta_{nm} = 1$ if $n = m$.

$\psi_0 = \exp\left[-\frac{y^2}{2}\right]$ is clearly symmetric about $y = 0$. All ψ_n with even n have even symmetry about $y = 0$ while ψ_n with odd n have odd symmetry. As can be seen directly from Equation A2.1, ψ_n is oscillatory between the two turning points $y = \pm\sqrt{2n+1}$ and decays as $y^n \exp\left[-\frac{y^2}{2}\right]$ poleward of each turning point.

The solutions have the following properties:

$$y\psi_n = \sqrt{\frac{n+1}{2}}\,\psi_{n+1} + \sqrt{\frac{n}{2}}\,\psi_{n-1} \tag{A2.3a}$$

and

$$\frac{d\psi_n}{dy} = -\sqrt{\frac{n+1}{2}}\,\psi_{n+1} + \sqrt{\frac{n}{2}}\,\psi_{n-1}. \tag{A2.3b}$$

The following two integrals prove useful in taking projections:

$$\int_{-\infty}^{\infty} \psi_{2n}(y)\,dy = \frac{\sqrt{2}\pi^{\frac{1}{4}}}{2^n}\frac{[(2n)!]^{\frac{1}{2}}}{n!} \tag{A2.4}$$

$$\int_{-\infty}^{\infty} y\psi_{2n+1}(y)\,dy = \frac{2\pi^{\frac{1}{4}}}{2^n}\frac{[(2n+1)!]^{\frac{1}{2}}}{n!}. \tag{A2.5}$$

The source of the square roots of π can be seen by considering the integral:

$$\int_{-\infty}^{\infty} \exp[-y^2]\,dy = 2\int_{0}^{\infty} \exp[-y^2]\,dy = \int_{0}^{\infty} t^{-\frac{1}{2}}\exp[-t]\,dt = \Gamma\left(\frac{1}{2}\right) = \sqrt{\pi},$$

$$\tag{A2.6}$$

where Γ is the gamma function (see Gradshteyn and Rizhik, 1965, Section 8.31):

$$\Gamma(z+1) = \int_{0}^{\infty} e^{-t}t^z\,dt. \tag{A2.7}$$

Appendix 3: Modal and non-modal growth

A3.1 Context

Assume we are dealing with a linear evolution equation:

$$\frac{d\mathbf{x}}{dt} = \mathbf{Ax}, \tag{A3.1}$$

where $\mathbf{x}$ is a n row vector which contains all the state variables of the problem and $\mathbf{A}$ is the n by n linearized evolution matrix. We will also take $\mathbf{A}$ to be independent of time (i.e. "autonomous"). If $\mathbf{A}$ is the result of discretizing a system of differential equations, $\mathbf{A}$ may be very large.

The formal solution to Equation A3.1 is:

$$\mathbf{x}(t) = [\exp \mathbf{A}(t - t_0)] \, \mathbf{x}(t_0), \tag{A3.2}$$

where the expression in brackets is the exponent of a matrix. This can be written $\mathbf{x}(t) = \mathbf{R}(t_0, t)\mathbf{x}(t_0)$ where $\mathbf{R}$ is the propagator, which takes the system from the initial state $\mathbf{x}(t_0)$ at time t_0 to the current state $\mathbf{x}(t)$ at time t.

If $\mathbf{e}_i$ is an eigenvector of $\mathbf{A}$ and λ_i the corresponding eigenvalue,

$$\mathbf{A}\mathbf{e}_i = \lambda_i \mathbf{e}_i, \tag{A3.3}$$

so that if we choose an eigenvector ("mode") for the initial state, $\mathbf{x}(t_0) = \mathbf{e}_i$ then the solution is

$$\mathbf{x}(t) = \exp[\lambda_i t] \, \mathbf{e}_i,$$

so that the solution grows (or decays) with rate $\mathrm{Re}[\lambda_i]$ and oscillates with period $\mathrm{Im}[\lambda_i]$. The shape of the solution stays the same as the solution evolves – i.e. it stays the shape of the original mode. Any initial vector can be expanded as an eigenvector expansion and each eigenvector separately will evolve to the final time without change of shape. The final state is then a sum of evolved modes (see Equation A3.9).

341

A3.2 Matrix background

A3.2.1 *Normal matrices*

A p by p square matrix $\mathbf{B}$ is said to be normal if $\mathbf{BB}^+ = \mathbf{B}^+\mathbf{B}$ where superscript "+" refers to the adjoint (complex conjugate transpose). If $\mathbf{BB}^+ \neq \mathbf{B}^+\mathbf{B}$ the matrix $\mathbf{B}$ is non-normal.

☼**Theorem** (Noble and Daniel, 1988 p. 329):

All normal matrices $\mathbf{B}$ can be diagonalized by a unitary transformation:

$$\mathbf{D} = \mathbf{U}^+\mathbf{BU} \text{ or } \mathbf{B} = \mathbf{UDU}^+ \tag{A3.4}$$

where $\mathbf{D}$ is diagonal and $\mathbf{U}$ is unitary: $\mathbf{UU}^+ = \mathbf{U}^+\mathbf{U} = 1$, i.e. $\mathbf{U}^+ = \mathbf{U}^{-1}$.

Since $\mathbf{U}$ is unitary,

$$\mathbf{BU} = \mathbf{UD}$$

and we see that $\mathbf{U}$ is composed of columns whose elements are the normalized eigenfunctions of $\mathbf{B}$:

$$\mathbf{U} = [\mathbf{e}_1, \mathbf{e}_2, \ldots, \mathbf{e}_p]$$

so that

$$\mathbf{UU}^+ = [\mathbf{e}_1, \mathbf{e}_2, \ldots, \mathbf{e}_p] \begin{bmatrix} \mathbf{e}_1^* \\ \mathbf{e}_2^* \\ \vdots \\ \mathbf{e}_p^* \end{bmatrix} = 1,$$

and the eigenvectors are orthonormal:

$$\sum_{ij} e_i e_j^* = \delta_{ij}.$$

The eigenvectors of a normal matrix are orthogonal.

If $\mathbf{B}$ is Hermitian, $\mathbf{B} = \mathbf{B}^+$, then the eigenvalues are real. All real symmetric matrices are Hermitian.

A3.2.2 *General matrices*

☼*Any p by q matrix $\mathbf{A}$ can be decomposed by a singular-value decomposition:*

$$\mathbf{A} = \mathbf{U\Sigma V}^+ \tag{A3.5}$$

where:

$\mathbf{U}$ is a p by p unitary matrix,

$\mathbf{V}$ is a q by q unitary matrix and,

Σ is a p by q "diagonal" matrix in the sense that the "diagonal" elements are

$$\Sigma_{ij} = \sigma_i \delta_{ij}$$

and the σ_i are real and $\sigma_i \geq 0$.

Because $\mathbf{U}$ and $\mathbf{V}$ are unitary,

$$\mathbf{A}^+\mathbf{A} = \mathbf{V}\mathbf{D}\mathbf{V}^+ \text{ or } \mathbf{D} = \mathbf{V}^+\mathbf{A}^+\mathbf{A}\mathbf{V} \qquad \text{(A3.6a)}$$

where $\mathbf{D} = \Sigma^+\Sigma$ is a q by q diagonal matrix whose elements are σ_i^2.

Similarly,

$$\mathbf{A}\mathbf{A}^+ = \mathbf{U}\mathbf{D}^+\mathbf{U}^+ \qquad \text{(A3.6b)}$$

where $\mathbf{D}^+ = \Sigma\Sigma^+$ is a p by p diagonal matrix with elements σ_i^2. When $p > q$, the additional matrix elements are zero.

It is easy to verify that both $\mathbf{A}\mathbf{A}^+$ and $\mathbf{A}^+\mathbf{A}$ are normal even when $\mathbf{A}$ is not. We can then use the Theorem to recognize that Equations A3.6a and A3.6b are unitary transformations. We can identify the columns of $\mathbf{U}$ with the eigenvectors of $\mathbf{A}\mathbf{A}^+$ (these are called the left singular vectors) and eigenvalues σ_i^2 and the columns of $\mathbf{V}$ with the eigenvectors of $\mathbf{A}^+\mathbf{A}$ (these are called the right singular vectors) with eigenvalues σ_i^2. (Note that if $\mathbf{A}$ is square and $\mathbf{U} = \mathbf{V}$, then $\mathbf{A}$ is normal.)

We use the notation

$$\mathbf{V} = [\mathbf{r}_1, \mathbf{r}_2, \ldots, \mathbf{r}_q],$$

and

$$\mathbf{U} = [\mathbf{l}_1, \mathbf{l}_2, \ldots, \mathbf{l}_p].$$

By Equation A3.5,

$$\mathbf{A}\mathbf{V} = \mathbf{U}\Sigma$$

so that

$$\mathbf{A}\mathbf{r}_i = \sigma_i\mathbf{l}_i \qquad \text{(A3.7)}$$

and we see that the matrix $\mathbf{A}$ takes the right singular vector into the left singular vector multiplied by a singular value. Recall that $\mathbf{A}^+\mathbf{A}\mathbf{r}_i = \sigma_i^2\mathbf{r}_i$.

Similarly, by Equation A3.5,

$$\mathbf{A}^+\mathbf{U} = \mathbf{V}\Sigma^+$$

so that

$$\mathbf{A}^+\mathbf{l}_i = \sigma_i \mathbf{r}_i, \tag{A3.8}$$

so that the adjoint of $\mathbf{A}$ takes the left singular vector into the right singular vector multiplied by the complex conjugate of the singular value. Recall that $\mathbf{A}\mathbf{A}^+\mathbf{l}_i = \sigma_i^2\mathbf{l}_i$.

Since $\mathbf{A}\mathbf{A}^+$ and $\mathbf{A}^+\mathbf{A}$ are separately self-adjoint, the left and right singular vectors separately form a complete and orthonormal basis set. For any matrix $\mathbf{B}$ with an eigenfunction $\mathbf{u}$ corresponding to eigenvalue λ, and for the adjoint matrix with an eigenfunction $\mathbf{v}$ corresponding to eigenvalue λ', we can see that $<\mathbf{v},\mathbf{Bu}> = \lambda<\mathbf{v},\mathbf{u}> = <\mathbf{B}^+\mathbf{v},\mathbf{u}> = \lambda'<\mathbf{v},\mathbf{u}>$ so that if $\lambda \neq \lambda'$, then $<\mathbf{u},\mathbf{v}> = 0$: any eigenfunction of a matrix $\mathbf{B}$ is orthogonal to all eigenfunctions of the adjoint $\mathbf{B}^+$ that does not correspond to the same eigenvalue. The eigenfunctions of $\mathbf{B}$ are said to be bi-orthogonal to the eigenfunctions of $\mathbf{B}^+$. The eigenfunctions of $\mathbf{B}$ are not themselves orthogonal. Further, all eigenvalues of $\mathbf{B}$ are also eigenvalues of $\mathbf{B}^+$. (☼ The proof is surprisingly hard: see Friedman, 1956.) Note that if $\mathbf{B}$ is self adjoint (Hermitian), the eigenfunctions *are* orthogonal.

A3.2.3   System evolution

The solution to the normal problem (Equation A3.1) can always be written in terms of eigenvectors as:

$$\mathbf{x}(t) = \sum_i \mathbf{e}_i \alpha_i \exp[-i\sigma_i t] \tag{A3.9}$$

where

$$\mathbf{A}\mathbf{e}_j = -i\sigma_j \mathbf{e}_j$$

and α_i are the elements of the matrix $\alpha = \mathbf{E}^{-1}\mathbf{x}(t=0)$ and $\mathbf{E}$ is the matrix whose columns are the eigenvectors. The eigenvalue with the largest imaginary part will eventually dominate the solution: at any finite time the solution is given by Equation A3.9.

We can use the bi-orthogonality relation to illustrate the profound difference between normal and non-normal systems. Suppose we have an initial disturbance $\mathbf{x}$ (t_0): we can expand this disturbance in the eigenvectors of $\mathbf{A}$ (which are complete but not orthogonal). The coefficients are given by the bi-orthogonality relation as proportional to:

$$\alpha_i = \frac{<\mathbf{f}_i, \mathbf{x}(t_0)>}{<\mathbf{f}_i, \mathbf{e}_i>}, \tag{A3.10}$$

so that the projection onto a given mode depends on the inner product with the adjoint mode, not the mode itself.

If we want the final state at long time to be the most rapidly growing eigenvector of $\mathbf{A}$ which will dominate the series, e_1, then it is clear that $\mathbf{x}(t_0)$ must be taken as proportional to the eigenvector of $\mathbf{A}^+$, $\mathbf{f}_1$, rather than the final state, the eigenvector of $\mathbf{A}$, $\mathbf{e}_1$. For any *mode*, the largest final mode will be obtained if the initial structure is the adjoint. The adjoint then evolves by changing its structure into the mode, contrary to the growing invariant structure that is characteristic of normal evolution. The final state in the mode is larger than if the same initial amplitude was put into the mode.

An alternate way of looking at the solutions to Equation A3.1 is at a finite time $t = \tau$. According to Equation A3.2,

$$\mathbf{x}(t_0 + \tau) = \mathbf{R}(t_0, t_0 + \tau)x(t_0) = \exp[\mathbf{A}\tau]\mathbf{x}(t_0).$$

The ratio of the amplitude at time $t_0 + \tau$ to that at the initial time t_0 is

$$\gamma = \frac{|\mathbf{x}(t_0 + \tau)|}{|\mathbf{x}(t_0)|} = \frac{<\mathbf{x}(t_0 + \tau), \mathbf{x}(t_0 + \tau)>^{\frac{1}{2}}}{<\mathbf{x}(t_0), \mathbf{x}(t_0)>^{\frac{1}{2}}},$$

where we take the norm measuring the square amplitude to simply be $<\mathbf{x},\mathbf{x}> = \mathbf{x}^+\mathbf{x}$, i.e. the usual L_2 norm.

$$\gamma = \frac{<\mathbf{R}(t_0, t_0 + \tau)\mathbf{x}(t_0), \mathbf{R}(t_0, t_0 + \tau)\,\mathbf{x}(t_0)>^{\frac{1}{2}}}{<\mathbf{x}(t_0), \mathbf{x}(t_0)>^{\frac{1}{2}}} = \frac{<\mathbf{x}(t_0), \mathbf{R}^+\mathbf{R}\,\mathbf{x}(t_0)>^{\frac{1}{2}}}{<\mathbf{x}(t_0), \mathbf{x}(t_0)>^{\frac{1}{2}}}.$$

Since by the singular vector decomposition of $\mathbf{R}$: $\mathbf{R}^+\mathbf{R} = \mathbf{V}\Sigma^+\Sigma\mathbf{V}^+$ and since the right singular vectors of $\mathbf{R}$ are the eigenvectors of $\mathbf{R}^+\mathbf{R}$ (all evaluated at the time τ) it is clear that the largest value γ at time τ will be attained when $\mathbf{x}(t_0)$ is chosen as the right singular vector $\mathbf{r}_1$ corresponding to the largest singular value σ_1^2. When $\mathbf{x}(t_0)$ is so chosen,

$$\gamma_1 = \frac{<\mathbf{r}_1, \sigma_1^2\mathbf{r}_1>^{\frac{1}{2}}}{<\mathbf{r}_1, \mathbf{r}_1>^{\frac{1}{2}}} = <\sigma_1^2>^{\frac{1}{2}} = \sigma_1.$$

The initial state $\mathbf{r}_1$ is taken into the left singular vector by Equation A3.7:

$$\mathbf{R}(t_0, t_0+\tau)\mathbf{r}_1 = \sigma_1\mathbf{l}_1.$$

We see that for a specific value of τ, we can find the initial state (the right singular vector of $\mathbf{R}(t_0, t_0 + \tau)$) that gives the largest final state (the left singular vector of $\mathbf{R}(t_0, t_0 + \tau)$). The amplification factor is σ_1 so that the initial disturbance grows only when there is a largest singular value larger than 1. We should note that this method gives the largest value of the amplitude that the system can grow to at time τ, recognizing that this value may not be the largest the system can grow to at some other time.

A3.3 A simple example

Note that all calculations in this section were easily performed by MATLAB.

A3.3.1 *Symmetric matrix*

Let us consider the symmetric matrix

$$\mathbf{A} = \begin{pmatrix} -1.12 & 0.025 \\ 0.025 & -0.027 \end{pmatrix}.$$

The eigenvalues of this matrix are -0.0264 and -1.1206 with corresponding normalized eigenvectors:

$$\mathbf{e}_1 = \begin{pmatrix} -0.9997 \\ +0.0229 \end{pmatrix}, \text{ and } \mathbf{e}_2 = \begin{pmatrix} -0.0229 \\ -0.9997 \end{pmatrix}.$$

It is clear that the eigenvectors are orthogonal so that any initial vector $\mathbf{x}(t = 0)$ can be expressed as:

$$\mathbf{x}(t = 0) = a_1\mathbf{e}_1 + a_2\mathbf{e}_2,$$

where

$$a_1 = \mathbf{e}_1^+\mathbf{x}(t = 0) \text{ and } a_2 = \mathbf{e}_2^+\mathbf{x}(t = 0).$$

And the solution to the evolution Equation A3.1 is given by:

$$\mathbf{x}(t) = a_1\mathbf{e}_1 \exp[\lambda_1 t] + a_2\mathbf{e}_2 \exp[\lambda_2 t].$$

Since the eigenvalues are negative, *any* initial conditions will decay. Since λ_2 is far more negative than λ_1, the second term will decay rapidly, leaving the relatively slow decay of the first term. Note that there are no initial conditions that will do anything but decay.

A3.3.2 *Asymmetric matrix*

We take an asymmetric matrix very close to the one used in the previous section:

$$\mathbf{B} = \begin{pmatrix} -1.12 & 0.500 \\ 0.025 & -0.027 \end{pmatrix}.$$

This matrix was the basis of a simple model of midlatitude atmosphere–ocean interactions in Bretherton and Battisti (2000). Similar 2×2 examples of non-normal evolution were given by Lacara and Talagrand (1988), Blumenthal (1991) and Ioannou and Farrell (2006).

The eigenvalues of matrix **B** are again both negative: $\lambda_1 = -1.1313$ and $\lambda_2 = -0.0157$ so that, again in this case, the asymptotic long-term behavior will always give decay – the system is clearly stable.

The normalized eigenvectors of **B** are:

$$\mathbf{e}_1 = \begin{pmatrix} -0.9997 \\ +0.0226 \end{pmatrix} \text{ and } \mathbf{e}_2 = \begin{pmatrix} -0.4127 \\ -0.9110 \end{pmatrix}$$

which are clearly *not* orthogonal. The eigenvalues of the adjoint matrix $\mathbf{B}^+$ are the same as those for **B** but the corresponding normalized eigenvectors are:

$$\mathbf{f}_1 = \begin{pmatrix} -0.9110 \\ +0.4127 \end{pmatrix} \text{ and } \mathbf{f}_2 = \begin{pmatrix} -0.0226 \\ -0.9997 \end{pmatrix}.$$

We see immediately that $\mathbf{f}_2^+ \mathbf{e}_1 = 0$ and $\mathbf{f}_1^+ \mathbf{e}_2 = 0$.

The fact that the eigenvectors are not orthogonal allows there to be initial conditions that will transiently grow despite the fact that asymptotically all disturbances decay. Let us ask that for the linear evolution equation

$$\frac{d\mathbf{x}}{dt} = \mathbf{Bx}. \tag{A3.11}$$

We find initial conditions to give maximum growth at $t = 2$. We perform a singular value decomposition on the propagator at time $t = 2$ (again, matrix exponentials are easily calculated in MATLAB):

$$\mathbf{R}(t = 2) = \exp[\mathbf{B}(t = 2)] = \begin{pmatrix} 0.1129 & 0.3877 \\ 0.0194 & 0.9602 \end{pmatrix},$$

with singular value decomposition:

$$\mathbf{R}(t = 2) = \begin{pmatrix} 0.3794 & 0.9252 \\ 0.9252 & -0.3794 \end{pmatrix} \begin{pmatrix} 1.0374 & 0 \\ 0 & 0.0972 \end{pmatrix} \begin{pmatrix} 0.0586 & 0.9983 \\ 0.9983 & -0.0586 \end{pmatrix}.$$

Since the singular value is 1.0374 and therefore greater than 1, we can expect the value, as measured by the square value in the regular L_2 norm, to be greater than 1. So we start with the right singular vector $\mathbf{x}(t = 0) = \begin{pmatrix} 0.0586 \\ 0.9983 \end{pmatrix}$ and plot the square size of the amplitude of the solution $\mathbf{x}(t) = \mathbf{R}(t)\,\mathbf{x}(0)$.

The size does indeed maximize at $t = 2$ and reaches the value of the square of the first singular value 1.076. The initial condition, the first right singular vector of

$\mathbf{R}(t = 2)$ is fairly close to that of the first adjoint eigenfunction $\mathbf{f}_1$ so that the maximum at $t=2$ is close to the maximum that $\mathbf{x}(t)$ reaches overall.

We can see more precisely what is going on by expanding in terms of the non-orthogonal eigenfunctions using the projections of the optimal initial condition $\mathbf{x}(0)$ given in Equation A3.11:

$$\mathbf{x}(t) = \alpha_1 \, \mathbf{e}_1 \, \exp[-1.1313t] + \alpha_2 \, \mathbf{e}_2 \, \exp[-0.0157], \qquad \text{(A3.12a)}$$

where $\alpha_1 = 0.3895$ and $\alpha_2 = -1.0862$. It can easily be checked that

$$\mathbf{x}(0) = \alpha_1 \, \mathbf{e}_1 + \alpha_2 \, \mathbf{e}_2 = \begin{pmatrix} 0.0586 \\ 0.9983 \end{pmatrix}. \qquad \text{(A3.12b)}$$

This modal decomposition allows us to see precisely what is happening in terms of the modes. (Note that we could have plotted the square size of the solution Equation A3.12a and gotten precisely the same as Figure A3.1.) The initial state is the sum of the two non-orthogonal modes that add up to something relatively small as in Figure A3.2. The smallest sum of initial modes is at right angles to $\mathbf{e}_1$ and is simply the adjoint mode $\mathbf{f}_2$.

As times goes on, the more rapidly decaying mode $\mathbf{e}_1$ gets smaller while the slowly decaying mode hardly changes, which allows the solution $\mathbf{x}(t)$ to get larger (Figure A3.3). Eventually the most rapidly decaying mode is mostly gone and the solution then simply decays as the less rapidly decaying mode $\mathbf{e}_2$.

While we have concentrated on non-modal growth in stable systems, non-normal unstable systems can also support non-modal growth – this growth may be greater than the growth rate of the unstable modes. One important property of non-normal systems, whether unstable or stable, is that the level of variance supported under random forcing is larger than that of a normal system under the same forcing (Ioannou, 1995).

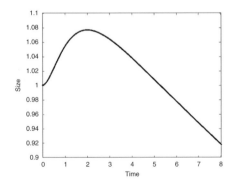

Figure A3.1. Evolution of squared size of solution starting from the first right singular vector.

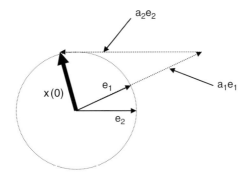

Figure A3.2. Schematic of construction of initial state $x(0)$ according to Equation A3.12b. The unit sphere is shown dotted.

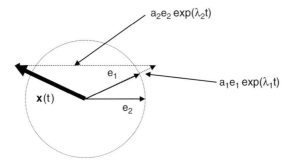

Figure A3.3. Schematic of evolution of state $x(t)$ at time t.

A3.4 Error evolution

In predicting the future state of a coupled system from a given initial state (Chapter 8), there are two types of errors that can grow to contaminate the forecast. The first is the errors growing from initial errors in the specification of the initial state. We know that deterministic chaotic systems exhibit sensitive dependence on initial conditions (Lorenz, 1963) and require arbitrarily small errors in the initial state to grow. Non-normal systems also exhibit sensitive dependence on initial conditions (at least for a finite time). Starting from a isotropic distribution of error, Figure A3.4 shows the evolution of the error in a stable non-normal system: the initial error ball decays in the more rapidly decaying mode and grows in the less rapidly decaying mode.

Although the Figure does not show it, it is relatively clear that if the initial error is carefully shaped and the size of the original errors changed, the error along the final (least decaying) mode can be controlled. The final error can therefore be arbitrary. This goes a long way to explaining the different results in the literature: unstable

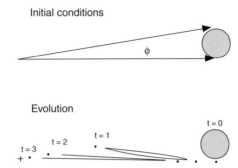

Figure A3.4. Evolution of initial error ball as a function of time in a stable non-normal system. (From Blumenthal, 1991.)

coupled models tend to have the initial error dominate the stochastic error induced during the transient growth (Karspeck *et al.*, 2006; Stan and Kirtman, 2008) while stable coupled models forced by stochastic noise during transient growth tend to have the effects of the initial error smaller than the effects of the continuing stochastic noise (Kleeman and Moore, 1997). It is clear that the actual distribution of error in the initial state projected onto the various modes and the actual stochastic noise as the system evolves need to be known in order to be able to gauge their relative importance for the predictability of the system.

References

AchutaRao, K., and K. R. Sperber, 2006: ENSO simulations in coupled ocean–atmosphere models: are the current models better? *Clim. Dyn.*, **27**, 1–16.

Anderson, D. L. T., 1984: An advective mixed layer model with applications to the diurnal cycle of the low-level East African jet. *Tellus*, **36A**, 278–91.

Anderson, D. L. T., and J. P. McCreary, 1985: Slowly propagating disturbances in a coupled atmosphere–ocean model. *J. Atmos. Sci.*, **42**, 615–29.

Anderson, D. L. T., and P. B. Rowlands, 1976a: The role of inertia-gravity and planetary waves in the response of a tropical ocean to the incidence of an equatorial Kelvin wave on a meridional boundary. *J. Mar. Res.*, **34**, 295–312.

Anderson, D. L. T., and P. B. Rowlands, 1976b: The Somali current response to the southwest monsoon: the relative importance of local and remote forcing. *J. Mar. Res.*, **34**, 395–417.

Asmerom, Y., V. Polyak, S. Burns, and J. Rasmussen, 2007: Solar forcing of Holocene climate: new insights from a speleothem record, southwestern United States. *Geology*, **35**, 1–4, doi:10.1130/G22865A.

Barnett, T. P., M. Latif, E. Kirk, and E. Roeckner, 1991: On ENSO physics. *J. Climate*, **4**, 487–515.

Barnston, A. G., M. H. Glantz, and Y. He, 1999: Predictive skill of statistical and dynamical climate models in SST forecasts during the 1997–98 El Niño episode and the 1998 La Niña onset. *Bull. Am. Met. Soc.*, **80**, 217–43.

Battisti, D. S., 1988: Dynamics and thermodynamics of a warming event in a coupled tropical atmosphere–ocean model. *J. Atmos. Sci.*, **45**, 2889–919.

Battisti, D. S., and A. C. Hirst, 1989: Interannual variability in a tropical atmosphere–ocean model: influence of the basic state, ocean geometry and nonlinearity. *J. Atmos. Sci.*, **46**, 1687–712.

Battisti, D. S., A. C. Hirst, and E. S. Sarachik, 1989: Instability and predictability in coupled atmosphere–ocean models. *Phil. Trans. Roy. Soc. Lond.*, **A329**, 237–47.

Battisti, D. S., E. S. Sarachik, and A. C. Hirst, 1999: A consistent model for the large-scale steady surface atmospheric circulation in the tropics. *J. Climate*, **12**, 2956–64.

Biondi, F., A. Gershunov, and D. R. Cayan, 2001: North Pacific decadal climate variability since 1661. *J. Climate*, **14**, 5–10.

Bjerknes, J., 1966: A possible response of the atmospheric Hadley circulation to equatorial anomalies of ocean temperature. *Tellus*, **18**, 820–9.

Bjerknes, J., 1969: Atmospheric teleconnections from the equatorial Pacific. *Mon. Weather Rev.*, **97**, 163–72.

Bjerknes, J., 1972: Large-scale atmospheric response to the 1964–65 Pacific equatorial warming. *J. Phys. Oceanogr.*, **2**, 212–7.

Blandford, R., 1966: Mixed gravity-Rossby waves in the ocean. *Deep. Sea. Res.*, **13**, 941–60.

Blumenthal, M. B., 1991: Predictability of a coupled ocean–atmosphere model. *J. Climate*, **4**, 766–84.

Bove, M. C., J. B. Elsner, C. W. Landsea, X. Niu, and J. J. O'Brien, 1998: Effect of El Niño on U.S. landfalling hurricanes, revisited. *Bull. Am. Met. Soc.*, **79**, 2477–82.

Bretherton, C. J., and D. S. Battisti, 2000: An interpretation of the results from atmospheric general circulation models forced by the time history of the observed sea surface temperature distribution. *Geophys. Res. Lett.*, **27**, 767–70.

Callahan, B. M., E. L. Miles, and D. L. Fluharty, 1999: Policy implications of climate forecasts for water resources management in the Pacific Northwest. *Policy Sci.*, **32**, 269–93.

Cane, M. A., 1979: The response of an equatorial ocean to simple wind stress patterns. II. Numerical results. *J. Mar. Res.*, **37**, 253–99.

Cane, M. A., 1992: Tropical Pacific ENSO models: ENSO as a mode of the coupled system. In *Climate System Modeling*, K. E. Trenberth, ed. Cambridge University Press.

Cane, M. A., 2005: The evolution of El Niño, past and future. *Earth Plan. Sci. Lett.*, **230**, 227–40.

Cane, M. A., and Y. du Penhoat, 1982: The effect of islands on low-frequency equatorial motions. *J. Marine Research*, **40**, 937–62.

Cane, M. A., and D. W. Moore, 1981: A note on low frequency equatorial basin modes. *J. Phys. Ocean.*, **11**, 1578–84.

Cane, M. A., and R. J. Patton, 1984: A numerical model for low-frequency equatorial dynamics. *J. Phys. Oceanogr.*, **14**, 1853–63.

Cane, M. A., and E. S. Sarachik, 1976: Forced baroclinic ocean motion. I. The equatorial unbounded case. *J. Marine Res.*, **34**, 629–65.

Cane, M. A., and E. S. Sarachik, 1977: Forced baroclinic ocean motions. II. The equatorial bounded case, *J. Marine Res.*, **35**, 395–432.

Cane, M. A., and E. S. Sarachik, 1979: Forced baroclinic ocean motions. III. The equatorial basin case. *J. Marine Res.*, **37**, 355–98.

Cane, M. A., and E. S. Sarachik, 1981: The response of a linear baroclinic equatorial ocean to periodic forcing. *J. Marine Res.*, **39**, 651–93.

Cane, M. A., and E. S. Sarachik, 1983: Seasonal heat transports in a forced equatorial baroclinic model. *J. Phys. Oceanogr.*, **13**, 1744–6.

Cane, M. A., and S. E. Zebiak, 1985: A theory for El Niño and the Southern Oscillation. *Science*, **228**, 1085–7.

Cane, M. A., S. E. Zebiak, and S. C. Dolan, 1986: Experimental forecasts of El Niño. *Nature*, **321**, 827–32.

Cane, M. A., M. Münnich, and S. E. Zebiak, 1990: A study of self-excited oscillations of the tropical ocean–atmosphere system. 1. Linear analysis. *J. Atmos. Sci.*, **47**, 1562–77.

Cane, M. A., G. Eshel, and R. W. Buckland, 1994: Forecasting Zimbabwean yield using eastern equatorial Pacific sea surface temperatures. *Nature*, **370**, 204–5.

Cane, M. A., S. E. Zebiak, and Y. Xue, 1995: Model studies of the long-term behavior of ENSO. In *Natural Climate Variability on Decade-to-Century Time Scales*, D. G. Martinson, K. Bryan, M. Ghil, M. M. Hall, T. R. Karl, E. S. Sarachik, S. Sorooshian, and L. D. Talley, eds. National Academy Press, Washington, D.C., DEC-CEN Workshop, Irvine, CA, pp. 442–57.

Cane, M. A., A. C. Clement, A. Kaplan, *et al.*, 1997: 20th century sea surface temperature trends. *Science*, **275**, 957–60.

Cane, M. A., P. Braconnot, A. Clement, *et al.*, 2006: Progress in paleoclimate modeling. *J. Climate*, **19**, 5031–57.

Capotondi, A., A. Wittenberg, and S. Masina, 2006: Spatial and temporal structure of tropical Pacific interannual variability in 20th century coupled simulations. *Ocean Modeling*, **15**, 274–98.

Chang, P., R. Saravanan, T. DelSole, and F. Wang, 2004: Predictability of linear coupled systems. Part I. Theory. *J. Climate*, **17**, 1474–86.

Changnon, D., 2000: Who used and benefited from El Niño forecasts? In *El Niño, 1997–1998: The Climate Event of the Century*, S. A. Changnon, ed. Oxford University Press, 215 pp.

Charney, J. G., 1948: On the scale of atmospheric motions. *Geofysiske Publikasjoner*, **17**, 1–17. Reprinted in: *The Atmosphere – A Challenge: The Science of Jule Gregory Charney*, R. S. Lindzen, E. N. Lorenx, and G. W. Platzman, eds. American Meteorological Society, 1990.

Chen, D., and M. A. Cane, 2008: El Niño prediction and predictability. *J. Computational Phys.*, **227**, 3625–40, doi:10.1016/j.jcp.2007.05.014.

Chen, D., S. E. Zebiak, A. J. Busalacchi, and M. A. Cane, 1995: An improved procedure for El Niño forecasting. *Science*, **269**, 1699–702.

Chen, D., M. A. Cane, A. Kaplan, S. E. Zebiak, and D. J. Huang, 2004: Predictability of El Niño over the past 148 years. *Nature*, **428**, 733–6.

Chen, Y.-Q., D. S. Battisti, T. N. Palmer, J. Barsugli, and E. S. Sarachik, 1997: A study of the predictability of tropical Pacific SST in a coupled atmosphere/ocean model using singular vector analysis: the role of the annual cycle and the ENSO cycle. *Mon. Wea. Rev.* **125**, 831–45.

Chiang, J. C. H., S. E. Zebiak, and M. A. Cane, 2001: Relative roles of elevated heating and surface temperature gradients in driving anomalous surface winds over tropical oceans. *J Atmos. Sci.*, **58**, 1371–94.

Clement, A. C., R. Seager, M. A. Cane, and S. E. Zebiak, 1996: An ocean dynamical thermostat. *J. Climate*, **9**, 2190–6.

Clement, A. C., R. Seager, and M. A. Cane, 1999: Orbital controls on tropical climate. *Paleoceanography*, **14**, 441–56.

Clement, A. C., R. Seager, and M. A. Cane, 2000: Suppression of El Niño during the mid-Holocene by changes in the earth's orbit. *Paleoceanography*, **15**, 731–7.

Clement, A. C., M. A. Cane, and R. Seager, 2001: An orbitally driven tropical source for abrupt climate change. *J. Climate*, **14**, 2369–75.

Cobb, K. M., C. D. Charles, R. L. Edwards, H. Cheng, and M. Kastner, 2003: El Niño–Southern Oscillation and tropical Pacific climate during the last millennium. *Nature*, **424**, 271–6.

Cole, J. E., J. T. Overpeck, and E. R. Cook, 2002: Multiyear La Niña events and persistent drought in the contiguous United States. *Geophys. Res. Lett.*, **29**, doi:10.1029/2001GL013561.

Crowley, T. J., 2000: Causes of climate change over the past 1000 years. *Science*, **289**, 270–7.

D'Arrigo, R., R. Villalba, and G. Wiles, 2001: Tree-ring estimates of Pacific decadal climate variability. *Climate Dynamics*, **18**, 219–24.

Davis, M., 2001: *Late Victorian Holocausts: El Niño Famines and the Making of the Third World*. Verso Press, London/New York, 465 pp.

DEMETER, 2005: *Tellus*, **57A**, 217–512.

Denman, K. L., 1973: A time-dependent model of the upper ocean. *J. Phys. Oceanogr.*, **3**, 173–84.

Deser, C., A. S. Phillips, and J. W. Hurrell, 2004: Pacific interdecadal climate variability: linkages between the tropics and North Pacific during boreal winter since 1900. *J. Climate*, **17**, 3109–24.

de Szoeke, S. P., and S. P. Xie, 2008: The tropical eastern Pacific seasonal cycle: assessment of errors and mechanisms in IPCC AR4 coupled ocean–atmosphere general circulation models. *J. Climate*, **21**, 2573–90.

Dima, I. M., and J. M. Wallace, 2003: On the seasonality of the Hadley cell. *J. Atmos. Sci.*, **60**, 1522–7.

Doblas-Reyes, F. J., R. Hagedorn, and T. N. Palmer, 2005: The rationale behind the success of multi-model ensembles in seasonal forecasting. II. Calibration and combination. *Tellus*, **57A**, 234–52.

du Penhoat, Y., and M. A. Cane, 1991: Effects of low-latitude western boundary gaps on the reflection of equatorial motions. *J. Geophys. Res.*, **96**, 3307–22.

Ekman, V. W., 1905: On the influence of earth's rotation on ocean currents. *Ark. Math. Astron. Fys.*, **2**, 1–53.

Emile-Geay, J., M. A. Cane, R. Seager, A. Kaplan, and P. Almasi, 2007: El Niño as a mediator of the solar influence on climate. *Paleoceanography*. **22**, PA3210, doi:10.1029/2006PA001304.

Emile-Geay, J., R. Seager, M. A. Cane, E. R. Cook, and G. H. Haug, 2008: Volcanoes and ENSO over the last millennium. *J. Climate*, **21**, 3134–48.

Evans, M. N., A. Kaplan, and M. A. Cane, 2000: Intercomparison of coral oxygen isotope data and historical SST: potential for coral-based SST field reconstructions. *Paleoceanography*, **15**, 551–63.

Evans, M. N., M. A. Cane, D. P. Schrag, *et al.*, 2001: Support for tropically-driven Pacific decadal variability based on paleoproxy evidence. *Geophys. Res. Lett.*, **28**, 3689–92.

Evans, M. N., A. Kaplan, and M. A. Cane, 2002: Pacific sea surface temperature field reconstruction from coral $\delta^{18}O$ data using reduced space objective analysis. *Paleoceanography*, **17**, doi:10.1029/2000PA000590.

Farrell, B. F., and P. J. Ioannou, 1993: Stochastic forcing of the linearized Navier-Stokes equations. *Phys. Fluids*, **A5**, 2600–9.

Fedorov, A. V., 2007: Net energy dissipation rates in the tropical ocean and ENSO dynamics. *J. Climate*, **20**, 1108–17.

Federov, A. V., and S. G. Philander, 2000: Is El Niño changing? *Science*, **288**, 1997–2002.

Fedorov, A. V., and S. G. Philander, 2001: A stability analysis of tropical ocean–atmosphere interactions: bridging measurements and theory for El Niño. *J. Climate*, **14**, 3086–101.

Fedorov, A. V., P. S. Dekens, M. McCarthy, *et al.*, 2006: The Pliocene paradox (mechanisms for a permanent El Niño). *Science*, **312**, 1485–9.

Friedman, B., 1956: *Principles and Techniques of Applied Mathematics*. John Wiley & Sons, 315 pp.

Garreaud, R. D., and D. S. Battisti, 1999: Interannual ENSO and interdecadal ENSO-like variability in the southern hemisphere tropospheric circulation. *J. Climate*, **12**, 2113–23.

Gedalov, Z., and D. J. Smith, 2001: Interdecadal climate variability and regime-scale shifts in Pacific North America. *Geophys. Res. Lett.*, **28**, 1515–8.

Gershunov, A., and T. P. Barnett, 1998: Interdecadal modulation of ENSO teleconnections. *Bull Am. Met. Soc.*, **79**, 2715–25.

Gill, A. E., 1980: Some simple solutions for heat induced tropical circulation. *Q. J. Roy. Met. Soc.*, **106**, 447–62.

Gill, A., and P. Niiler, 1973: The theory of the seasonal variability in the ocean. *Deep-Sea Res.*, **20**, 141–77.

Goddard, L., and M. Dilley, 2005: El Niño: catastrophe or opportunity. *J. Climate*, **18**, 651–65.

Goddard, L., and N. E. Graham, 1999: Importance of the Indian Ocean for simulating rainfall anomalies over eastern and southern Africa. *J. Geophys. Res.*, **104**, 19099–116.

Goddard, L., S. J. Mason, S. E. Zebiak, *et al.*, 2001: Current approaches to seasonal-to-interannual climate predictions. *Int. J. Climatol.*, **21**, 1111–52.

GCOS (Global Climate Observing System), 2004: *Implementation Plan for the Global Observing System for Climate in Support of the UNFCCC*. World Meteorological Organization TD No. 1219, 136 pp. Available at www.wmo.ch/web/gcos/gcoshome.html.

Gradshteyn, I. S., and I. M. Rhyzhik, 1965: *Table of Integrals, Series and Products*, 4th edn. Academic Press, 1086 pp.

Guilyardi, E., 2006: El Niño–mean state–seasonal cycle interactions in a multi-model ensemble. *Clim. Dyn.*, **26**, 329–48.

Guilyardi, E., A. Wittenberg, A. Fedorov, *et al.*, 2009: Understanding El Niño in ocean–atmosphere general circulation models: progress and challenges. *Bull. Am. Met. Soc.*, **90**, 325–40.

Hagedorn, R., F. J. Doblas-Reyes, and T. N. Palmer, 2005: The rationale behind the success of multi-model ensembles in seasonal forecasting-I. Basic concept. *Tellus*, **57A**, 219–33.

Hamlet, A. F., and D. Lettenmaier, 2006: *West-Wide Seasonal Hydrologic Forecast System*. www.hydro.washington.edu/forecast/westwide/index.shtml.

Hammer, G., 2000: A general systems approach to applying seasonal forecasts. In *Applications of Seasonal Climate Forecasting in Agricultural and Natural Ecosystems*, G. L. Hammer, N. Nicholls, and C. Mitchell, eds. Kluwer Academic Publishers, 469 pp.

Harrison, D. E., and N. K. Larkin, 1998: El Niño–Southern Oscillation sea surface temperature and wind anomalies, 1946–1993. *Rev. Geophys.*, **36**, 353–99.

Hartmann, D. L., 1994: *Global Physical Climatology*. Academic Press, 411 pp.

Haywood, A. M., P. Dekens, A. C. Ravelo, and M. Williams, 2005: Warmer tropics during the mid-Pliocene? Evidence from alkenone paleothermometry and a fully coupled ocean–atmosphere GCM. *Geochem. Geophys. Geosyst.*, **6**, doi:10.1029/2004GC000799.

Haywood, A. M., P. J Valdes, and V. L. Peck, 2007: A permanent El Niño-like state during the Pliocene? *Paleoceanography*, **22**, doi:10.1029/2006PA001323.

Hewitt, C. D., A. J. Broccoli, J. F. B. Mitchell, and R. J. Stouffer, 2001: A coupled model study of the last glacial maximum: was part of the North Atlantic relatively warm? *Geophys. Res. Lett.*, **28**, 1571–4.

Hide, R., 1969: Dynamics of the atmospheres of the major planets with an Appendix on the viscous boundary layer at the rigid bounding surface of an electrically-conducting rotating fluid in the presence of a magnetic field. *J. Atmos. Sci.*, **26**, 841–53.

Hirst, A. C., 1986: Unstable and damped equatorial modes in simple coupled models. *J. Atmos. Sci.*, **43**, 606–30.

Hirst, A. C., 1988: Slow instabilities in tropical ocean basin global–atmosphere models. *J. Atmos. Sci.*, **45**, 606–30.

Hoerling, M. P., and A. Kumar, 2002: Atmospheric response patterns associated with tropical forcing. *J. Climate*, **15**, 2184–203.

Hoerling, M. P., A. Kumar, and M. Zhong, 1997. El Niño, La Niña, and the nonlinearities of their teleconnections. *J. Climate*, **10**, 1769–86.

Hough, S. S., 1898: On the applications of harmonic analysis to the dynamical theroy of tides. II. On the general integration of Laplace's tidal equations. *Phil. Trans. Roy. Soc. London*, **A191**, 139–85.

Hughen, K. A., D. P. Schrag, S. B. Jacobsen, and W. Hantoro, 1999: El Niño during the last interglacial period recorded by a fossil coral from Indonesia. *Geophys. Res. Lett.*, **26**, 3129–32.

Hurrell, J., 1995: Decadal trends in the North-Atlantic oscillation – regional temperatures and precipitation. *Science*, **269**, 676–9.

IPCC, 2007: *Climate Change 2007: The Physical Science Basis*. Contribution of Working Group I to the Fourth Assessment Report of the Intergovernmental Panel on Climate Change [S. Solomon, D. Qin, M. Manning, Z. Chen, M. Marquis, K. B. Averyt, M. Tignor, and H. L. Miller, eds.]. Cambridge University Press, Cambridge, UK, and New York, 996 pp.

Ioannou, P., 1995: Nonnormality increases variance. *J. Atmos. Sci.*, **52**, 1155–8.

Ioannou, P. J., and B. F. Farrell, 2006: Application of generalized stability theory to deterministic and statistical prediction. In *Predictability of Weather and Climate*, T. Palmer, and R. Hagedorn, eds. Cambridge University Press, pp. 181–216.

Jin, E. K., J. L. Kinter, B. Wang, *et al.*, 2008: Current status of ENSO prediction skill in coupled ocean–atmosphere models. *Clim. Dyn.*, **31**, 647–64.

Jin, F.-F., 1997a: An equatorial ocean recharge paradigm for ENSO. Part I. Conceptual model. *J. Atmos. Sci.*, **54**, 811–29.

Jin, F.-F., 1997b: An equatorial ocean recharge paradigm for ENSO. Part II. A stripped-down coupled model. *J. Atmos. Sci.*, **54**, 830–847.

Jin, F.-F., and J. D. Neelin, 1993a: Modes of interannual tropical ocean–atmosphere interaction – a unified view. Part I. Numerical results. *J. Atmos. Sci.*, **50**, 3477–502.

Jin, F.-F., and J. D. Neelin, 1993b: Modes of interannual tropical ocean–atmosphere interaction – a unified view. Part III. Analytical results in fully coupled cases. *J. Atmos. Sci.*, **50**, 3523–40.

Jin, F.-F., J. D. Neelin, and M. Ghil, 1994: El Niño on the devil's staircase: annual subharmonic steps to chaos. *Science*, **264**, 70–2.

Jones, P. D., T. J. Osborn, and K. R. Briffa, 2001: The evolution of climate over the last millennium. *Science*, **292**, 662–7.

Josey, S. A., E. C. Kent, and P. K. Taylor, 1998: *The Southampton Oceanography Centre (SOC) Ocean–Atmosphere Heat, Momentum and Freshwater Flux Atlas*. Southampton Oceanography Centre Report No. 6, Southampton, UK, 30 pp.

Kållberg, P., P. Berrisford, B. Hoskins, *et al.*, 2005. *The ERA-40 Atlas*. ECMWF ERA-40 Report Series No. 19, 191 pp.

Kalnay, E., 2003: *Atmospheric Modeling, Data Assimilation and Predictability*. Cambridge University Press, 341 pp.

Kaplan, A., M. A. Cane, Y. Kushnir, *et al.*, 1998: Analyses of global sea surface temperature, 1856–1991. *J. Geophys. Res.*, **103**, 18567–89.

Karspeck, A. R., A. Kaplan, and M. A. Cane, 2006: Predictability loss in an intermediate ENSO model due to initial error and atmospheric noise. *J. Climate*, **19**, 3572–88.

Kiehl, J. T., and K. E. Trenberth, 1997: Earth's annual global mean energy budget. *Bull. Am. Met. Soc.*, **78**, 197–208.

Kleeman, R., 2008: Stochastic theories for the irregularity of ENSO. *Phil. Trans. Roy. Soc.*, **A366**, 2511–26.

Kleeman, R., and A. M. Moore, 1997: A theory for the limitation of ENSO predictability due to stochastic atmospheric transients. *J. Atmos. Sci.*, **54**, 753–67.

Koutavas A., and J. Lynch-Stieglitz, 2005: Variability of the marine ITCZ over the eastern Pacific during the past 30,000 years: regional perspective and global context. In *The Hadley Circulation: Present Past and Future*, R. Bradley, and H. Diaz, eds. Springer, pp. 347–69.

Koutavas, A., J. Lynch-Stieglitz, T. M. Marchitto, and J. P. Sachs, 2002: El Niño-like pattern in ice age tropical Pacific sea surface temperature. *Science*, **297**, 226–30.

Kraus, E. B., and J. A. Businger, 1994: *Atmosphere-Ocean Interactions*, 2nd edn. Oxford University Press, 362 pp.

Kraus, E. B., and J. S. Turner, 1967: A one-dimensional model of the seasonal thermocline. II. The general theory and its consequences. *Tellus*, **19**, 98–106.

Kumar, K., B. Rajagopalan, and M. A. Cane, 1999: On the weakening relationship between the Indian Monsoon and ENSO. *Science*, **284**, 2156–9.

Lacara, J.-F., and O. Talagrand, 1988: Short-range evolution of small perturbations in a barotropic model. *Tellus*, **40A**, 81–95.

Large, W. G., and G. Danabasoglu, 2006: Attribution and impacts of upper-ocean biases in CCSM3. *J. Climate*, **19**, 2325–46.

Large, W. G., J. C. McWilliams, and S. C. Doney, 1994: Oceanic vertical mixing: a review and a model with a nonlocal boundary layer parameterization. *Rev. Geophys.*, **32**, 363–403.

Latif, M., D. Anderson, T. Barnett, *et al.*, 1998: A review of the predictability and prediction of ENSO. *J. Geophys. Res.*, **103**, C7, 14375–93.

Legler, D. M., and J. J. O'Brien, 1988: Tropical Pacific wind stress analysis for TOGA. In *IOC Time Series of Ocean Measurements*, IOC Technical Series 33, Vol. 4, UNESCO.

Lewis, H. W., 1997: *Why Flip a Coin? The Art and Science of Good Decisions*. John Wiley & Sons, 206 pp.

Lighthill, M. J., 1969: Dynamical response of the Indian Ocean to the onset of the Southwest Monsoon. *Phil. Trans. Roy. Soc. London*, **A265**, 45–92.

Lindzen, R. D., 1967: Planetary waves on beta-planes. *Mon. Wea. Rev.*, **95**, 441–51.

Lindzen, R. S., 1966: On the theory of the diurnal tide. *Mon. Wea. Rev.*, **94**, 295–301.

Lindzen, R. S., 1970: Atmospheric tides. In *Mathematical Problems in the Geophysical Sciences*. American Mathematical Society.

Lindzen, R. S., and S. Nigam, 1987: On the role of sea surface temperature gradients in forcing low-level winds and convergence in the tropics. *J. Atmos. Sci.*, **44**, 2418–36.

Liu, Z., and M. Alexander, 2007: Atmospheric bridge, oceanic tunnel, and global climatic teleconnections. *Rev. Geophys.*, **45**, RG2005, doi:10.1029/2005RG000172.

Liu, Z, J. Kutzbach, and L. Wu, 2000: Modeling the climatic shift of El Niño variability in the Holocene. *Geophys. Res. Lett.*, **27**, 2265–8.

Longuet-Higgins, 1968: The eigenfunctions of Laplace's tidal equations on a sphere. *Phil. Trans. Roy. Soc.*, **A262**, 511–607.

Lorenz, E. N., 1963: Deterministic nonperiodic flow. *J. Atmos. Sci.*, **20**, 130–41.

Lunt, D. J., G. L. Foster, A. M. Haywood, and E. J. Stone, 2008: Late Pliocene Greenland glaciation controlled by a decline in atmospheric CO_2 levels. *Nature*, **454**, 1102–5.

Lyon, B., and A. G. Barnston, 2005: ENSO and the spatial extent of interannual precipitation extremes in tropical land areas. *J. Climate*, **18**, 5095–109.

Malkus, J. S., 1958: *On the Structure of the Trade-Wind Moist Layer*. Papers in Physical Oceanography and Meteorology, Vol. XIII, No. 2, MIT and Woods Hole Oceanographic Institution, 47 pp.

Malkus, J. S., 1962: Interactions of properties between sea and air. 4. Large scale interactions. In *The Sea*, Vol. 1, M. N. Hill, ed. John Wiley and Sons, pp. 88–294.

Manabe, S., and F. Möller, 1961: On the radiative equilibrium and heat balance of the atmosphere. *Mon. Wea. Rev.*, **89**, 503–32.

Mann, M. E., R. S. Bradley, and M. K. Hughes, 2000: Long-term variability in the El Niño/ southern oscillation and associated teleconnections. In *El Niño and the Southern Oscillation: Multiscale Variability and Global and Regional Impacts*, H. F. Diaz, and V. Markgraf, eds. Cambridge University Press.

Mann, M. E., 2002: The value of multiple proxies. *Science*, **297**, 1481–2.

Mann, M. E., M. A. Cane, S. E. Zebiak, and A. C. Clement, 2005: Volcanic and solar forcing of El Niño over the past 1000 years. *J. Climate*, **18**, 447–56.

Mantua, N. J., and D. S. Battisti, 1994: Evidence for the delayed oscillator mechanism for ENSO: the "observed" oceanic Kelvin wave in the far western Pacific. *J. Atmos. Sci.*, **48**, 1238–48.

Mantua, N. J., and D. S. Battisti, 1995: Aperiodic variability in the Zebiak–Cane coupled atmosphere-ocean model: air–sea interactions in the western equatorial Pacific. *J. Climate*, **8**, 2897–927.

Mantua, N. J., S. R. Hare, Y. Zhang, J. M. Wallace, and R. C. Francis, 1997: A Pacific interdecadal oscillation with impacts on salmon production. *Bull. Am. Met. Soc.*, **78**, 1069–79.

Matsuno, T., 1966: Quasi-geostrophic motions in the equatorial area. *J. Met. Soc. Japan*, **44**, 25–43.

McPhaden, M. J., A. J. Busalacchi, R. Cheney, *et al.*, 1998: The tropical ocean–global atmosphere observing system. *J. Geophys. Res.*, **103**, 14169–240.

Mellor, G. L., and T. Yamada, 1982: Development of turbulence closure models for geophysical fluid problems. *Rev. Geophys. Space Phys.*, **4**, 851–75.

Merryfield, W. J., 2006: Changes to ENSO under CO_2 doubling in the IPCC AR4 coupled climate models. *J. Climate*, **19**, 4009–27.

Molnar, P., and M. A. Cane, 2002: El Niño's tropical climate and teleconnections as a blueprint for pre–Ice-Age climates. *Paleoceanography*, **17**, 2, doi:10.1029.2001PA000663.

Molnar, P., and M. A. Cane, 2008: Early Pliocene (pre-Ice Age) El Niño-like global climate: which El Niño? *Geosphere*, **3**, 337–65, doi:10.1130/GES00103.1.

Moore, A. M., and R. Kleeman, 1999: Stochastic forcing of ENSO by the intraseasonal oscillation. *J. Climate*, **12**, 1199–220.

Moore, A. M., J. Zavala-Garay, Y. Tang, *et al.*, 2006: Optimal forcing patterns for coupled models of ENSO. *J. Climate*, **19**, 4683–99.

Moore, D. W., 1968: Planetary-gravity waves in an equatorial ocean. Ph.D. Dissertation, Harvard University, 207 pp.

Moore, D. W., and S. G. H. Philander, 1976: Modeling of the tropical oceanic circulation. In *The Sea*, Vol. 6, Chap. 8. Interscience.

Moura, A. D., 1976: The eigensolutions of the linearized balance equations on a sphere. *J. Atmos. Sci.*, **33**, 877–907.

Moy, C. M., G. O. Seltzer, D. T. Rodbell, and D. M. Anderson, 2002: Variability of El Niño/southern oscillation activity at millennial timescales during the Holocene epoch. *Nature*, **420**, 162–5.

Munk, W. H., and E. R. Anderson, 1948: Notes on a theory of the thermocline. *J. Mar. Res.*, **7**, 276–95.

Münnich, M., M. A. Cane, and S. E. Zebiak, 1991: A study of self-excited oscillations of the tropical ocean atmosphere system. 2. Nonlinear cases. *J. Atmos. Sci.*, **48**, 1238–48.

National Research Council, 1999: *Making Climate Forecasts Matter*. P. C. Stern, and W. E. Easterling, eds. National Academy Press, 175 pp.

Neelin, J. D., 1989: On the interpretation of the Gill model. *J. Atmos. Sci.*, **46**, 2466–8.

Neelin, J. D., 1991: The slow sea surface temperature mode and the fast-wave limit: analytic theory for tropical interannual oscillations and experiments in a hybrid coupled model. *J. Atmos. Sci.*, **48**, 584–606.

Neelin, J. D., and F.-F. Jin, 1993: Modes of interannual tropical ocean–atmosphere interaction – a unified view. Part II. Analytical results in the weak-coupling limit. *J. Atmos. Sci.*, **50**, 3504–22.

Neelin J. D., D. S. Battisti, A. C. Hirst, *et al.*, 1998: ENSO theory. *J. Geophys. Res.*, **103**, C7, 14261–90.

Nicholls, N., 2000a: Cognitive illusions, heuristics and climate prediction. *Bull. Am. Met. Soc.*, **80**, 1385–97.

Nicholls, N., 2000b: Opportunities to improve the use of seasonal climate forecasts. In *Applications of Seasonal Climate Forecasting in Agricultural and Natural Ecosystems*, G. L. Hammer, N. Nicholls, and C. Mitchell, eds. Kluwer Academic Publishers, 469 pp.

Niiler, P. P., and E. B. Kraus, 1977: One-dimensional models of the upper ocean. In *Modelling and Prediction of the Upper Layers of the Ocean*, E. B. Kraus, ed. Pergamon Press.

Noble, B., and J. W. Daniel, 1988: *Applied Linear Algebra*, 3rd edn. Prentice Hall, 521 pp.

Oberhuber, J. M., 1988: *An Atlas Based on the COADS Data Set: The Budgets of Heat, Buoyancy and Turbulent Kinetic Energy at the Surface of the Global Ocean.* MPI Report 15.

Otto-Bleisner, B. L., E. C. Brady, S.-I. Shin, Z. Liu, and C. Shields, 2003: *Modeling El Niño and its tropical teleconnections during the last glacial-interglacial cycle.* Geophys. Res. Lett, **30**, doi:10.1029/2003GL018553.

Pacanowski, R. C., and S. G. H. Philander, 1981: Parameterization of vertical mixing in numerical models of tropical oceans. *J. Phys. Oceanogr.*, **11**, 1443–51.

Palmer, T. N., A. Alessandri, U. Andersen, *et al.*, 2004: Development of a European multimodel ensemble system for seasonal-to-interannual prediction (DEMETER). *Bull. Am. Met. Soc.*, **85**, 853–72.

Pearson, P. N., and M. R. Palmer, 2000: Atmospheric carbon dioxide concentrations over the past 60 million years. *Nature*, **406**, 695–9.

Pedlosky, J., 1965: A note on the western intensification of the oceanic circulation. *J. Mar. Res.*, **23**, 207–10.

Penland, C., and T. Magorian, 1993: Prediction of Niño3 sea surface temperature using linear inverse modeling. *J. Climate*, **6**, 1067–76.

Penland, C., and P. Sardeshmukh, 1995: The optimal growth of tropical sea surface temperature anomalies. *J. Climate*, **8**, 1999–2024.

Persson, A., and F. Gravini, 2005: *User Guide to ECMWF Forecast Products*. European Centre for Medium Range Weather Forecasts, 154 pp. Available at www.ecmwf.int/products/forecasts/guide/user_guide.pdf.

Philander, S. G. H., 1978: Forced oceanic waves. *Rev. Geophys.*, **16**, 15–46.

Philander, S. G. H., 1990: *El Niño, La Niña, and the Southern Oscillation*. Academic Press, 293 pp.

Philander, S. G. H., and A. Fedorov, 2003: Is El Niño sporadic or cyclic? *Ann. Rev. Earth Planet. Sci.*, **31**, 579–94.

Philander, S. G. H., and A. V Fedorov, 2003: Role of tropics in changing the response to Milankovich forcing some three million years ago. *Paleoceanography*, **18**, doi:10.1029/2002PA000837.

Philander, S. G. H., T. Yamagata, and R. C. Pacanowski, 1984: Unstable air–sea interactions in the tropics. *J. Atmos. Sci.*, **41**, 604–13.

Phillips, N. A., 1957: A coordinate system having some special advantages for numerical forecasting. *J. Atmos. Sci.*, **14**, 184–5.

Phillips, N. A., 1968: Reply. *J. Atmos. Sci.*, **25**, 1155–7.

Phillips, N. A., 1973: Principles of large scale numerical weather prediction. In *Dynamical Meteorology*, P. Morel, ed. D. Reidel Publishing Co., 622 pp.

Picaut, J., C. Menkes, J.-P. Boulanger, and Y. du Penhoat, 1993: Dissipation in a Pacific equatorial long wave model. *TOGA Notes*, **10**, 11–15.

Pielke, R. A., Jr., 2000: Policy responses to El Niño 1997–1998. In *El Niño, 1997–1998: The Climate Event of the Century*, S. A. Changnon, ed. Oxford University Press, 215 pp.

Pielke, R. A., and C. N. Landsea, 1999: La Niña, El Niño, and Atlantic hurricane damages in the United States. *Bull. Am. Met. Soc.*, **80**, 2027–33.

Pollard, R. T., P. B. Rhines, and R. O. R. Y. Thompson, 1973: The deepening of the wind mixed layer. *Geophys. Fluid. Dyn.*, **4**, 381–404.

Power, S., T. Casey, C. Folland, A. Colman, and V. Mehta, 1999: Inter-decadal modulation of the impact of ENSO on Australia. *Clim. Dyn.*, **15**, 319–24.

Price, J. F., 1979: On the scaling of stress-driven entrainment experiments. *J. Fluid Mech.*, **90**, 509–29.

Price, J. F., R. A. Weller, and R. Pinkel, 1986: Diurnal cycling: observations and models of the upper ocean response to diurnal heating, cooling, and wind mixing. *J. Geophys. Res.*, **91**, 8411–27.

PROVOST, 2000: *Q. J. Roy. Met. Soc.*, **126B**, 1989–2351, July 2000 issue.

Pulwarty, R. S., and K. T. Redmond, 1997: Climate and salmon restoration in the Columbia River Basin: the role and suability of seasonal forecasts. *Bull. Am. Met. Soc.*, **78**, 381–7.

Qian, J. H., 2008: Why precipitation is mostly concentrated over islands in the maritime continent. *J. Atmos. Sci.*, **65**, 1428–41.

Rajagopalan, B., U. Lall, and M. A. Cane, 1997: Anomalous ENSO occurrences: an alternate view. *J. Climate*, **10**, 2351–57.

Rasmusson, E. M., and T. H. Carpenter, 1982: Variations in tropical sea surface temperature and surface wind fields associated with the Southern Oscillation/El Niño. *Mon. Wea. Rev.*, **110**, 354–84.

Rayner, S., D. Lach, and H. Ingram, 2005: Weather forecasts are for wimps: why water resource managers do not use climate forecasts. *Climatic Change*, **69**, 197–227.

Reed, R. J., and E. E. Recker, 1971: Structure and properties of synoptic-scale wave disturbances in the equatorial western Pacific. *J. Atmos. Sci.*, **28**, 1117–33.

Reynolds, R. W., N. A. Rayner, T. M. Smith, D. C. Stokes, and W. Wang, 2002: An improved in situ and satellite SST analysis for climate. *J. Climate*, **15**, 1609–25.

Rodbell, D., G. Seltzer, D. Anderson, *et al.*, 1999: A 15,000 year record of El Niño-driven alluviation in southwestern Ecuador. *Science*, **283**, 516–20.

Rogers, E. M., 2003: *Diffusion of Innovations*, 5th edn. Free Press, 551 pp.

Ropelewski, C. F., and M. S. Halpert, 1987: Global and regional scale precipitation patterns associated with the El Niño/Southern Oscillation. *Mon. Wea. Rev.*, **115**, 1606–26.

Ropelewski, C. F., and M. S. Halpert, 1996: Quantifying Southern Oscillation–precipitation relationships. *J. Climate*, **9**, 1043–59.

Sandweiss, D., J. Richardson, E. Reitz, and H. Rollins, 1996: Geoarchaeological evidence from Peru for a 5000 year B.P. onset of El Niño. *Science*, **273**, 1531–3.

Sarachik, E. S., 1978: Tropical sea surface temperature: an interactive one-dimensional atmosphere–ocean model. *Dyn. Atmos. Oceans*, **2**, 455–69.

Sarachik, E. S., 1985: A simple theory for the vertical structure of the tropical atmosphere. *Pure Appl. Geophys.*, **123**, 261–71.

Sarachik, E. S., 1990: Predictability of ENSO. In *Climate–Ocean Interaction*, M. E. Schlesinger, ed. Kluwer Academic Publishers, 161–71.

Sarachik, E. S., 1999: The application of climate information. *Consequences*, **5**, 27–36. Available at www.gcrio.org/CONSEQUENCES/index.htm.

Schneider, E. K., 1977: Axially symmetric steady-state models of the basic state for instability and climate studies. Part II. Nonlinear calculations. *J. Atmos. Sci.*, **34**, 280–97.

Schneider, E. K., and R. S. Lindzen, 1976: A discussion of the parameterization of momentum exchange by cumulus convection. *J. Geophys. Res.*, **81**, 3158–60.

Schneider, E. K., and R. S. Lindzen, 1977: Axially symmetric steady-state models of the basic state for instability and climate studies. I. Linearized calculations. *J. Atmos. Sci.*, **34**, 263–79.

Schopf, P. S., and M. A. Cane, 1983: On equatorial dynamics, mixed layer physics and sea-surface temperature. *J. Phys. Oceanogr.*, **13**, 917–35.

Schopf, P. S., and M. J. Suarez, 1988: Vacillations in a coupled ocean–atmosphere model. *J. Atmos. Sci.*, **45**, 549–66.

Shaman, J., M. Stieglitz, S. Zebiak, and M. Cane, 2003: A local forecast of land surface wetness conditions derived from seasonal climate predictions. *J. Hydromet.*, **4**, 611–26.

Shaman, J., J. F. Day, M. Stieglitz, S. Zebiak, and M. Cane, 2006: An ensemble seasonal forecast of human cases of St. Louis encephalitis in Florida based on seasonal hydrologic forecasts. *Climatic Change*, **75**, 495–511.

Shukla, J., 1981: Predictability of the tropical atmosphere. NASA Tech., Memo 83829.

Smith, N. R., 1995: An improved system for tropical ocean subsurface temperature analyses. *J. Atmos. Ocean. Technol.*, **12**, 850–70.

Stan, C., and B. P. Kirtman, 2008: The influence of atmospheric noise and uncertainty in ocean initial conditions on the limit of predictability in a coupled GCM. *J. Climate*, **21**, 3487–503.

Stephens, D., D. Butler, and G. Hammer, 2000: Using seasonal climate forecasts in forecasting the Australian wheat crop. In *Applications of Seasonal Climate Forecasting in Agricultural and Natural Ecosystems*, G. L. Hammer, N. Nicholls, and C. Mitchell, eds. Kluwer Academic Publishers, 469 pp.

Suarez, M. J., and P. S. Schopf, 1988: A delayed action oscillator for ENSO. *J. Atmos. Sci.*, **45**, 3283–7.

Sun, D.-Z., T. Zhang, C. Covey, *et al.*, 2006: Radiative and dynamical feedbacks over the equatorial cold tongue: results from nine atmospheric GCMs. *J. Climate*, **19**, 4059–74.

Sverdrup, H. U., 1947: Wind driven currents in a baroclinic ocean with application to the equatorial currents of the eastern Pacific. *PNAS*, **33**, 318–26.

Tennekes, H., 1973: A model for the dynamics of the inversion above a convective boundary layer. *J. Atmos. Sci.*, **30**, 558–67.

Thompson, C. J., and D. S. Battisti, 2000: A linear stochastic dynamical model of ENSO. I. Model development. *J. Climate*, **13**, 2818–32.

Thompson, C. J., and D. S. Battisti, 2001: A linear stochastic dynamical model of ENSO. II. Analysis. *J. Climate*, **14**, 445–66.

Trenberth, K. E., and T. J. Hoar, 1997: El Niño and climate change. *Geophys. Res. Lett.*, **24**, 3057–60.

Trenberth, K. E., and J. M. Caron, 2001: Estimates of meridional atmosphere and ocean heat transports. *J. Climate*, **14**, 3433–43.

Trenberth, K. E., G. W. Branstator, D. Karoly, *et al.*, 1998: Progress during TOGA in understanding and modeling global teleconnections associated with tropical sea surface temperatures. *J. Geophys. Res.*, **103**, C7, 14291–324.

Tudhope, A. W., C. P. Cilcott, M. T. McCulloch, *et al.*, 2001: Variability in the El Niño-Southern Oscillation through a glacial-interglacial cycle. *Science*, **291**, 1511–7.

Tziperman, E., L. Stone, M. Cane, and H. Jarosh, 1994: El Niño chaos: overlapping of resonances between the seasonal cycle and the Pacific ocean–atmosphere oscillator. *Science*, **264**, 72–4.

Tziperman, E., M. A. Cane, and S. Zebiak, 1995: Irregularity and locking to the seasonal cycle in an ENSO prediction model as explained by the quasi-periodicity route to chaos. *J. Atmos. Sci.*, **50**, 293–306.

Tziperman, E., S. E. Zebiak, and M. A. Cane, 1997: Mechanisms of seasonal-ENSO interaction. *J. Atmos. Sci.*, **52**, 61–71.

Uppala, S. M., Kållberg, P. W., Simmons, *et al.*, 2005: The ERA-40 re-analysis. *Q. J. Roy. Met. Soc.*, **131**, 2961–3012, doi: 10.1256/qj.04.176.

USGS, 1994: *Climatic Fluctuations, Drought, and Flow in the Colorado River Basin.* USGS Fact Sheet 2004-3062 version 2. Available at http://pubs.usgs.gov/fs/2004/3062.

USGS, 2004: *Climatic Fluctuation, Drought and Flow in the Colorado River Basin.* USGS Fact Sheet 2004-3062 Version 2.

Uvo, C. B, C. A. Repelli, S. E. Zebiak, and Y. Kushnir, 1998: The relationships between tropical Pacific and Atlantic SST and northeast Brazil monthly precipitation. *J. Climate*, **11**, 551–62.

van Oldenborgh, G. J., S. Philip, and M. Collins, 2005: El Niño in a changing climate: a multi-model study. *Ocean Science*, **1**, 81–95.

Vecchi, G. A., and N. A. Bond, 2004: The Madden–Julian oscillation (MJO) and northern high latitude wintertime surface air temperatures. *Geophys. Res. Lett.* **31**, L04104, doi: 10.1029/2003GL018645.

Vecchi, G. A., B. J. Soden, A. T. Wittenberg, *et al.*, 2006: Weakening of tropical Pacific atmospheric circulation due to anthropogenic forcing. *Nature*, **441**, 73–6.

Veronis, 1973: Large scale ocean circulation. *Adv. Appl. Mech.*, **13**, 1–92.

Villalba, R. R., D. D'Arrigo, E. R. Cook, G. C. Jacoby, and G. Wiles, 2001: Decadal-scale climate variability along the extratropical western coast of the Americas: evidence from tree-ring records. In *Interhemispheric Climate Linkages*, V. Markgraf, ed., Academic Press, San Diego, pp. 155–72.

Vimont, D. J., 2005: The contribution of the interannual ENSO cycle to the spatial pattern of ENSO-like decadal variability. *J. Climate*, **18**, 2080–92.

Wakata, Y., and E. S. Sarachik, 1991: Unstable coupled atmosphere ocean basin modes in the presence of a spatially varying basic state. *J. Atmos. Sci.*, **48**, 2060–77.

Wallace, J. M., E. M. Rasmusson, T. P. Mitchell, *et al.*, 1998: On the structure and evolution of ENSO-related climate variability in the tropical Pacific: lessons from TOGA. *J. Geophys. Res. Oceans*, **103**, C7, 14241–60.

Wang, B., 1994: On the annual cycle in the tropical eastern central Pacific. *J. Climate*, **7**, 1926–42.

Wang, C., 1994: Understanding ENSO physics – a review. In the *Ocean–Atmosphere Interaction*, Geophysical Monograph Series, **147**, 21–48.

Wang, C., 2001: A unified oscillator model for the El Niño-Southern Oscillation. *J. Climate*, **14**, 98–115.

Wara, M. W., A. C. Ravelo, and M. L. Delaney, 2005: Permanent El Niño-like conditions during the Pliocene warm period. *Science*, **309**, 758–61.

White, W. B., G. A. Meyers, J. R. Donguy, and S. E. Pazan, 1985: Short-term climatic variability in the thermal structure of the Pacific Ocean during 1979–82. *J. Phys. Oceanogr.*, **15**, 917–35.

Wiley, M. *et al.*: Available at www.tag.washington.edu/projects/midrange.html.

Woodhouse, C. A., S. T. Gray, and D. M. Meko, 2006: Updated streamflow reconstructions for the upper Colorado river basin. *Water Resources Res.*, **42**, W05415, doi:10.1029/2005WR004455.

Woodruff, S. D., R. J. Slutz, R. L. Jenne, and P. M. Steurer, 1987: A comprehensive ocean–atmosphere data set. *Bull. Am. Met. Soc.*, **68**, 1239–50.

Woodruff, S. D., S. J. Lubker, K. Wolter, S. J. Worley, and J. D. Elms, 1993: Comprehensive ocean-atmosphere data set (COADS) release 1a: 1980–92. *Earth Syst. Monitor*, **4**, 1–8.

Wu, Z., 2003: A shallow CISK, deep equilibrium mechanism for the interaction between large-scale convection and large-scale circulations in the tropics. *J. Atmos. Sci.*, **60**, 377–92.

Wu, Z., E. S. Sarachik, and D. S. Battisti, 1999: Thermally forced surface winds on an equatorial beta-plane. *J. Atmos. Sci.*, **56**, 2029–37.

Wu, Z., D. S. Battisti, and E. S. Sarachik, 2000a: Rayleigh friction, Newtonian cooling, and the linear response to steady tropical heating. *J. Atmos. Sci.*, **57**, 1937–57.

Wu, Z., E. S. Sarachik, and D. S. Battisti, 2000b: The vertical structure of convective heating and the three-dimensional structure of the forced circulation in the tropics. *J. Atmos. Sci.*, **57**, 2169–87.

Wu, Z., E. S. Sarachik, and D. S. Battisti, 2001: Thermally driven tropical circulations under Rayleigh friction and Newtonian cooling: analytic solutions. *J. Atmos. Sci.*, **58**, 724–41.

Wunsch, C., 1999: The interpretation of short climate records, with comments on the North Atlantic and Southern Oscillations. *Bull. Am. Met. Soc.*, **80**, 245–55.

Wyrtki, K., 1975: El Niño – the dynamic response of the ocean to atmospheric forcing. *J. Phys. Oceanogr.*, **5**, 572–84.

Wyrtki, K., 1979: The response of sea surface topography to the 1976 El Niño. *J. Phys. Oceanogr.*, **11**, 1205–31.

Wyrtki, K., 1985a: Water displacements in the Pacific and the genesis of El Niño cycles. *J. Geophys. Res.*, **90**, 7129–32.

Wyrtki, K., 1985b: Sea level fluctuations in the Pacific during the 1982–83 El Niño. *Geophys. Res. Lett.*, **12**, 125–8.

Yamagata, T., and S. G. H. Philander, 1985: The role of damped equatorial waves in the oceanic response to winds. *J. Oceanogr. Soc. Japan*, **41**, 345–57.

Yanai, M., S. Esbensen, and J. Chu, 1973: Determination of bulk properties of tropical cloud clusters from large-scale heat and moisture budgets. *J. Atmos. Sci.*, **30**, 611–27.

Yuan, D., 2005: Role of the Kelvin and Rossby waves in the seasonal cycle of the equatorial Pacific Ocean circulation. *J. Geophys. Res.*, **110**, doi:10.1029/2004JC002344.

Yulaeva, E., and J. M. Wallace, 1994: The signature of ENSO in global temperature and precipitation fields derived from the Microwave Sounding Unit. *J. Climate*, **7**, 1719–36.

Zebiak, S. E., 1990: Diagnostic studies of Pacific surface winds. *J. Climate*, **3**, 1016–31.

Zebiak, S. E., 1982: A simple atmospheric model of relevance to El Niño. *J. Atmos. Sci.*, **39**, 2017–27.

Zebiak, S. E., 1986: Atmospheric convergence feedback in a simple model for El Niño. *Mon. Wea. Rev.*, **114**, 1263–71.

Zebiak, S. E., and M. A. Cane, 1987: A model El Niño–Southern Oscillation. *Mon. Wea. Rev.*, **115**, 2262–78.

Zebiak, S. E., and M. A. Cane, 1991: Natural climate variability in a coupled model. In *Greenhouse Gas-Induced Climatic Change: Critical Appraisal of Simulations and Observations*, M. E. Schlesinger, ed., Elsevier, 457–70.

Zhang, C., 2005: Madden–Julian oscillation. *Rev. Geophys.*, **43**, 2004RG000158.

Index